ESSENTIALS OF MODERN HOSPITAL SAFETY
VOLUME 3

EDITED BY

WILLIAM CHARNEY, DOH
Director of Environmental Health
San Francisco General Hospital
San Francisco, California

Library of Congress Cataloging-in-Publication Data
(Revised for vol. 3)

Essentials of modern hospital safety.

Includes bibliographical references.
1. Hospitals--Safety measures. 2. Hospitals--Staff--Health and hygiene. I. Charney, William, 1947–
II. Schirmer, Joseph.
RA969.9.E77 1990 363.15 89-13704
ISBN 0-87371-198-X (v. 1)
ISBN 0-87371-584-5 (v. 2)
ISBN 1-56670-083-3 (v. 3)

Lewis Publishers is an imprint of CRC Press

No claim to original U.S. Government works
International Standard Book Number 1-56670-083-3
Library of Congress Card Number 89-13704
Printed in the United States of America 1 2 3 4 5 6 7 8 9 0
Printed on acid-free paper

Acknowledgments

I would like to thank Jane Doe for writing the introduction to Volume 3. Her spirit is the energy behind this volume. Her needlestick symbolizes, for me, the daily battle being waged in health care between health and safety and "other priorities." It is often recorded in the history of the occupational health and safety movement, that front line workers have had to pay a great price due to the hazards confronted in the workplace; Karen Silkwood (exposed safety violations in the Nuclear Laboratory), Herve Rousseau (Québec asbestos miner who now suffers from irreversible pleural plaque disease, has helped organize other stricken miners), Melissa Campbell, Jean Roe, Dr. Patty Wessell, all who seroconverted to HIV positive after a needlestick and who have spoken out about the conditions faced by front line health care workers, 250 health care workers a year die from hepatitis B, nine health care workers who died from multiple drug resistant TB, seven of whom were HIV positive. In that spirit, I dedicate this volume to Jane Doe, RN.

The hard work and good humor of Sharon Ray, of Lewis Publishers, have been greatly appreciated on Volumes 2 and 3 of this series.

I acknowledge the work of the SFGH Engineering Department, Eric Miller, Director, Terry Saltz, Chief Engineer and Max Bunuan, Senior Engineer for their tireless efforts implementing the SFGH Tuberculosis Engineering Plan. Their creative response to the need for engineering controls for TB and the applicability of the systems that they installed influenced the models behind the TB chapter in this volume.

I would also like to take this opportunity to acknowledge the work of all the contributing authors of Volumes 1–3. Their work, I believe, has set a good standard for the approaches to health care safety and has helped create an ongoing dialogue for the continuing need for this type of writing, analysis and solutions to the multiple risks facing health care professionals.

About the Editor

William Charney, DOH, is an industrial hygienist who specializes in hospital systems. For five years Mr. Charney was the Director of the Industrial Hygiene and Occupational Health and Safety Department at the Jewish General Hospital, a 600-bed tertiary, teaching, and research hospital in Montreal, Québéc. He is presently Director of Environmental Health at San Francisco General Hospital, San Francisco, California. Mr. Charney is a member of the American Conference of Governmental Industrial Hygienists and the Canadian Society of Safety Engineers.

About the Authors

Beth Blackwell, BA, Statistical Analyst, Department of Neurosurgery, University of Virginia Health Sciences Center, 202 Park Place, Box 407 Medical Center, Charlottesville, VA 22908

Murray Cohen, PhD, MPH, CIH, Chief, Medical Device Evaluations, Centers for Disease Control and Prevention, 1600 Clifton Road NE, Atlanta, GA 30333

Lenora S. Colbert, Vice President and Director, Occupational Safety and Health, National Health & Human Service Employees Union, 310 W. 43rd St., New York, NY 10036-6407

Lorraine M. Conroy, ScD, CIH, University of Illinois at Chicago, School of Public Health, Environmental and Occupational Health Sciences (M/C 922), 2121 W. Taylor, Chicago, IL 60612

Edward W. Finucane, PE, CSP, CIH, QEP, High Technology Enterprises, P.O. Box 7835, Stockton, CA 95267

June Fisher, MD, Associate Clinical Professor of Medicine, University of California, San Francisco, and Director, Training for Development of Innovative Control Technology, 1180 Dolores Street, San Francisco, CA 94110

John E. Franke, PhD, CIH, PE, University of Illinois at Chicago, School of Public Health, Environmental and Occupational Health Sciences (M/C 922), 2121 W. Taylor, Chicago, IL 60612

Arun Garg, PhD, University of Wisconsin–Milwaukee, Industrial and Systems Engineering, P.O. Box 784, Milwaukee, WI 53201

Michael L. Garvin, Safety Engineer, University of Iowa Hospitals and Clinics. President, Garvin Consulting Services, Iowa City, IA

Derrick Hodge, MS, Safety and Security Specialist, F.O.J.P. Service Corp., 130 E. 54th St., New York, NY 10022

Janine Jagger, MPH, PhD, Associate Professor of Neurosurgery, Director, Health Care Worker Safety Project, Department of Neurosurgery, University of Virginia Health Sciences Center, 202 Park Place, Box 407 Medical Center, Charlottesville, VA 22908

Kathleen Kahler, MPH, Kaiser Foundation Hospital, 2241 Geary Boulevard, San Francisco, CA 94115

Daniel Kass, MSPH, Director, The Center for Occupational and Environmental Health, Hunter College, 425 East 25th Street, New York, NY 10010

Lindsey V. Kayman, MS, CIH, UMDNJ, Department of EOHSS, 675 Hoes Lane, Trailer #1, Piscataway, NJ 08854

Jane Lipscomb, RN, PhD, University of California, School of Nursing, DMHAN, San Francisco, CA 94143-0608

David Makofsky, PhD, Department of Public Health, AIDS Office, 25 Van Ness Avenue, Suite 500, San Francisco, CA 94102-6033

Marian McDonald, DrPH, Assistant Professor of Health Communication and Education, Tulane University School of Public Health and Tropical Medicine, 1501 Canal Street, New Orleans, LA 70112

John Mehring, BA, Service Employees International Union, Western Region Health and Safety Department, P.O. Box 19360, Seattle, WA 98109

Bernice D. Owen, PhD, RN, University of Wisconsin–Madison, School of Nursing, Madison, WI 53792-2455

Jacob D. Paz, PhD, President, J&L Environmental Service, Inc., P.O. Box 33036, Las Vegas, NV 89133

Jolie Pearl, RN, MPH, Clinical Research Coordinator, E.P.I. Center, San Francisco General Hospital, 1001 Potrero Avenue, Room 5H22, San Francisco, CA 94110

Anthony Schapera, Department of Anesthesiology, San Francisco General Hospital, 1001 Potrero Avenue, Room 5H22, San Francisco, CA 94110

Wendy Shearn, MD, MPH, Kaiser Permanente Medical Center, 2200 O'Farrell Street, San Francisco, CA 94115

David A. Sterling, PhD, CIH, School of Public Health, St. Louis University Health Sciences Center, St. Louis University, 3663 Lindell Boulevard, St. Louis, MO 63108

Bernadette Stringer, RN, MSc (A) Previously employed full time as an occupational health and safety staff person for the BC Nurses Union. Now 2nd year PhD in "Occupational Health," Faculty of Medicine, McGill University, Montréal, Québec.

Byron S. Tepper, PhD, CSP, Director, Office of Safety and Environmental Health, 2024 East Monument Street, Suite 2-700, Baltimore, MD 21205

Introduction

In September 1987, I learned that a recent blood exposure had rendered me HIV-positive. The only "dirty" needlestick of my career, this accident led me to become San Francisco General Hospital's first health care worker "known" to have acquired HIV occupationally. Nationally, I was the 13th. The years following would challenge me to attempt to comprehend the broader context in which such an accident could have occurred.

I came to San Francisco General Hospital a young but experienced nurse and received comprehensive training in universal precautions during a hospital orientation. My perception of occupational risk at that time was influenced by multiple factors: a still youthful sense of immortality, an uncritical assumption that the institution and the larger regulatory agencies overseeing it had employed every measure to provide a safe working environment, and a rudimentary understanding of health and safety tenets (my most salient memory of health and safety training during my BSN education is being told not to bite my nails and thus create open sores). My perceptions were politicized as well; problematically, the voices conveying alarm over occupational transmission were often tinged with homophobia, disdain for IV drug users, and proposals for withholding care. In the absence of a reasoned and comfortable forum for discussion of risk, a health care worker was left keeping concerns over occupational transmission to and from herself.

But I did not incur my injury because of an underestimation of risk or the inappropriate handling of a needle. I acquired HIV because I had no adequate means with which to discard a used, blood-filled, unsheathed needle. At the moment of my needlestick, I found myself across a crowded room from the bathroom containing the sharps disposal container. In retrospect, I would recall a "gadget" I had used at another San Francisco hospital earlier that year. Marketed as the "Click Lock," this engineering device provided a cylindrical sheath over a needle that would have prevented my accident. Ironically, the Click Lock was being utilized in the other institution for the purposes of patient safety with central IV lines.

In December 1991, federal OSHA issued its final standard on bloodborne pathogens. Effective in March 1992, the standard set a hierarchy of controls for bloodborne hazards in the workplace, emphasizing that engineering controls be used in preference to other control measures when possible. This emphasis was a victory for health care workers who historically have had the onus for our safety placed on our own behavior. The standard called for each institution to develop an exposure control plan and required the employer to provide the hepatitis B vaccine to employees at risk free of cost.

It is known that prior to the development of the hepatitis B vaccine in 1982, 6000 to 8000 U.S. health care workers contracted the hepatitis B virus occupationally each year, resulting in two hundred deaths annually and leaving hundreds open to cirrhosis and liver cancer in years to come. As late as 1992, statistics issued by CDC and federal OSHA indicated no significant reduction in the incidence of occupationally transmitted hepatitis B. Yet it took federal OSHA a decade to mandate that the hepatitis B vaccine be made available to every health care worker at risk. As safety devices and needleless systems are implemented in our institutions at present, I am left to wonder where these devices have been as thousands of my co-workers have been dying of hepatitis B. Why has it taken a stigmatized, seemingly 100% lethal and ironically much less virulent disease to elicit a federal OSHA standard and to provoke an industry to design, manufacture, and market the devices Dr. Jagger tells us could prevent at least 85% of all needlesticks? (Jagger, congressional testimony, February 7, 1992.)

In August 1990, the nation responded with shock and horror as the CDC reported a young woman's acquisition of HIV as a patient in a Florida dental practice. Later it would be strongly suggested that inadequate sterilization techniques and infection control practices were responsible for the transmission of HIV to the young woman and four other of the dentist's patients. In July 1991, the U.S. Senate voted 81 to 18 for Jesse Helm's amendment requiring mandatory disclosure of HIV status by HIV-positive health care providers who perform invasive procedures, with criminal penalties applying to those health care workers who do not comply. No definition of invasive procedures was provided by the amendment.

Though the amendment did not become law, the message of its overwhelming vote was clear. While infection control practices, the crux of the transmission issue, went unaddressed, the first legislative responses to this tragedy were punitive of health care workers.

At the time of this vote, 22 health care workers were "known" to have seroconverted occupationally, a statistic considered to be underreported. Health care workers still waited for a standard on bloodborne pathogens, still worked with unsafe devices, and in some areas continued to be blamed for their injuries. As we waited, federal funding for research on the efficacy of needlestick prevention technology remained inadequate. Clearly, the federal agencies charged by Congress through the OSHAct of 1970 had failed to demand that safety devices become an industry standard in a timely fashion. (The FDA had not developed performance standards for needle devices.) While the media inflamed public fear about contracting HIV from one's dentist, the risk to health care workers was dramatically underplayed.

Paradoxes exist as well in the responses of an institution renowned for its provision of AIDS care. Following my seroconversion, SFGH would create a leading needlestick response program within two years. Providing 24-hour counseling, evaluation, Zidovudine therapy when appropriate, and blood exposure data collection, this program is notable for its management of occupational blood exposures. Prevention measures, however, would be implemented more gradually. The most obvious and facile prevention measure, the relocation of the sharps disposal container from the bathroom to the bedside (the point of use), would take three years. In 1991, when SFGH employed a safety intravenous stylet solely in its emergency department, workers in other units, together with SEIU, would launch a campaign to get the device distributed hospitalwide. It would take a class-action grievance to overcome institutional resistance and make the device available to all workers. Since 1991, a Needlestick Prevention Committee, with representation of occupational infectious disease personnel, management and frontline workers, has reviewed safety products, evaluated data on needlesticks and planned trainings. As of the end of 1993, the exposure rate has fallen to half the rate reported in 1989.

While any decrease in needlesticks is positive, the overall statistics are alarming and beg immediate action. With 1,000,000 needlesticks occurring nationally each year, approximately 20,000 or 2% are contaminated with HIV. If surveillance data indicating a 1/250 to 1/400 chance of contracting HIV from an HIV-positive needlestick are accurate, up to 50 to 80 health care workers a year are contracting HIV on the job (Jagger, Congressional Testimony, February 7, 1992) one health care worker each week. And thousands are plummeted into the shadows of worrying, deciding whether to embark on AZT therapy and undergoing testing over a period of six months. The cost of managing these needlesticks (counseling, blood testing, treatment) is estimated to be $750 million annually.

In the face of these compelling statistics made real to me by my own seroconversion, I attempt to understand what forces can account for the often fragmented, delayed and obstructionist approaches to health and safety exhibited by our institutions and regulatory agencies. The health care industry is one of the most profitable industries in the nation. When profit is a motive, or resources are limited, institutions have not prioritized worker safety and have generally responded after tragedies occurred. This pattern is consistent with our health care delivery system, which is based on a disease management model rather than a disease prevention model. Our regulatory agencies are supported in their insufficient response by the political climate of antiregulatory sentiment and the resultant underfunding.

The vast majority of health care workers are women and people of color, and our well-being as workers has long been ignored. In addition, the socialization of care givers has taught us to prioritize the interests of the patient and the institution; a sense of self-advocacy is not cultivated or encouraged. Our sense of responsibility toward our patients is often exploited by an institution choosing to abdicate its responsibilities by placing the onus for safety on the worker rather than itself.

WORKERS' COMPENSATION?

What happens to the health care worker after an occupational needlestick transmission? In the wake of my seroconversion, SFGH went on record as upholding my rights to remain employed and to have my confidentiality protected. In pursuing workers' compensation from the City and County of San Francisco, I requested a confidential means of processing my claim that would ensure minimal disclosure of my identity. The Workers' Compensation Division and City Attorney's Office did not agree to such a procedure for 19 months, and did so only after the involvement of my attorney, Service Employees International Union, ACT UP, numerous hospital administrators, the Board of Supervisors, the media, and finally the Mayor. That procedure (disclosure to two high-level personnel only) was later negotiated into the RNs' collective bargaining agreement, putting it into place for all other city RNs.

Workers' compensation will be my sole source of monetary compensation for becoming HIV-positive on the job. Its benefits include financing ongoing medical and mental health costs. In the event of disabling disease, it will provide no more than the cost of state disability as income maintenance. By law, my employer is protected from a suit for damages from my injury. In order to glean an additional 50% in workers' compensation benefits in the form of penalties, I would have to prove serious and willful misconduct on the part of my employer. One other option for the infected health care worker is to bring a product liability suit against the needle manufacturer for the absence of a safer device in the workplace. While the statute of limitations has passed for me, I know of a few health care workers who have won settlements in such cases. I am familiar as well with health care workers who have been obstructed by their employers or employer's insurers in the process of applying for job relocation, workers' compensation and death benefits. Reimbursement of medical costs will be my only true compensation, and in the absence of a dependent at the time of my injury, my death benefit will go to the state of California rather than to my family or a partner.

ZERO RISK?

Who determines what is an "acceptable" rate of needlesticks and at what pace an institution can afford to implement safer devices and needleless systems? Dr. Jagger's preliminary research tells us that at least 85% of needlesticks could be preventable; clearly a rate in excess of 15% of current needlesticks is unacceptable. Federal agencies and health care workers must intensify cooperative efforts to bring the rate to as close to zero as possible.

Health care workers, historically distanced from the health and safety infrastructures of our institutions, bring to the evaluation process the hands-on wisdom of how a device or work practice control may or may not decrease the risk of exposure and may or may not affect patient care. Active participation and leadership by frontline health care providers in the decision-making process about health and safety programs provides a sense of urgency and a more ethical time frame for needed changes. Effective participation will be contingent upon the development of a creative forum for discourse about hazards in the workplace and perceptions of risk, such as educational trainings based on interactive learning rather than brief, formal instruction. Workers and management must learn to talk freely about these issues without a sense of antagonism and cross-purposes. Health care worker involvement should be facilitated on work time or with compensation and at no risk of negative sanctions.

CONCLUSION

Six years after my needlestick, I understand more clearly the personal, institutional and national politics as to how a needle intersected with my life, changing it irrevocably. I have witnessed the creative perseverance of health and safety advocates and union activists in demanding and shaping change from a recalcitrant industry. I have watched infected health care workers speak the truth with courage. I have heard, as well, administrative doublespeak in response to workplace hazards demanding urgent resolution. Sadly, this resistance exists despite the example of what has happened and continues to happen to human lives when health and safety is not prioritized.

What I have learned has provided me with valuable perspective, but it is with great sorrow that I realize that this understanding cannot restore to me and other infected health care workers the tremendous losses that result from this type of injury. My hope is that the lessons wrought from our experiences will be heeded so that further tragedies are prevented. Inherent in these lessons is the imperative to err on the side of caution and precaution, to harness strength and power from knowledge, and to righteously strive to save each others' lives.

Jane Doe, RN

Table of Contents

CHAPTER 1

Overview of Health and Safety in the Health Care Environment

David A. Sterling

INTRODUCTION

In addition to the well documented air quality problems encountered in building structures, both workers and patients in health care settings may be exposed to special environmental contaminant problems. Health care workers, by the very nature of their work, have both unique and common health hazards in the workplace environment. Many of these hazards are obvious while others are difficult to recognize and include physical risk factors and agents such as communicable diseases, exposure to chemical and biological toxins, carcinogens, ionizing and non-ionizing radiation and ergonomic/human factor hazards; as well as psychological risks, stress-induced disorders, chemical dependency, marital dysfunction and suicide. Paradoxically, these same workers are responsible for the health care of others. Many of these health problems may also be brought home and be transmitted to their families. Meanwhile, the cost of health insurance for health care practitioners and their employers in the United States is increasing.

Health care facilities themselves represent a situation where the facility and operational needs to satisfy health care requirements serve as the source of occupational hazards. The health care environment is diverse and includes hospitals of all types, outpatient clinics and other provider services, emergency medical treatment centers (stationary and mobile), dental clinics, pharmacies, testing and research laboratories, patient rehabilitation (physical therapy), as well as veterinary facilities.

Selected examples of health care facility operations that may have a health impact on personnel include operating rooms, patient and treatment areas, sterilization areas, pharmacies and support laboratory facilities.[37] These areas may have sources of contaminants such as: toxic gases and vapors (anesthetic gases—nitrous oxide, ethane, halothane; sterilant gases—ethylene oxide; and antineoplastics); physical hazards (radiation, noise, ergonomic); and infectious microorganisms.[25] Investigations of operating room, sterilization and other specific exposed personnel have shown symptoms ranging from acute effects of fatigue, headaches and skin

1-56670-083-3/94/$0.00+$.50

irritation, to adverse reproductive and cancer outcomes.[1–13,29–30,35,77] Maintenance and deterioration of the facility and equipment can expose non-health care staff to toxic[69] and infectious agents.[14] Disposal of hazardous chemicals, radioactive, and infectious waste is a proven potential health hazard to health care personnel, non-health care staff, the community and the environment.[15,19]

The stress of providing health care also has psychological impacts, which may be associated with increased chemical dependency, lifestyle and family difficulties.[16] High stress situations may also lead to increased accident rates as well as physiological upsets, which may reduce tolerance to other toxic exposures.[16,31,36] Violence toward health care workers is well documented in psychiatric wards, but is becoming more prevalent under other routine health care settings such as the emergency room, pediatric clinics, medical surgery units and long-term care facilities.[78]

An aspect, which in the past has not typically been a concern, although is now often a major consideration, is the building design decisions in new buildings and renovations. Proper venting of work areas such as laboratories,[28,65] operating rooms and pharmacies, as well as specific equipment, is crucial to the health of the general patient population and the employees. Improper design, installation, operation or maintenance may be responsible for insufficient removal or spread of toxic and infectious agents. With ongoing energy conservation and cost-cutting measures, building ventilation and maintenance is often the first to suffer, but may have one of the greatest consequences.[26–27,65]

Energy conservation in buildings, hospitals included, has concentrated on reducing ventilation. Hospitals, because of special potentially hazardous conditions, are even more prone to problems than other buildings.[25] Any energy conservation strategy must seek to guarantee reasonable air quality in hospitals. Other design aspects of growing concern are properly designed, or ergonomically acceptable, work stations and appropriate lighting.[22–23]

The issue of health and safety for health care providers has been growing in not only professional concerns, as shown by the abundance of articles in the professional and scientific journals[2–11,14–15,25–27,29–33,35–37] and books,[16–20,28,34] but also by the increasing regulatory requirements.[21]

Yet, despite the ubiquitousness documentation concerning hazards of work in hospitals, hospital workers, especially those considered well educated, have been identified as one of the workforces ranked with the greatest difference between proportion of workers exposed to workplace hazards and the number perceiving risk to themselves and co-workers.[44] Hospital workers, who have been shown to be highly exposured to various hazards, such as radiologic technicians, have low awareness of their own exposure.[45] The diversity and complexity of hospital exposures may be one reason why awareness of hazard is low. Hospital workers may be well informed concerning their specific responsibilities, but may be unaware of other exposures that may be directly related to their work.

Although continuing education and training awareness is a requirement for most workers either under the Occupational Safety and Health Administration (OSHA) Hazard Communication Standard and other specific standards, various health care facilities must comply with, the level and quality is variable and unfortunately, is typically ineffective. Programs are most often brief and infrequent and tend to be routine and without hands-on components. Individuals in technically oriented skills are often, by virtue of their training, considered already educated in the recognition of such hazards, and instruction is passive. In certain hospital services participants may have poor education, be minimally literate and/or speak English as a second language. The utilization of education and training programs not designed or does not address appropriately the problems to specific worker populations is of questionable value.[46]

The early years of hospital administration in the 1900s centered on ways to attract middle class (paying) patients, keeping the census high and the costs low. To this end visual appearances were important and led to health and safety improvements primarily in infectious control by improved housekeeping, use of flooring materials in wards and operating rooms, which were easy to clean and, by chance alone, were aseptic. On the other hand, cost-cutting techniques contrary to health and safety were also promoted, such as: cleaning and re-use of soiled gauze and reduction in surgical instrument use to reduce cleaning and sterilization costs.[47,61] The cost benefit approach to hospital management has not changed, particularly in these times of rapidly rising health costs. Continuing education and training of employees in a manner that would ensure appropriate awareness of hazards and engineering methods of control are examples of numerous areas that have suffered.

SELECTED HAZARDS IN THE HEALTH CARE PROFESSION

Health care workers have traditionally been viewed as accepting of certain risks as part of patient care. The worker compensation acts, which all states have, were to eliminate the concept of acceptance of risk as part of employment, and because the health care services are thought of as part of the service industry, health care is considered less hazardous than the manufacturing industry. In contrast, the incidence of lost workdays per 100 full-time workers per year for health service workers in 1987 was 66.7, compared with 45.8 for all service workers combined.[24] In addition, the Centers for Disease Control reports that among the ten highest statistically significant proportionate mortality ratios (PMRs) by occupation, industry and cause of death are: malignant lymphomas (PMR of 216 with $p<0.01$) for males over 20 years of age employed in health diagnosis and treatment occupations; and malignant melanoma of the skin (PMR of 188 with $p<0.05$) for females over 20 years of age employed in health service occupations.[40]

The highest incidents of injuries are reported in custodial/housekeeping personnel,[67] followed by food services/nutrition employees, nurses and laboratory technicians, as compared to all job classifications. Needlestick injuries, typically by used needles, were the most prevalent type of injury followed by: strains and sprains, over half of which are related to the back from lifting and twisting; lacerations; and contusions. Chemical and biologic exposures are also a large component of reported injuries.[38–39,49]

Needlestick injuries expose personnel to bloodborne pathogens and other body fluids with the increased potential for such serious infections as Human Immunodeficiency Virus (HIV), hepatitis C virus (HCV), hepatitis B virus (HBV), and others. It is also estimated that there may be underreporting of injuries ranging from 40 to 60%.[39,49,62,66,73]

Housekeeping and custodial personnel are routinely exposed to cleaning solvents and disinfectant agents of industrial strength, as well as refuse that may contain biologically contaminated items such as needles, gauze and broken glass improperly disposed. Because of the work environment and varied tasks, typical industrial engineering controls are rarely feasible, resulting in regular exposure to dermal irritants as well as inhalation of suspect and known systemic toxins, placing this group at high risk to chemically induced skin diseases and chronic illnesses.

Maintenance workers are also found to have similar exposures as housekeeping and custodial personnel and be at increased risk.[38–39] For example, the simple repair of broken blood pressure machines, which is seldom performed, has been shown to lead to elevated exposure to mercury of maintenance workers and those in the vicinity.

Musculoskeletal injuries, the sprains and strains, rank second among all work-related injuries, with the greatest prevalence, according to the National Institute for Occupational and Safety and Health (NIOSH), among those employed in the health care industry.[63] This is no surprise given the requirements of lifting, pulling, sliding and turning of patients, transfer of patients, moving of equipment and standing for long hours.

Potential exposures and adverse health outcomes of personnel in the operating rooms have been an area of concern and controversy. Although exposure to anesthetics has been recognized for years, the methods of control can be costly, and exposure to anesthetics cannot be regarded as a problem solved.[34] Percutaneous injuries during surgical procedures occur regularly and increase the risk of infection to surgical personnel.[73] Electrocautery surgery produces aerosols and smoke, which may be mutagenic,[74] as well as blood-containing aerosols,[75] and pose a potential respiratory infection hazard to operating room personnel. The electrosurgical devices, as well as other operating room equipment, produce microwave and other non-ionizing radiation, which may be of concern, particularly for ocular exposure to surgeons.[70]

Antineoplastics, although of therapeutic value to patients, have mutagenic, carcinogenic and teratogenic potential. Health care personnel such as pharmacists, physicians, nurses and others handling these drugs have been shown to be potentially at risk.[71–72,81] Often used laminar flow hoods are more useful in protecting the integrity of the drug than personnel. Antineoplastics have been measured within rooms outside of the hoods where they are handled,[71,80] posing a risk to the users and other occupants in the room.

Elevated levels of noise capable of producing noise-induced hearing loss are common in large hospital facilities. Noise levels over 80 dBA and up to 110 dBA have been measured in food preparation areas, medical laboratories, mechanical and power plant rooms, medical records offices, floor nursing units, print shops and maintenance areas.[76]

Use of high technology medical devices for diagnosis and treatment is common. These devices, depending on the conditions of use, have potential health risks associated. For example, the use of magnetic resonance imaging (MRI) as a diagnostic procedure is increasing in use. MRI devices produce magnetic fields, radio frequency radiation, noise and vibration. Although at this time there is no evidence on adverse health effects of technicians associated with the proper use of MRI devices, it has been suggested that surveillance continue with the continued increase use.[68]

GOVERNMENT AND PROFESSIONAL ASSOCIATION IMPACT ON THE HEALTH CARE PROFESSION

Government regulatory agencies are in a large part responsible for the manner in which health care is managed and priorities are set.[47] Acceptance of federal funds allowed for acquisition and use of state-of-the-art diagnostic equipment, financing for uninsured patient care and other services. However, also acquired, were regulatory standards with which to comply, ranging from burdensome documentation and reporting paperwork, to requirements associated with building construction and equipment specifications for hospitals and medical facilities.[48] Selected governmental agencies and standards are discussed below.

Governmental Agencies

Diversity of the health care environment brings it under the compliance requirements of many governmental regulatory agencies. The primary agency responsible for employee health and safety is the Occupational Safety and Health Administration (OSHA), established in 1970 in the U.S. Department of Labor. OSHA is responsible for the promulgation and enforcement of workplace safety and health standards. Almost all of the OSHA standards are not workplace specific, but apply to all workplaces, except where exempt, to protect the majority of healthy workers. Examples of relevant standards associated with health care include: the list of permissible exposure levels (PELs) to gases, vapors and particles, as well as specific requirements for ethylene oxide, mercury, formaldehyde, noise, bloodborne pathogens, hazard communication and laboratory standard.

Probably the OSHA standard that has had the greatest impact in all work places has been the OSHA Hazard Communication Standard (HCS), which came into full effect as of March 1989. Among the requirements, employers are responsible for performing inventory and labeling all hazardous chemicals, maintaining, reviewing and updating Material Safety Data Sheets of all hazardous chemicals, and training/educating all employees as to the hazards, health effects and protective measures for all hazardous chemicals they may contact. The requirement alone of educating all employees, when performed appropriately, has brought forth both an increase in awareness that many workers did not have of the hazards associated with their work, as well as a decrease in anxiety over work conditions wrongly perceived as hazardous due to lack of information.

A related standard that also applies to most hospital facilities is the Laboratory Standard, which was specifically designed to deal with the use of small quantities of multiple chemicals and procedures that are not part of a production process. In the hospital environment this would apply to the medical laboratory diagnostic support facilities, pharmacy and all chemical research activities.

The Bloodborne Pathogen Standard is the first standard OSHA has promulgated specifically for the protection of health care workers. Compliance with this standard became effective as of March 1992. OSHA's Bloodborne Pathogen requirements are based on the guidelines from the Centers for Disease Control (CDC), which were first available in 1982 and consolidated with recommended practices for universal precautions in 1987.[49–50]

Discharges to the environment, such as through incineration, wastewater discharge, hazardous and biological waste disposal and fugitive (uncontrolled) emissions are regulated under standards by the U.S. Environmental Protection Agency (EPA). The Resource, Conservation and Recovery Act (RCRA) is administered by the EPA and is basically a permit system requiring inventorying and tracking of all hazardous chemicals from the time they are considered waste products through their storage, treatment and disposal. The OSHA Hazard Communication Standard and Laboratory Standard requiring hazardous chemical documentation and worker education, along with the relevant EPA standards such as RCRA, have been instrumental in the recognition of potentially hazardous substances and situations. Regulation is the first step in hazard control.

The CDC is charged with the surveillance and investigation of infectious disease outbreaks in hospitals, and as such, the regulations are mostly in the form of reporting requirements. The CDC is also responsible for preparing recommendations and guidelines for the control of infectious disease.[51] The National Institute for Occupational Safety and Health (NIOSH) is a similar agency within the U.S. Department of Health and Human Services (USDHHS). This agency conducts research and investigates workplace hazards toward the preparation of recommendations and guidelines for their evaluation and control.[52]

The Nuclear Regulatory Commission (NRC) regulates sources and isotopes that produce ionizing radiation. These impact on areas such as diagnostic X-rays, nuclear medicine and research utilizing radioactive sources. The Federal Drug Agency (FDA) impacts in the hospital environment through regulation of drug products, medical devices, on food services handling and certain food products related to infectious disease.

State and local governments are typically involved with the enforcement of federal regulations through equivalent state programs, such as a state OSHA, state and local health departments and other state and local agencies with varying titles in areas such as: radiation, infectious disease control, biological and hazardous waste and food handling. State and local governments may adopt other agency and association guidelines and standards and/or generate their own in regards to regulatory standards and licensing requirements.[52,60]

Hospital Accreditation and the JCAHO

Accreditation is a voluntary procedure and, on its own, it is an overhead few administrators would put forth the time and expense to comply with. With the passage of the Medicare Act in 1965, accreditation by the Joint Commission on Accreditation of Hospitals (JCAH) (known as the Joint Commission on Accreditation of Healthcare Organizations, JCAHO, as of 1987), among others, was promoted as a means of compliance with the Medicare and Medicaid program reimbursement requirements.[53–57] In addition, many state licensing programs accept accreditation by the JCAHO for hospital certification. Insurance companies have also added incentive toward accreditation. Certain insurers require accreditation for reimbursement certification requirements and offer discounts for physician, and other staff, for malpractice, liability and other hazard insurance.[55]

Requirements for accreditation have been, and still are, however, primarily in support of improving and maintaining the quality of health services that hospitals and other health care organizations provide. Health and safety concerns of the patient are at the forefront, and those of the employees (the health care workers and supportive services personnel), are a secondary concern.

Professional Associations

Professional associations play a large role in the development of recommendations and guidelines, which often are adopted as regulatory requirements and/or accreditation and licensing standards by federal, state and local agencies. For example, sections of the National Fire Protection Association (NFPA) code for safety to life from fire in buildings and structures, and other NFPA codes, have been adopted by OSHA, JCAHO and HRSA for hospitals. The first OSHA PELs were mostly adopted from the ACGIH 1968 TLVs, and much of the OSHA safety standards sections were adopted from National Safety Council (NSC) recommendations. Table 1 contains a short list of selected associations, among many more, which have a relevant impact on hospital and health care worker health and safety.

Unions

Employee unions are not typically given credit for health and safety reform. However, in many cases they have been a driving force through membership on

Table 1. Selected professional associations with health care worker health and safety interest.

American Association of Occupational Health Nurses
American College of Occupational and Environmental Medicine
American College of Physicians
American Conference of Governmental Industrial Hygienists
American Hospital Association
American Industrial Hygiene Association
American Medical Association
American Nursing Association
American Society for Heating, Refrigeration and Air Conditioning Engineers
American Society for Safety Engineers
American Standards for Testing and Materials
National Fire Protection Association
National Safety Council

health and safety committees, through grievances and grievance committees and contract negotiations. The Bloodborne Pathogen Standard was initiated by petitioning of OSHA to issue an emergency temporary standard to protect workers from bloodborne pathogens by several hospital employee unions.[50] Although an emergency standard was never issued, this led ultimately to the revision of the CDC guidelines of 1988 and development of the present OSHA Bloodborne Pathogen Standard. Probably the greatest union activities with the most success have come from working women, which make up approximately 46% of the U.S. workforce and 80% of all hospital employees, represented by such unions at the Service Employees International Union and the Coalition of Labor Union Women, which advocates on behalf of women's issues and rights.[44,58]

GROWTH OF THE HEALTH CARE PROFESSION

The demographics of health care workers are unclear. A 1988 report by the Bureau of Labor Statistics (BLS) reported there to be approximately 4.5 million hospital employees out of 8 million health care workers in the U.S., about 4% of the total U.S. workforce.[41] Another report from NIOSH only gave the number of health care workers in the U.S. to be 6.5 million in 1983.[64] More recent data from the BLS using only Standard Industrial Code (SIC) 806 for hospitals (Table 2) indicate approximately 3.8 million employees in hospitals as of 1993, where 80% of these individuals are women and 92% of all employees serving in non-supervisory positions.[42] This may leave out up to 1 million support employees who fall under different SIC numbers. Also, according to these statistics, in 1981 there were only 2.9 million individuals employed in hospitals. This is an increase of almost one million people over the past ten years employed in hospitals alone.

Table 2. Number of employees in U.S. hospitals for selected years*

Year	Number of Employees ($\times10^3$)	Number of Females Employed ($\times10^3$)	Non-Supervisory Positions ($\times10^3$)	Number in General Med. & Surg. Hospital ($\times10^3$)
1981	2,904.2	2,348.7	2,661.9	2,856.4**
1991	3,655.1	2,956.1	3,352.6	3,359.2
1993 (Feb.)	3,806.7	3,056.2	3,492.8	3,500.2

* Derived from BLS 1993 Employment, Hours, and Earnings for SIC number 806.[42]
** For year 1982.

The American Hospital Association (AHA), using statistics only from member hospitals (Table 3), estimated approximately 4.2 million hospital employees as of 1991, with 22% as registered nurses and 4.6% as licensed practical nurses. A total expenditure for hospital operations of over 250 billion dollars per year for 1991 was estimated.[43] The data from the AHA over the previous four decades show an increase of almost one million additional hospital employees every ten years since 1950. This corresponds with the statistics from the BLS. The hospital expenditures indicate an increase of 30 to 40% each decade. It is expected that the health care workforce in the U.S. may exceed 10 million by the year 2000.[42]

CONCLUSIONS

Health care work and the facilities in which the work is performed pose an adverse health risk to the health care worker and support personnel. These individuals deserve, and should demand, healthy and safe conditions in which to work. Necessarily, because of the type of work performed, flexible controls need to be instituted to achieve and maintain a healthy and safe work environment. The majority of health and safety controls and management procedures observed in the health care profession and facilities are for the protection of the patient, not the worker. In many instances the health and safety practices serve to protect both the patient and worker. However, this is not always the case. For example, surgical masks, intended to protect the patient have been used by the health care workers to protect against airborne aerosols, such as droplet nuclei tuberculosis, for which it is not effective.[84] The CDC and NIOSH have proposed guidelines for more effective respirator use. However, the respirators recommended are not designed for health care worker-patient interaction in terms of size, convenience and looks, and have met resistance.[83] This is one example of the numerous issues, only some of which have been included in this chapter, which must be addressed to properly protect workers in the health care environment.

Table 3. Number of hospitals, employees and total expenses of selected years for member hospitals of the American Hospital Association*

Year	Number of Hospitals	Full Time Equivalent Employees ($\times 10^3$)	Expenses ($\times 10^6$)	Number of RNs	Number of LPNs
1950	6,788	1,058	3,651	---	---
1960	6,876	1,598	8,421	---	---
1970	7,123	2,537	25,556	---	---
1980	6,965	3,492	91,886	---	---
1991	6,634	4,165	258,508	925,947	193,507

*Derived from AHA 1991 Annual Survey of Member Hospitals[43]
---Indicates information is not available

Other issues of concern that need to be addressed as well include the lack of health and safety education and training within the certificate and degree-granting programs for doctors, nurses, medical laboratory technicians and others in the health care environment who will be exposed to potential hazards on a daily basis; the impact of the Americans for Disability Act; and health care reform. These issues need to be addressed on a pro-active, not reactive, basis.

REFERENCES

1. NIOSH: Criteria for a recommended standard: Occupational exposure to waste anesthetic gases and vapors. DHHS(NIOSH), Publ. No. 77-140, Cincinnati, OH. 1977.
2. Knill-Jones R, Newman B, Spence A: Anesthetic practice and pregnancy: Controlled survey of male anesthetists in the United Kingdom. *The Lancet,* 25:1326–1328, 1975.
3. Knill-Jones R, Rodrogues L, Moir D, Spence A: Anesthetic practice and pregnancy: Controlled survey of women anesthetists in the United Kingdom. *The Lancet,* 17:1326–1328, 1972.
4. Layzer R: Myeloneuropathy after prolonged exposure to nitrous oxide. *The Lancet,* 9: 1227–1230, 1978.
5. Haas J, Schottenfeld D: Risk of the offspring from parental occupational exposure. *J of Occupational Medicine,* 21(9):21–31, 1979.
6. Cohen E, et al.: Occupational disease in dentistry and chronic exposure to trace anesthetic gases. *J American Dental Association,* 101:21–31, 1980.
7. Tarmenbaum T, Goldbery R: Exposure to anesthetic gases and reproductive outcomes. *J Occupational Medicine,* 27(9):659–668, 1985.
8. Choi-Lao A: Trace anesthetic vapors in hospital operating room environments. *J Nursing Research,* 30(3):156–161, 1981.
9. Vainio H: Inhalation anesthetics, anticancer drugs and sterilants as chemical hazards in hospitals. *Scand J Work Environmental Health,* 8:94–107, 1982.

10. Nguyen T, Theiss J, Matney T: Exposure of pharmacy personnel to mutagenic antineoplastic drugs. *Cancer Research,* 42:4792–4796, 1982.
11. Anderson R, et al.: Risk of handling injectable antineoplastic agents. *American J Hospital Pharmacy,* 39:1881–1887, 1982.
12. USDOL(OSHA): Guidelines for cytotoxic (antineoplastic) drugs. OSHA Instruction Publ. No. 8-1.1, Washington, DC. 1986.
13. Proceedings 26th Interscience Conference on Antimicrobial Agents and Chemotherapy. *American Society Microbiology.*
14. Goldberg M, et al.: Mercury exposure from the repair of blood pressure machines in medical facilities. *Applied Occupational and Environmental Hygiene,* 5(9):604, 1990.
15. Block S, Netherton J: Infectious hospital wastes: Their treatment and sanitary disposal. In *Disinfection, Sterilization, and Preservation,* 2nd edition. S. Block Editor, Philadelphia, PA: Lea & Febiger. 1977.
16. Pelleier K: *Healthy People in Unhealthy Places: Stress and Fitness at Work.* Dell Publishing Company, Inc., New York, NY. 1985.
17. Brune D, Edling C (Eds.): *Occupational Hazards in the Health Profession.* CRC Press, Inc., Boca Raton, FL. 1989.
18. Charney W, Shirmer J (Eds.): *Essentials of Modern Hospital Safety. Volume 1.* Lewis Publishers, Boca Raton, FL. 1990.
19. Reinhardt P, Gordan J: *Infectious and Medical Waste Management.* Lewis Publishers, Boca Raton, FL. 1990.
20. Lewy R: *Employees at Risk: Protection and Health of the Health Care Worker.* Van Nostrand Reinhold, Florence, KY. 1990.
21. 29CFR 1910. 1450: Occupational exposure to hazardous chemicals in laboratories. DOL(OSHA). Federal Register, 55(21):3327–3335, 1990.
22. 2nd International Conference on AIDS, Paris, France, June 23–25, 1986.
23. 3rd International Conference on AIDS, Paris, France, June 1–5, 1987.
24. BLS: Occupational injuries and illnesses in US by industry, 1987. USDOL, Bureau of Labor Statistics, Bulletin No. 2328, Washington, DC. 1989.
25. Sterling E, Sterling D: Air quality in hospitals and health care facilities. Proceedings of the 3rd International Conference on Indoor Air Quality, Stockholm, Sweden, August 20–24, 5:209–213, 1984.
26. Bleckman J, Albrecht R, Bertz E: Hospital air quality. Proceedings of the 3rd International Conference on Indoor Air Quality, Stockholm, Sweden, August 20–24, 5:215–220, 1984.
27. Sterling D, Clark C, Bjornson S: The effects of air control systems on the indoor distribution of viable particles. *Environment International,* 8:559–571, 1982.
28. USDHHS: *Biosafety in Microbiological and Biomedical Laboratories,* 2nd ed. US Department of Health and Human Services (CDC/NIH), Washington, DC. 1988.
29. Rubin R, et al.: Neurobehavioural effects of the on-call experience in housestaff physicians. *J Occupational Medicine,* 33(1):13–18, 1991.
30. Sarri C, Eng E, Rungan C: Injuries among medical laboratory housekeeping staff: Incidence and workers perception. *J Occupational Medicine,* 33(1):52–56, 1991.
31. Gauch R, Feeney K, Brown J: Attitudes and behaviours of medical technologists as a result of AIDS. *J Occupational Medicine,* 33(10):74–79, 1991.
32. Putz V, Anderson (Eds.): *Cumulative trauma disorders: A manual for musculoskeletal disease of the upper limbs.* Philadelphia, PA: Taylor & Francis, Inc. 1988.
33. Rodgers S (Ed.): *Ergonomic Design for People at Work,* volumes 1 and 2. New York, NY: Van Nostrand Reinhold. 1986.
34. Halsey M: Occupational health and pollution from anesthetics. *Anaesthesia,* 46:486–488, 1991.

35. Keleher K: Occupational health: How work environment can affect reproductive capacity and outcome. *The Nurse Practitioner J,* 16(1):23–34, 1991.
36. Sechrist S, Frazer G: Identification and ranking of stressors in Nuclear Medicine Technology. *J Nuclear Medicine,* 18(1):44–48, 1990.
37. Schwartz J, et al.: The risk of radiation exposure to laboratory personnel. *J Laboratory Medicine,* 22(2):114–119, 1991.
38. Weaver V, McDiarmid M, Guidera J, Humphrey F, Schaefer J: Occupational chemical exposures in an academic medical center. *J Occupational Medicine,* 35(7):701–705, 1993.
39. Wilkinson W, Salazar M, Uhl J, Koepsell T, Dekoos R, Long R: Occupational injuries: A study of health care workers at a Northwestern health science center and teaching hospital. *American Association of Occupational Health Nurses J,* 40(6):287–293, 1992.
40. NCHS: Monthly vital statistics report, 42(4). Final draft from the US Department of Health and Human Services, Center for Disease Control and Prevention, and National Center for Health Statistics. 1993.
41. BLS: Employment and earnings, 35(3). US Department of Labor, Bureau of Labor Statistics, Office of Employment and Unemployment Statistics, Washington, DC. 1988.
42. BLS: Employment, hours, and earnings, United States, 1981–93: March 1992 benchmark revisions and historical corrections. US Department of Labor, Bureau of Labor Statistics, Washington, DC. 1993.
43. AHA: American Hospital Association hospital health statistics. AHA 1991 Annual Survey of Hospitals, Chicago, IL. 1993.
44. Sexton P: *The new nightingales: Hospital workers, unions, new women issues.* New York, NY: Enquiry Press. 1982.
45. Behrens V, Brackbill R: Worker awareness of exposure: Industry and occupations with low awareness. *American J Industrial Medicine,* 23:695–701, 1993.
46. LaMontagne A, Kelsey K, Ryan C, Christiani D: A participatory workplace health and safety training program for ethylene oxide. *American J Industrial Medicine,* 22:651–664, 1992.
47. Granshaw L, Porter R (Eds.): Managing medicine: Creating a profession of hospital administration in the U.S., 1895–1915, by Vogel M. In, *The Hospital in History,* New York, NY: Routledge. 1989.
48. HRA: Minimum requirements of construction and equipment for hospitals and medical facilities. US Department of Health and Human Services, Publication No: 79-14500, Hyattsville, MD. 1979.
49. Behling D, Gay J: Industry Profile: Hazards of the healthcare profession. *Occupational Health and Safety,* 62(2):54–57, 1993.
50. Goldstein L, Johnson S: OSHA Bloodborne pathogen standard: Implications for the occupational health nurse. *American Association Occupational Health Nurses,* 39(4): 182–188, 1991.
51. CDC: Universal precautions for prevention of transmission of human immunodeficiency virus, hepatitis B virus, and other bloodborne pathogens in health care settings. *Morbidity and Mortality Weekly Review,* Center for Disease Control, Washington DC. pp. 37: 377–382, 387–388, 1988.
52. NIOSH: *Guidelines for protecting the safety and health of health care workers.* US Department of Health and Human Services (NIOSH), Publication No. 88-119, 1988.
53. Martin F: *Fifty years of medicine and surgery: An autobiographical sketch,* Chicago, IL: Lakeside Press. pp 338, 1934.
54. Roberts J, et al.: A history of the Joint Commission on Accreditation of Hospitals. *J American Medical Association,* 258:936–940, 1987.

55. Yodaiken R, Zeitz P: Accreditation policies in occupational health care. *J Occupational Medicine,* 35(6):562–567, 1993.
56. Lynch J, Pendergrass: Occupational health and safety program accreditation commission. *American Industrial Hygiene Association J,* 7:387–390, 1976.
57. JCAHO: Accreditation manual for hospitals, Volume I, Standards. Joint Commission on Accreditation of Healthcare Organizations, 645 N. Michigan Ave, Chicago, IL. 1992.
58. Stellman J, Stellman S, et al.: The role of the union health and safety committee in evaluating the health hazards of hospital workers: a case study. *Preventive Medicine,* 7(3):332–337, 1978.
59. Hanson K: Occupational dermatosis in hospital cleaning women. *Contact Dermatosis,* 9:343–351, 1983.
60. Monagle J: *Risk Management: A guide for healthcare professionals.* Rockville, MA: Aspen Systems Corp. 1985.
61. Vogel M: *The Invention of the Modern Hospital: Boston 1870–1930.* Chicago, IL: The University of Chicago Press. 1980.
62. Sellick J, Hazany P, Mylotte J: Influence of an educational program and mechanical opening needle disposal boxes on occupational needlestick injuries. *Infection Control Epidemiology,* 12(12):725–731, 1991.
63. Allen A: On the job injury: A costly problem. *J Post Anesthesia Nursing,* 5(5):367–368, 1993.
64. Martin L: Prevention of occupational transmission of bloodborne pathogens. Abstract from the American Occupational Health Conference, April 28–30, Atlanta, GA. pp 434, 1993.
65. Burton J: Choice, location of lab fume hood will effect performance, protection. *Occupational Health and Safety,* 61(9):50, 1992.
66. Hoffman K, Weber D, Rutala W: Infection control strategies relevant to employee health. *American Association Occupational Health Nurses,* 39(4):167–181, 1991.
67. Toivanen H, Helin P, Hanninien O: Impact of regular relaxation training and psychosocial working factors on neck-shoulder tensions and absenteeism in hospital cleaners. *Journal Occupational Medicine,* 35(11):1123–1130, 1993.
68. Evans J, et al.: Infertility and pregnancy outcomes among magnetic resonance imaging workers. *Journal Occupational Medicine,* 35(12):1191–1195, 1993.
69. Goldberg M, et al.: Mercury exposure from the repair of blood pressure machines in medical facilities. *Applied Occupational and Environmental Hygiene Journal,* 5(9):604, 1990.
70. Paz J, et al.: Potential ocular damage from microwave exposure during electrosurgery: Dosimetric survey. *Journal Occupational Medicine,* 29(7):580–583, 1987.
71. McDevitt J, Lees P, McDiarmid M: Exposure of hospital pharmacists and nurses to antineoplastic agents. *Journal Occupational Medicine,* 35(1):57–60, 1993.
72. McDiarmid M, Garley H, Arrington D: Pharmaceuticals as hospital hazards: Managing the risk. *Journal Occupational Medicine,* 33(2):155–158, 1991.
73. Tokars J, et al.: Percutaneous injuries during surgical procedures. *Journal American Medical Association,* 267(21):2899–2904, 1992.
74. Gatti J, et al.: The mutagenicity of electrocautery smoke. *Plastic and Reconstructive Surgery,* 89(5):781–786, 1992.
75. Heinsohn P, Jewett D: Exposure to blood-containing aerosols in the operating room: A preliminary study. *American Industrial Hygiene Association Journal,* 54(8):446–453, 1993.
76. Yassi A, Gaborieau D, Gillespie I, Elias J: The noise hazard in a large health care facility. *Journal Occupational Medicine,* 33(10):1067–1070, 1991.

77. Schulte P, et al.: Biologic markers in hospital workers exposed to low levels of ethylene oxide. *Mutation Research,* 278:237–251, 1992.
78. Lipsomb J, Love C: Violence toward health care workers. *American Association Occupation Health Nursing Journal,* 40(5):219–227, 1992.
79. Saurel-Cubizolles M, et al.: Neuropsychological symptoms and occupational exposure to anesthetics. *British Journal Industrial Medicine,* 49:276–281, 1992.
80. Harlow V: Occupational exposure to antineoplastics. Thesis for Master of Science in Community Health, emphasis in Industrial Hygiene, Old Dominion University, Norfolk, VA. May, 1990.
81. Balanis B, et al.: Antineoplastic drug handling protection after OSHA guidelines: Comparison by profession, handling activity, and work site. *Journal Occupational Medicine,* 34(2):149–155, 1992.
82. Harber P, Hsu P, Fedoruk M: Personal risk assessment under the Americans with disability act. *Journal Occupational Medicine,* 35(10):1000–1010, 1993.
83. Nelson H: Objections to protective respirators. *Lancet,* 340(8827):1088, 1992.
84. Charney W: The inefficiency of surgical masks for protection against droplet nuclei tuberculosis. *Journal Occupational Medicine,* 33(9):943–944, 1991.

CHAPTER 2

Epidemiological Analysis of Occupational Injury and Illness in Hospital and Health Care Workers, 1980–1990

David Makofsky

OVERVIEW

It is possible now to get some insight on the current (1980–1990) epidemiological patterns of occupational health and safety as they apply to hospital and health care workers. The data are in—the discussion concerns the interpretation.

Issues of occupational injury and illness are embedded in the history of the American labor movement. There have been volumes written about miner's black lung disease, lung impairment due to asbestos exposure among shipyard workers, mercury poisoning in the nineteenth century hat industry and loss of life and limb in logging, construction and agricultural work. Industrial accidents such as the Triangle Shirtwaist fire in New York City at the turn of the twentieth century played a major role in the development of protective legislation.

The effort here is focused entirely on the two-digit SIC code for health care workers (80) and the three-digit SIC code for hospital workers (806). For federal and state reporting standards, hospital work is simply a subdivision of health care work, and it is important to put hospital work in the context of the more general health care environment.

First we will look on the national level, where the data are more comprehensive, but subject to different reporting standards. Then we will look at the California data, where the reporting is relatively uniform, and the categories of occupational injury are broken out in more detail.

On both a national and state (California) level (see Tables 1 and 3), rates of occupational illness and injury have been rising for the past decade. There are some that feel that this is a function of legal acknowledgment of compensation: now that there is widespread recognition of the availability of disability benefits, there are many more claims. Some feel that the increase in rates is a function of better surveillance and reporting: now that competent surveillance systems are in place, more of these injuries are being recorded.

1-56670-083-3/94/$0.00+$.50

Table 1. U.S. Total: Incidence rates of occupational injury and illness per 100 full-time workers.

	1980	1981	1982	1983	1984	1985	1986	1987	1988	1989	1990
Health care workers, private sector (SIC 80)	6.4	6.1		6.3	6.3		6.6	7.2	7.3	7.3	8.4
Health care workers, private nursing homes (SIC 806)	10.7	10.5					13.5	14.2	15	15.5	15.6
Health care workers, private hospitals (SIC 806)	7.9	7.2		7.4	7.3		7.8	8.5	8.7	8.5	10.6
Health care workers, private sector (SIC 80), employment (000)		5555.1			6104.1			6827.8			7844
Health care workers, private sector (SIC 806), employment (000)		1029.4			2993.5			3153.9			3547

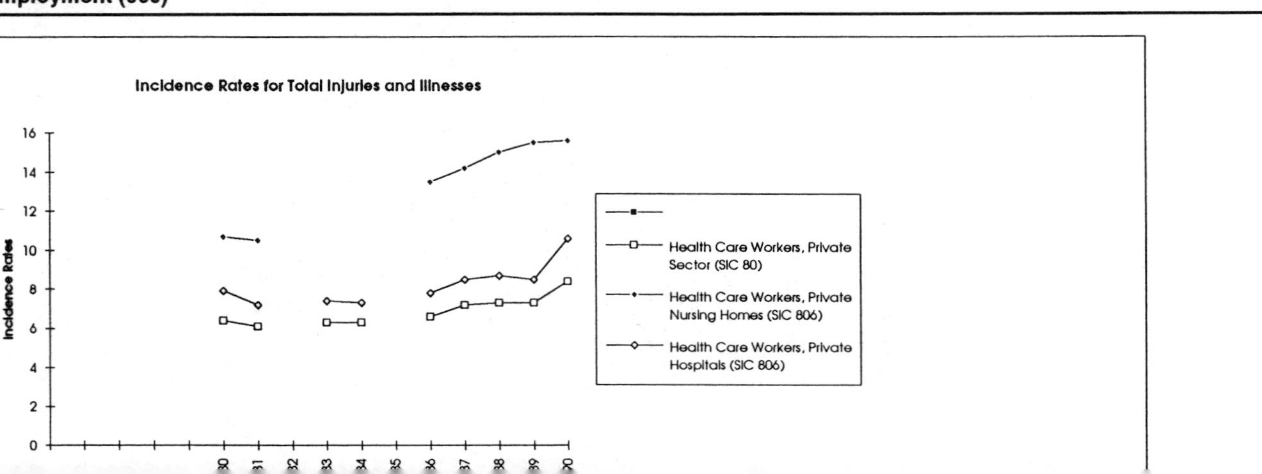

Although both of the preceding statements are true, it is also evident that there are other underlying forces at work. There are increasing occupational injury and illness rates among highly paid professionals (physicians, surgeons), who are certainly not doing this to claim workers' compensation benefits. There are increasing rates of fatal occupational injuries and illness, and these are not cases of fraudulent claims or better reporting. The work is more dangerous, and working conditions may be inadequate to meet new threats to worker health and safety. A work environment that was relatively safe prior to the AIDS and tuberculosis epidemic is now inadequate.

Finally, we will review the data on the most common causes of illness and injury in the health care and hospital workplace.

INCREASING OCCUPATIONAL INJURY AND ILLNESS RATES: THE NATIONAL DATA[1]

The national data on occupational injury and illness are summarized in Table 1. Incidence rates are presented for occupational injury* and illness**, 1980–1990. The figures are annual incidence rates, that is, they represent the number of new cases controlling for the number of hours of employed labor***.

The data are presented here for the SIC two-digit industrial division: 80 represents all health care workers, and for the SIC three-digit more specific industrial levels, 806 represents all hospital workers.

The national data indicate that occupational injury and illness rates have been increasing for health care workers, and hospital workers have even higher rates. In the 1990 Overview of the Department of Labor Report,[1] it is noted that more cases were reported in hospitals than in manufacturing sites, which have historically been considered more dangerous in the country. Among the three-digit SIC

---

*From the Overview of Reference 1, supra: An occupational injury is one that results from a work-related event or from a single instantaneous explosion in the work environment. Injuries are reported if they result in death, lost work time, medical treatment other than first aid, loss of consciousness, restriction of work or motion or transfer to another job.

**From the Overview of Reference 1: Occupational illness is any abnormal condition or disorder other than one resulting from an occupational injury, caused by exposure to factors associated with occupational employment. It includes acute and chronic illness or disease that may be caused by inhalation, absorption, ingestion or direct contact. Typical examples include occupational skin diseases, dust diseases of the lungs, respiratory conditions due to toxic agents, poisoning (systematic effects of toxic material) and disorder due to physical agents (heat stroke, freezing, ionizing and non-ionizing radiation, disorder associated with repeated trauma).

***Specifically, incidence rates represent the number of lost workday injuries and/or illnesses per 100 full-time workers. Incidence rates are calculated as $N/EH \times 200{,}000$ where N = the number of lost workday injuries or lost workdays; EH = total hours worked by all employees during the calendar year; 200,000 = the base for 100 full-time workers, working 40 hours per week, 50 weeks per year.

Table 2. Disabling nonfatal work injuries and illnesses under workers' compensation, California, 1991.

	State Health Services	Local Hospitals
Total	3668	4360
Amputations	—	—
Burns and scalds	24	44
Contusions, crushing	512	360
Cuts, punctures	176	180
Abrasions, scratches	36	88
Fractures	80	108
Strains, sprains	2064	2556
Other injuries	20	56
Occupational illnesses	420	704
Nature not stated	336	28

code designations, hospitals, with 282,400 injuries, were second only to eating and drinking establishments, with 353,600 injuries. Nursing and personal care facilities workers (also within the 80 SIC category) were in fifth place, with 168,200 injuries. These figures reflect the relative decline of the importance of manufacturing and the rise of health care in the economy. The incidence rates for manufacturing are much higher than for health care, but the absolute numbers are cited to demonstrate the relative importance of the problem.

Based on data reported to the Department of Labor, the incidence rate for 1990 was 32% higher for health care workers (SIC 80) than it was in 1980, and 34% higher for hospital workers (SIC 806) in 1990 when compared to 1980.

For a more technically precise analysis: when a linear time series regression is created for incidence rates for the decade, the slope of the regression line (B) is positive and significantly greater than zero. The average incidence rate was 6.87 cases of occupational injury or illness for every one hundred full-time health care workers. The rate of increase was 1.82/1000 cases for one thousand full-time health care workers.

EXPLAINING THE INCREASING RATES

One cannot dispute the fact that occupational injury and illness rates for health care workers have increased. The major issue deals with the analysis of "cause." As professionals in the field, we support the first and second of the following explanations, (1) even though many workplace problems have been "solved," (2) more serious ones have emerged.

Table 3. Incidence rates for total occupational injuries and illness, California, per 100 full-time workers.

	1975	1976	1977	1978	1979	1980	1981	1982
Health care workers, private sector (SIC 80)	8	8.8	9.4	9	9.1	9	8.6	8.3
Health care workers, private nursing homes (SIC 806)	15.2	19						
Health care workers, private hospitals (SIC 806)	10.5	10.3						
Health care workers, state setting (SIC 80)			11.9	11.9	11.5	10	10.5	9
Health care workers, county and local (SIC 806)			10.8	10.7	10.6	10	10.1	12.1

Table 3. Continued.

	1983	1984	1985	1986	1987	1988	1989	1990
Health care workers, private sector (SIC 80)	8.3	8.7	9.3	9.1	9.1	9	8.6	8.9
Health care workers, private nursing homes (SIC 806)								
Health care workers, private hospitals (SIC 806)								
Health care workers, state setting (SIC 80)	8.8	9.7	11	14.4	13.5			
Health care workers, county and local (SIC 806)	11.2	11.2	12.7	14.8	14.9			

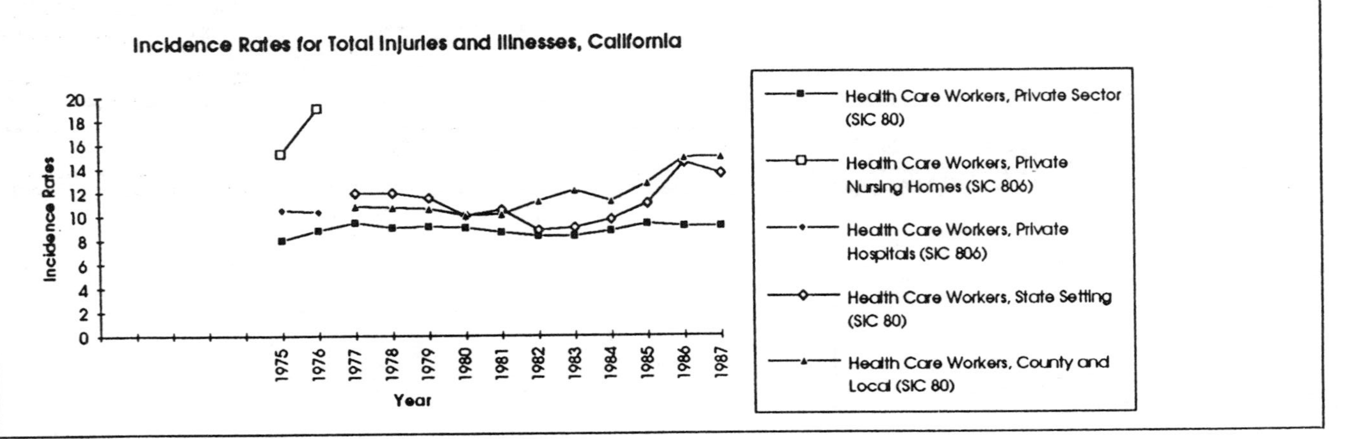

This does not mean that there should not be some room for other points of view. Surveillance systems for injury reporting have most certainly been improved (explanation number three). This simply means that occupational injury and illness rates have always been as they are at the current time, and that there were no "good old days." Finally, many critics assert that some of this increase can be explained by an increase in fraudulent claims. An example: in a discussion of the issues in the local police department, a reporter for the San Jose Mercury News mentioned that over 80% of the police force had retired with disability benefits.[2] While we are in no position to challenge or verify that fact, it is one more instance of the "public's belief that fraud is rampant." Law enforcement is historically a field with many occupational injuries. Can any cynic devise a means to verify what is the true proportion of disabled police officers and what is the true proportion of fraudulent claims? Does anyone have a way of measuring "fraud," a charge that throws suspicion on all legitimate claims? In any case, there are four explanations for this rise:

1. Health care institutions, and hospitals in particular, have not kept pace with job-related threats to the health and safety of their employees.
2. A rapid expansion of the health care and hospital work force and the introduction of potentially dangerous new machinery and toxic substances have made health care workers more susceptible to injury and illness than they have been in the past.

   We will develop the analysis of needlestick injuries at greater length, but some comments should be made at this point. A decade ago, needlestick injuries to physicians, registered nurses, laboratory technicians and phlebotomists occurred on a regular basis, and barely any attention was paid to this occupational injury. At the current time, needlestick injuries are one of the most common types of exposures resulting in the documented transmission of both human immunodeficiency virus (HIV) and hepatitis B virus (HBV) among health care workers.[3] Of the documented cases of occupationally transmitted HIV, 87% involved percutaneous exposure, cuts and needle injuries. The largest occupational groupings for the 101 cases reported as of 1992 were nurses (26 cases), clinical laboratory technicians (23) and physicians (13).[4] According to the Centers for Disease Control's (CDC) surveillance of health care workers exposed to blood from patients known to be seropositive, risk of HIV seroconversion is estimated at 0.5%, or 1 out of every 200 needlesticks.[5]
3. The rise in injury and illness rates is due to increased awareness and better reporting

   A major epidemiological study of sharps injuries in hospitals was undertaken as part of the effort against occupational HIV and hepatitis B infection. The authors note:[6] "Despite greatly increased institutional efforts to prevent sharps injuries, the annual incidence has increased more than threefold (60.4 to 187.0 per 1000 full-time health care workers), reflecting better reporting and increased exposure.
4. Protective legislation has lead to an increase in fraudulent claims.

   Summarizing the situation,[7] an author (Leavitt) notes: "Studies of the industrially injured worker invariably suggest that compensation reinforces pain and prolongs disability." The researcher in this case is attempting to demonstrate that this conclusion has some hidden flaws, but many in the field acknowledge its underlying validity. Leavitt continues: "Patients on compensation are more likely involved in heavy physical exertion on their jobs... The results suggest that injury on the job operates both independent of the level of physical exertion, as well as in interaction with it."

HEALTH AND SAFETY PROFESSIONALS VIEW THE EVIDENCE

National figures do not indicate the nature of occupational injuries and illnesses suffered by hospital workers. The state of California has broken down this data into nine categories (Table 2). We take a specific and representative injury or illness from each of three major categories and explain why the focus of attention should be on occupational problems rather than on fraud. The focus will be on needlestick injuries (cuts and punctures), tuberculosis and pneumonia (occupational illness) and back injuries (strains, sprains). These subjects will be covered with some attention in this edition, but this brief review offers only an overall context for analysis.

Cuts and Punctures—Needlestick Injuries

In all likelihood, there has always been much underreporting of cuts and puncture wounds. The incidence rates are only now beginning to reflect how many injuries took place. The increased reporting is a function of the fact that the entire health care profession now takes these injuries more seriously, and the injuries themselves are potentially life-threatening.

The first study to describe the epidemiology of needlestick injuries in health care workers was published in 1981 by McCormick.[****] In this first effort, it was found that the housekeeping staff most frequently reported these injuries.

Within a few years, the occupational impact of the AIDS epidemic was felt by a wide variety of health care workers. For example, statistical samples of surgeons,[9] emergency medical technicians[10] and hospital-based nurses and housestaff[11–12] revealed much higher rates of needlestick injury that was previously reported. Some physicians and nurses reported more than five punctures with needles infected with HIV-contaminated blood.

With the aim of avoiding either complacency or hysteria, professionals in the field devised some responses to the problem. Initial preventive efforts involved changing needlestick containers and needle design. Educational programs were initiated. Room design was changed to enhance injury prevention.[13–14] This response involved a great deal of time and effort. In some cases, local criminal

---

[****]Many of these references can be found in Makofsky D, Cone JE. Work Environment and Needle Handling Injuries. pp. 117–134 in Charney W, Schirmer J. *Essentials of Modern Hospital Safety, Volume 2.*, Lewis, 1993. This particular reference is to McCormick RD, Maki DG. (See Reference 8.)

ordinances had to be changed to allow needle containers to be placed in the same room with patients. Educational campaigns involved hiring and retraining personnel. Changes in equipment are cost and budget decisions and are made in the context of cost containment. Those who rely on explanations involving fraud or better injury reporting reveal their ignorance of the full dimension of the problem.

Occupational Respiratory Illness: Tuberculosis and Pneumonia

Two decades ago there would have been no mention in a volume such as this of occupationally transmitted tuberculosis and pneumonia. By 1990 these have become major occupational hazards. One explanation for the rise in incidence rates is the fact that social changes give birth to new illnesses, which then affect health care workers.

Quoting Edward Nardell: "The transmission of tuberculosis from patient to physicians and nurses, and from patient to patient, is a problem as old as the disease itself. *The rise in nosocomial TB transmission is associated with the HIV epidemic.* After years of steady decline, TB cases are increasing in many areas of this and other developed countries, especially among the homeless intravenous drug users, and others with risk factors for HIV and TB co-infection. Among the earliest reports of nosocomial TB transmission in the HIV era is a case that occurred in a South Florida clinic where 17 workers were infected."[15] Although tuberculosis can be treated and cured, it is also more contagious than HIV disease, and potentially affects more health care workers.

A similar case can be made for the threat of occupationally transmitted pneumonia. As Quinlan notes: "Pentamidine...is used as a treatment for patients with *Pneumocystis carinii* (PC). Pentamidine use has increased in recent years in preventive therapy and treatment of PC pneumonia, as there is a high frequency of the occurrence of this organism in HIV-infected individuals. Because of the sense of urgency which accompanied the development of effective treatments of HIV-infected patients, the potential consequences to health care workers who administer the drug in aerosolized form did not receive critical attention until recently."[16]

It is true that the discussion of occupational illness and injury has been on HIV disease and AIDS. The current epidemic has captured public attention to a situation that has existed for a long time: the danger of infectious disease and injury to those who work with those in poor health. This danger may take the form of tuberculosis, pneumonia and AIDS, or hepatitis B or some other infectious epidemic that will strike in the coming decades.

In other fields new threats to occupational health and safety have replaced older problems. In a study of fatal occupational injuries in California, it was discovered that 40% of fatalities were the result of motor vehicle accidents and another 25% were the result of homicide and suicide.[17] Two decades ago, these would not have been considered occupational hazards. The fact that they are recognized as such at the current time reflects the many changes that have taken place in service and industrial organizations, and in society as a whole.

Strains and Sprains: Health Care Workers and Back Injuries

The data in Table 2 leave no doubt that the most frequently occurring occupational injuries are sprains and strains. Although there is no way to directly tie these injuries to lack of employee training, there are some indications that this may be the cause. Health care and hospital employment increased greatly over the decade.

Writing about the problem of back injury, Charney, Zimmerman and Walara note: "Occupational low back injury continues to be a significant problem in nursing. These injuries are the most expensive in terms of workers' compensation claims, and in many hospitals low back injury accounts for the greatest amount of lost work time. One author states that 12% of the nurses have decided to leave nursing because of low back pain."[17]

The authors believe that the cause of these injuries is a lack of ergonomic prevention. There has been little attention paid to ergonomic design in every area of the hospital from the laundry to patient intake. With respect to back injuries, untrained personnel engage in uncoordinated lifts, and orderlies lifting with nurses create height and weight differentials. Typically, there is no use of mechanical lifting devices.

At the hospital under study, the net result is that the injuries cost the compensation system in excess of $200,000 per year, and that the injuries place a burden on the entire hospital. Temporary hires are needed for replacement personnel, overtime is paid to others who absorb the duties of injured workers and damage to equipment and legal costs for claims must also be absorbed by the system. The massive expansion of the health care system and the ergonomic deficiencies may be major contributors to the rise in incidence rates.

CONCLUSION

The information collected by the Bureau of Labor Statistics offers only a glimpse into the area of occupational injury and illness. Our protective agencies, which are only one or two decades old, have not yet effectively dealt with many of the occupational problems. Issues such as fatigue due to shift work have scarcely been addressed in the research literature. The Workers' Compensation program was designed to offer financial assistance for manual workers in industrial settings, and is not adequate to compensate nurses or physicians with occupationally transmitted AIDS. It makes no sense to wait for the courts to process these problems, as was done in the case of asbestos poisoning and exposed construction workers. The only professional response is to take these problems seriously, and suggest a preventive strategy.

REFERENCES

1. U.S. Department of Labor, Bureau of Labor Statistics. Occupational Injuries and Illnesses in the United States by Industry.
2. San Jose Mercury News, January 2, 1994, Magazine Section.
3. Department of Labor, Occupational Safety and Health Administration. Occupational exposure to bloodborne pathogens; proposed rule and notice of hearing. Federal Register, 54:23041–23139, 1989.
4. Ibid.
5. Centers for Disease Control, Guidelines for prevention of transmission of Human Immunodeficiency Virus and Hepatitis B Virus to health-care and public safety workers. *Morbidity and Mortality Weekly Report,* 38(S6):1-X, 1989.
6. McCormick RD, Meuch MG, Irunk IG, Maki DG: Epidemiology of hospital sharps injuries: a 14 year prospective study in the pre-AIDS and AIDS eras. *American Journal of Medicine,* (3B):301S–307S, Sept 16, 1991.
7. Leavitt F: The physical exertion factor in compensable work injuries: a hidden flaw in previous research. *Spine,* 17(3):307–310, March, 1992.
8. McCormick RD, Maki DG: Epidemiology of needlestick injuries in hospital personnel. *American Journal of Medicine,* 70:378–388, 1981.
9. Porteous MJ: Operating practices and precautions taken by orthopaedic surgeons to avoid infection with HIV and Hepatitis B virus during surgery. *British Journal of Medicine,* 301:167–169, 1990.
10. Tandberg D, Stewart KK, Doesema D: Underreporting of contaminated needlestick injuries in emergency health care workers. *Annals of Emergency Medicine,* 206:66–70, 1991.
11. Mangione CM, Gerberding JL, Cummings SR: Occupational exposure to HIV and rates of underreporting of percutaneous and mucocutaneous exposures by medical housestaff. *American Journal of Medicine* 90:85–90, 1991.
12. Cone J, Makofsky D, Mueller K, Ficarrotto T: Needle recapping as preventable accidents: estimations of rates. American Public Health Association Annual Meeting, 1990.
13. Ribner BS, Ribner BS: An effective educational program to reduce the frequency of needle recapping. *Infection Control and Hospital Epidemiology,* 11:635–638, 1988.
14. Makofsky D, Cone JE: Installing needle disposal boxes closer to bedside reduces needle recapping rates in hospital units. *Infection Control and Hospital Epidemiology,* 14:3, 140–144, 1993.
15. Nardell EA: Nosocomial tuberculosis transmission: biological, behavioral, and environmental factors. *Essentials of Modern Hospital Safety, Volume 2.* Charney W, Schirmer J (Eds.), pp. 1–9, 1993.
16. Quinlan P: Aerosolized pentamidine exposures in health care workers. *Essentials of Modern Hospital Safety, Volume 2,* Charney W, Schirmer J (Eds.), pp. 219–230, 1993.
17. Cone JE, Daponte A, Makofsky D, Reiter R, Becker C, Harrison R, Balmes J: Fatal injuries at work in California. *Journal of Occupational Medicine,* 33:813–818, July, 1991.
18. Charney W, Zimmerman K, Walara, E: A design method to reduce lost time back injury in nursing. *Essentials of Modern Hospital Safety, Volume 2,* Charney W, Schirmer J (Eds.), pp. 313–322, 1993.

CHAPTER 3

Violence in the Health Care Industry

INTRODUCTION

A study by the International Association for Hospital Security indicates that in 1990 virtually every hospital will experience on-site violence. One in four will experience an arson fire. One in five will experience an armed robbery, an on-site rape or an on-site homicide.

Two disturbing future trends are being noted. The first is an acceleration of lawsuits charging that health care facilities have failed to maintain a safe premises. The second trend is toward increased kidnappings, extortion and terrorism at health care facilities with prominent physicians and principals of hospitals becoming targets of threats and requiring protection.

Pressures That Cannot Be Overlooked

Even if the administrator of a health care facility were tempted to ignore their moral and ethical responsibilities to provide a safe premises, there are three overriding sets of pressures that cannot be overlooked—lawsuits, staff moral, and bad publicity.

With hospitals no longer immune from suits by patients, visitors, staff, vendors, or even outsiders, the number of lawsuits for failure to provide adequate security, as well as the size of the awards or settlements, has become a matter of concern second only to malpractice. For example:

- Two lawsuits seeking $42.5 million in damages have been filled in federal court by families of two people shot to death at a Kansas Medical Center. The lawsuit stems from a fatal shotgun attack in the facility's emergency room. The victims were a hospital visitor and a doctor who was completing his residency training at the center.
- A $13 million negligence lawsuit has been filed against a hospital in Utah by lawyers for a 12 year-old mother who claimed she had been raped at 3:00 a.m. after giving birth at the hospital.
- A lawsuit asking $1k million in damages from a Tennessee Medical Center has been filed on behalf of two children whose mother was stabbed to death at the hospital while visiting her grandmother who was a patient there.
- A medical center in Springfield, MA, it's chief of security, and others were named in a $44 million lawsuit for lack of security by the mother of a 7 year-old boy who was stabbed to death in the medical center's parking lot.

- In another parking lot incident, an anesthesiologist at a memorial hospital sued the hospital for $6 million for inadequate security following his being shot in the abdomen during a mugging.
- The director of a psychiatric center in Brooklyn, NY was sued for $16 million by the brother of a nurse who was killed by a patient there.

Many hospitals have quietly settled a number of lawsuits, especially in connection with rape, other types of assaults, or infant kidnapping—for big dollars. Furthermore, the current trend is for juries to substantially increase the amount awarded to a plaintiff through punitive damages when they find a hospital negligent in providing adequate security.

Whether it is true or not, the perception by nurses, physicians and other staff members, that security is inadequate at a hospital, frequently provokes protests and disputes. At a hospital in New Haven, CT, 700 employees signed a petition demanding tighter security following a mugging in the hospital's garage. At a medical center in Bakersfield, big headlines in local newspapers reported that employees were protesting the lack of security and indicated that some nurses were carrying revolvers to work. At a hospital in Brooklyn, interns and residents picketed the hospital demanding increased security at their residence hall. Of even greater consequence was the difficulty of a hospital in New Orleans, in attracting nurses to meet JCAH staffing requirements. In a local newspaper, a former nurse charged that the lack of security hampered the recruitment of nurses. When JCAH denied accreditation to the hospital, the hospital, among other improvements, added security personnel.

A violent crime in a hospital often brings with it newspaper, radio, and TV publicity, sometimes national as well as local, and tends to distort the true picture of security at the institution. Sometimes the very nature of the crime, especially infant kidnappings, encourages the press to have a field day at the expense of the institution. Bad publicity of this kind can undo years of excellent public relations work.

In order to minimize the potential for violent crime, it is of paramount importance to establish a viable access control program. Institutional officers must be able to regulate the flow of visitors and patients, and everyone's identity must be easily recognized. At Children's Memorial Hospital in Chicago, the color of visitor passes has been set up on each floor. The hospital learned the hard way, after a baby was kidnapped there in 1980. At the Presbyterian Hospital in New York City, uniformed officer issue color-coded adhesive passes which limit the visitor to a specific floor or area of the hospital. Visitors must state who they are visiting, and the name is checked on a computer printout. The two-visitors-at-a-time rule is also enforced. The result of such strict access controls? A 65% reduction in the number of reported crimes over an 18-month period.

The following chapter, in three parts, outlines through epidemiology and design programs some of those steps that can be implemented to reduce violence to health care workers.

PART 1

Violence in the Health Care Industry: Greater Recognition Prompting Occupational Health and Safety Interventions

Jane Lipscomb

INTRODUCTION

We are currently experiencing an epidemic of violence toward workers in the health care industry. In order to combat this epidemic, employers, front-line health care workers, occupational health and safety professionals and security experts from a variety of disciplines and constituencies, as well as state and federal regulatory agencies and legislators, must work together to develop a comprehensive approach to this problem.

Injuries and illnesses associated with employment in the health care industry have been of concern to health care workers and occupational safety and health professionals for at least the past two decades. However, violence in this setting has only recently been added to the list of workplace hazards facing health care workers. In response to the recognition of this hazard, regulatory agencies, health and safety professionals and organizations representing health care workers must act aggressively by designing and applying occupational health control strategies to this workplace hazard.

This chapter will provide an overview of the problem of violence in the health care industry and serve as an introduction to two additional chapters on the topic. The first of these chapters describes CAL/OSHA's "Guidelines for Security and Safety of Health Care and Community Service Workers" while the second chapter presents one labor union's (SEIU) approach to educating and organizing front-line health care workers around the issue of violence in the workplace.

BACKGROUND ON THE PROBLEM

The problem of violence in mental health care facilities has an extensive history, with the first documented case of a patient fatally assaulting a psychiatrist in 1849 (Bernstein, 1981). Psychiatric settings have been the subject of the most research to date, with the majority of studies examining the risk of violence to psychiatrists and other therapists and focusing on the victim's role, the assaultive patient's characteristics and contextual factors surrounding the assault. Unfortunately, until recently, violence in the psychiatric setting was viewed as a problem with psychiatrists and other therapists and their inability to predict and control the violent or dangerous patient rather than a problem with the workplace.

1-56670-083-3/94/$0.00+$.50

More recent studies have focused on others, in particular nurses as the victims of assault, and have begun to approach the problem from an occupational health perspective with a focus on environmental factors that place the health care worker at risk.

PREVALENCE OF ASSAULTS

The media has made us all aware of the fact that physical assaults, sometimes fatal attacks, occur in U.S. hospitals, but how often health care workers are assaulted and which are most often the victims of these assaults is much less understood. Over the past decade, a number of surveys have been conducted in an attempt to measure the magnitude of the problem. The following review provides some answers to these questions.

A number of studies of mental health care institutions, several surveys of emergency departments and a survey of registered nurses in Ontario, Canada have estimated the prevalence of assaultive patients and/or assaults on health care workers in these settings. Larkin et al. (1988) found that 37% of 600 patients engaged in assaultive behavior, defined as any behavior that could physically damage the individual him- or herself, another individual or property. They reported that very serious incidents (defined as life threatening, e.g., strangulation) occurred at a rate of five per month. Rossi et al. (1985) reported that the incidence of violent or "fear-inducing" behavior among the mentally ill was associated with 45% of patient admissions (n = 1687) to a general hospital's psychiatric unit from 1979–1982.

Lanza (1983) reported the assault experience of 40 nursing staff (60% response rate) in a Veteran's Administration (VA) neuropsychiatric hospital. The respondents, who had an average of six years of psychiatric nursing experience, reported being assaulted an average of seven times in their career. Prior assaults included being grabbed, hit, choked, knocked out or thrown to the floor.

The Occupational Safety and Health Administration (OSHA) defines an occupational injury as one that results in death or in "lost workdays, loss of consciousness, restriction of work or motion, termination of employment, transfer to another job, or medical treatment (other than first aid)." This definition was used by Carmel and Hunter (1989) to conduct a one-year study of injuries in a 973-bed maximum security forensic hospital in California. They found that nursing staff sustained 16 injuries per 100 staff. By comparison, work-related injuries defined as above and reported to OSHA during 1989 occurred at a rate of 8.3 per 100 full-time workers in all industries combined, with the highest rate of 14.2 per 100 full-time workers in the construction industry (Bureau of Labor Statistics, 1991). These data suggest that at this maximum security hospital, the rate of injuries from assaults alone put this group of workers at a higher risk than that of the most hazardous industry in the country, the construction industry.

Lavoie et al. (1988) surveyed the medical directors (75% response rate) of 127 large university-based hospital emergency departments and reported that 43% had at least one physical attack on a medical staff member per month and 7% described acts of emergency department violence that resulted in death. Keep et al.'s (1992) informal survey of nurse managers in five California metropolitan area emergency departments reported that over 40% of 103 managers surveyed reported physical threats to staff 5 or more times per month. Three managers reported threats with a weapon 5 to 15 times per month.

HIGH RISK SETTINGS AND TRENDS IN VIOLENCE

There are a number of settings that place health care workers at high risk of assault, including mental health care facilities, emergency departments, pediatric units, medical-surgical units and long-term care facilities. One study of a university hospital found that assaults were distributed throughout the hospital, with 41% occurring in the psychiatric unit, 18% in the emergency room, 13% on medical units, 8% in surgical units and 7% in the pediatric unit (Conn, 1983). In contrast, a 1989 survey conducted by the International Association for Healthcare Security and Safety found that 42% of reported assaults occurred in emergency departments and 23% occurred in psychiatric areas. Two years later the survey showed that general patient rooms had replaced psychiatric units as the second most frequent area for assaults (Worthington, 1993).

Although community mental health outpatient settings have not been the subject of systematic study of violence, at least one fatal attack on a psychologist in Southern California in 1989 has raised the level of concern among this group of health care workers. Bernstein (1981), in his survey of psychiatrists and therapists, found that 33% of incidents took place in an inpatient setting, 26% took place in an outpatient setting and 21% took place in private practice. In Canada, in community settings in general, physical attacks by patients were reported by 1.1 to 14.1% of nurses surveyed by the Manitoba Association of Registered Nurses (Liss, 1993). Nursing homes are undoubtedly a setting with a high prevalence of assaultive patients, but with the exception of one qualitative study of nurses aids in this setting (Lusk, 1992) the medical literature is devoid of reports of the problem in nursing homes.

As is apparent from this brief review, no health care setting is immune from the risk of violence and as such, each workplace must be assessed to identify high risk areas, activities and workers. Intervention strategies described in this and subsequent chapters must then be implemented to control all types of assaults to employees in the health care industry.

POSSIBLE EXPLANATIONS FOR INCREASING VIOLENCE

The root causes of violence in the health care industry are multifaceted. Undoubtedly, increases in poverty, substance abuse, unemployment, the portrayal of violence in the media and the increased availability of firearms (in particular handguns) all contribute to the problem. In addition, health care providers see individuals, families and communities in time of illness and injury, which invokes stress and involves a loss of control. In a small percentage of cases, patients with various types of brain dysfunction may be violent as a consequence of their condition.

In psychiatric settings, the increase in violence has been attributed to: 1) deinstitutionalization of the chronically mentally ill; 2) the right of involuntary patients to refuse psychotropic medications; 3) the changed involuntary commitment criteria from "in need of treatment" to "dangerousness" and 4) shorter hospital stays and frequent readmissions of violent patients. The blurred boundary between mental illness and criminality is also thought to have contributed to the problem of violence in hospitals. At times, law enforcement personnel will divert a violent individual who is acting "crazy" to a locked psychiatric setting rather than to jail (Jones, 1985).

Experts from the social sciences, criminal justice, law enforcement and public health must all work together to develop and implement strategies for intervening in the root causes of violence in the U.S. In the meantime though, all available means must be used to protect health care workers from the threat of violence in the workplace.

WEAPON-CARRYING IN HOSPITALS

Weapon-carrying patients are a concern in emergency departments throughout the country. The literature suggests that weapon-carrying in both psychiatric and general emergency services is not uncommon and that the ability of staff to predict which patients are carrying weapons is poor (McCulloch, 1986). Media coverage of murders and hostage situations in a number of emergency departments and other areas of hospitals across the country have brought the problem of violence in these settings to the attention of the general public.

Anderson et al. (1989) found that 8.4% of 287 inpatients admitted to a New Jersey hospital during the first six months in 1987 carried weapons. McCulloch et al. (1986) found that 8% of 175 patients searched in a psychiatric emergency department carried weapons, while McNiel and Binder (1987) found 4% of all patients seen in a psychiatric emergency room carried weapons. Wasserberger et al. (1989) reported that at one Los Angeles Level 1 trauma center, 4796 weapons were confiscated from 21,456 patients over a 9-year period and that at least 25% of the major trauma patients were carrying lethal weapons.

Lavoie et al. (1988), in their survey of 127 emergency department directors, found that 43% of departments confiscated at least one weapon per month; one hospital confiscated 300 weapons per month using a metal detector; 18% reported at least one threat with a weapon each month and two hostage incidents were reported to have occurred at knifepoint.

Goetz et al. (1991) conducted a retrospective study of 500 patients who were searched for weapons by security officers in a university emergency department during a 20-month period. Eleven percent of psychiatric patients were searched, with 17.3% found to be carrying weapons while 0.4% of medical patients were searched, with 15.7% found carrying weapons.

ENVIRONMENTAL RISK FACTORS

An occupational health approach to the problem of violence in the workplace requires a careful examination of the environmental factors that place health care workers at risk of assault. Risk factors identified to date include poor security, inadequate training, staffing patterns, time of day and containment activities.

Training is an important component of any risk reduction program. Three studies have shown that training in the management of assaultive behavior can reduce injury from assault (Carmel, 1990; Infantino, 1985; Lehman, 1988). Reports that inexperienced health care workers are at increased risk of assault suggests that additional training of novice health care workers may prevent assaults. Bernstein (1981) found that less experienced therapists were threatened or assaulted more frequently than seasoned clinicians by a ratio of 4:1. Hodgkinson et al. (1984) reported that student nurses were at greater risk of assault than nursing assistants.

Carmel and Hunter (1989) found that the majority of injuries (9.9/100 staff) were sustained while containing patient violence while the remainder (6.1/100 staff) were the result of battery-type assaults. Lion et al. (1981) also found that the largest percentage of patient assaults on staff occurred during seclusion or restraining activities.

Staffing patterns have been associated with increased assaults. Jones (1985) found that short staffing increased the likelihood of violent behavior. Fineberg et al. (1988) found a relationship between violence in a psychiatric ward and the use of agency nursing provision ($r = 0.56$) after a 12-month period during which permanent nursing provisions were halved.

Several studies have reported that assaults occur most frequently during times of high activity and interaction on patient wards. Ionno (1983) found that physical assaults were most common on visiting days, and suggested that the increased activity level associated with these days may lead to increased assaultiveness. Jones (1985) concluded that increased activity was the single most significant environmental factor in assaults in a VA medical center. She also found that the majority of violent incidents reported occurred during the day (59% during the day shift), with the largest number of incidents in a two-hour period occurring between 8:30 and 10:30 am. Carmel and Hunter (1989) reported that injuries were more likely to occur during three peak periods (meal hours) than at other times of day. Results from these studies should be used to develop and support comprehensive security and training programs.

Much effort has been directed toward predicting violent behavior among psychiatric patients. Bernstein (1981) found that 76% of those who had threatened or attacked a therapist had a history of violence. Kurlowicz (1990) found that a history of violent behavior within the past several hours was a strong predictor of assaults in the emergency department. He suggests that this history should be made available or elicited from all available sources. Drummond et al. (1989) reported that they reduced assaults against staff by 91% by flagging charts of patients with a history of assaultive or disruptive behavior. The fact that a number of studies have documented that a small percentage of patients are responsible for the majority of assaults supports the hypothesis that patients with a history of violent behavior place health care workers at greatest risk of future assaults. Health care institutions must develop and evaluate strategies for early identification of patients with a history of violent behavior.

Engineering controls such as the use of security guards, metal detectors, personal alarms, adequate lighting and access control in high risk settings such as emergency departments are an essential component of any comprehensive health and safety program. Administrative controls such as escort services, staffing patterns, assault prevention and control training for health care workers, self-defense training, patient education regarding assaults and its consequences and post-assault intervention (victim counseling) programs should all be included as a second line of defense. Clear policies and procedures regarding assault control must be a basis for all of these control measures. It should go without saying that input from front-line health care workers in the form of health and safety committees is the key to a successful assault prevention program.

LEGISLATIVE AND REGULATORY EFFORTS

To date three states have legislation enacted or in progress addressing violence in the health care setting. The California Legislature enacted AB 508, addressing violence in the emergency department which requires all hospitals to perform a safety assessment and correct problems identified as well as requiring mandatory education of direct care givers. This legislation also states that persons interfering with the reporting of any act of assault shall be guilty of a misdemeanor. The Mississippi Legislature enacted SB 2004, addressing increased penalties for conviction of assault on a health care worker. In the New Mexico Senate Judiciary Committee there is a bill changing the documentation of assault as a misdemeanor (Boucher, 1993).

To date, Federal OSHA has no standard specifically addressing violence in the workplace, but in 1992, the agency issued a "field memo" describing the problem and provided guidance for inspectors on citing workplaces under 5(a)1, the "General Duty Clause." Since that time at least three citations have been issued including one citation of a Chicago area hospital for failing to protect workers from patients' violent behavior. The two other citations involved settings other than health care settings that have a recognized hazard of violence.

The question of legal redress for assault and battery in the workplace is an unresolved issue. A number of employees have attempted to sue employers for damages suffered as a result of assault in the workplace and have been denied any recourse other than workers' compensation. Civil suit against the perpetrator, however, is a course of action that is allowed and is being encouraged by many advocates of worker protection in the health care industry.

CONCLUSION

Violence in the health care industry has reached epidemic proportions and as such requires immediate and aggressive control strategies. Mental health care workers continue to be at high risk of assault on the job, however, emergency department personnel and health care workers in general medical, community and long-term care settings are also at risk. Weapon-carrying, which is highly prevalent in emergency departments, increases the risk of serious injury from assaults for all health care workers.

The adequacy of existing hospital security, employee training and policies and procedures for preventing violence has rarely been examined, and when it has, it has been shown to be sorely inadequate, particularly in emergency settings. Identified risk factors for assaults include staffing patterns, time of day, and containment activities. Recent legislation and regulatory activity is beginning to address the problem, however, these efforts are clearly in their infancy. Front-line health care workers and occupational health professionals from many disciplines and constituencies must work together to advocate for and develop a comprehensive approach to the problem. In addition, health care institutions need to be educated that they have much to gain from efforts to identify and reduce the current epidemic of violence in these settings.

REFERENCES

Anderson AA, Ghali AY, Bansil RK: Weapon carrying among patients in a psychiatric emergency room. *Hospital and Community Psychiatry,* 40(8):845–847, 1989.

Bernstein HA: Survey of threats and assaults directed toward psychotherapists. *American Journal of Psychotherapy,* 35(4):542–549, 1981.

Boucher D: Recommendations for Legislative Approaches to Violence in the Health Care Setting. *Proceedings from the American Academy of Nursing Annual Conference,* Washington, DC, November, 1993.

Carmel H, Hunter M: Staff injuries from inpatient violence. *Hospital and Community Psychiatry,* 40(1):41–46, 1989.

Carmel H, Hunter M: Compliance with training in managing assaultive behavior and injuries from inpatient violence. *Hospital and Community Psychiatry,* 41(5):558–560, 1990.

Conn LM, Lion JR: Assaults in a university hospital. In: Lion JR, Reid WH (Eds.) *Assaults Within Psychiatric Facilities.* Philadelphia, PA: W.B. Saunders, pp. 61–70, 1983.

Fineberg NA, James DV, Shah AK: Agency nurses and violence in a psychiatric ward. *The Lancet,* 1, 474, 1988.

Goetz RR, Bloom JD, Chenell SL, Moorhead JC: Weapons possession by patients in a university emergency department. *Annals of Emergency Medicine,* 20(1):8–10, 1991.
Hodgkinson P, Hillis T, Russell D: Assaults on staff in a psychiatric hospital. *Nursing Times,* 80, 44–46, 1984.
Infantino JA, Musingo SY: Assaults and injuries among staff with and without training in aggression control techniques. *Hospital and Community Psychiatry,* 36(12):1312–1314, 1985.
Ionno JA: A prospective study of assaultive behavior in female psychiatric inpatients. In: Lion JR, Reid WH (Eds.) *Assaults Within Psychiatric Facilities.* Philadelphia, PA: W.B. Saunders, pp. 71–80, 1983.
Jones MK: Patient violence report of 200 incidents. *Journal of Psychosocial Nursing and Mental Health Services,* 23(6):12–17, 1985.
Keep N, Gilbert P: California Emergency Nurses Association's informal survey of violence in California emergency departments. *Journal of Emergency Nursing,* 18(5):433–439, 1992.
Kurlowicz L: Violence in the emergency department. *American Journal of Nursing,* 474, 35–40, 1990.
Lanza ML: The reactions of nursing staff to physical assault by a patient. *Hospital and Community Psychiatry,* 34(1):44–47, 1983.
Larkin E, Murtagh S, Jones S: A preliminary study of violent incidents in a special hospital (Rampton). *British Journal of Psychiatry,* 153, 226–231, 1988.
Lavoie F, Carter GL, Danzl DF, Berg RL: Emergency department violence in United States teaching hospitals. *Annals of Emergency Medicine,* 17(11):1227–1233, 1988.
Lehmann LS, Padilla M, Clark S, Loucks S: Training personnel in the prevention and management of violent behavior. *Hospital and Community Psychiatry,* 34, 40–43, 1983.
Lion JR, Snyder W, Merrill GL: Brief reports: Understanding of assaults of staff in a state hospital. *Hospital and Community Psychiatry,* 322(7):497–498, 1981.
Liss GM: *Examination of workers' compensation claims among nurses in Ontario for injuries due to violence.* Unpublished report, Health and Safety Studies Unit, Ministry of Labor, Ontario, Canada, 1993.
Lusk SL: Violence experienced by nurses' aides in nursing homes. *The Journal of the American Association of Occupational Health Nurses,* 40(5):237–241, 1992.
McCulloch LE, McNeil DE, Binder RL, Hatcher C: Effects of a weapon screening procedure in a psychiatric emergency room. *Hospital and Community Psychiatry,* 37, 837–838, 1986.
McNeil DE, Binder RL: Patients who bring weapons to the psychiatric emergency room. *Journal of Clinical Psychiatry,* 48, 230–233, 1987.
Registered Nurses' Association of Ontario, Psychiatric Nursing Interest Group, Nurse Assault Project Team. *Nurse assault survey,* 1992.
Rossi AM, Jacobs M, Monteleone M, Olson R, Surver RW, Winkler E, Wommack A: Violent or fear-inducing behavior associated with hospital admission. *Hospital and Community Psychiatry,* 36(6):643–647, 1985.
U.S. Department of Labor, Bureau of Labor Statistics. *Occupational injuries and illnesses in the United States by industry, 1989.* Bulletin 2379, 1991.
Wasserberger J, Ordog GJ, Kolodny M, Allen K: Violence in a community emergency room. *Archives of Emergency Medicine,* 6, 266–269, 1989.
Worthington K: Taking action against violence in the workplace. *American Nurse,* June, 1993.

PART 2

Guidelines for Security and Safety of Health Care and Community Service Workers

CAL/OSHA
Medical Unit Division of
Occupational Safety & Health
Department of Industrial Relations

PREFACE

Violence is an escalating problem in workplaces, and health care settings are not exempt from this trend. In the past, health care professionals have generally regarded themselves as immune to harm arising from their work. When workplace violence resulted in injuries, administrators and supervisors often expressed sentiments that the health care worker might have been at fault or that these incidents were "part of the job." These guidelines were developed in response to the increasing number of severe injuries, some resulting in death, experienced by health care and community service workers. A variety of individuals, organizations, unions and state and local government agencies have requested assistance from CAL/OSHA to control this serious occupational health hazard.

These guidelines are designed to assist all health care, community workers and support staff who may be exposed to violent behavior from patients, clients or the public. Because every work situation that could present a threat to worker safety cannot be covered, the focus is on major health care and community service locations, such as public and private medical services (acute-care hospitals, emergency rooms, long-term care facilities, public health and other clinics, home health services and pre-hospital care); psychiatric service (inpatient, clinic, residential and home visiting); alcohol and drug treatment facilities and social welfare agencies including unemployment and welfare eligibility offices, homeless shelters, parole and public defender services and child welfare services. The recommended courses of action, however, may also be applied to prevent violence in any facility.

Measures to prevent assaults should include engineered systems and administrative measures as well as training. Alarm systems are one of the most important protective measures for hospitals and clinics. Of course, back-up security staff to respond to the alarm must accompany any alarm system. Many of the basic problems leading to violence may be traced to inadequate staffing levels, therefore an effective administrative measure is to ensure that appropriate staffing is maintained at all times. Training of all personnel is a necessary preventive as well as protective measure. Although the temptation to place the responsibility for controlling violence on the trained employee is great, it will not be sufficient to prevent serious injuries. A concerned administration that implements and main-

1-56670-083-3/94/$0.00+$.50

tains a well-developed program should be able to succeed in reducing the incidence of assaults and injuries in the workplace.

This document does not cover public safety work, such as in police or corrections departments where exposure to violence is well recognized and already addressed by department guidelines or regulations. These guidelines do not specifically address conditions of service industry workers in the private sector, such as sales personnel or restaurant workers. Some of the guidelines, however, may be adapted to protect these workers from violent acts that may occur in their contacts with the general public.

I am grateful to all who have contributed to this document. I am aware that updates will be needed as needs are identified and technology is refined.

I want to specifically acknowledge Melody M. Kawamoto, MD, for her assistance in editing this document.

Joyce A. Simonowitz, RN, MSN
Occupational Consultant

I. INTRODUCTION: THE PROBLEM

During the past two decades, we have seen a sharp increase in violence in our cities, country and society. Estimates show that nearly one third of all Americans are victimized by crime each year (Poster and Ryan, 1989). Violence in the workplace is a manifestation of this problem, with homicide being the third leading cause of occupational death among all workers in the U.S. from 1980 to 1988 (Jenkins et al., 1992) and the leading cause of fatal occupational injuries among women from 1980 to 1985 (Levin et al., 1992).

Higher rates of occupational homicides were found in the retail and service industries, especially among sales workers (Jenkins et al., 1992). This increased risk may be explained by contact with the public and the handling of money (Kraus, 1987). Research into the causes of the increasing incidence of death and serious injury to health care workers has led to the theory that exposure to the public may be an important risk (Lipscomb and Love, 1992; Lavoie et al., 1988). The risk is increased particularly in emotionally charged situations with mentally disturbed persons or when workers appear to be unprotected.

During the past few years, violence resulting in the death of California health care and community workers occurred in emergency rooms, psychiatric hospitals, community mental health clinics and social service offices. Assaults, hostage taking, rapes, robbery and other violent actions are also reported at these and other health care and community settings. In a study by Conn and Lion (1983), assaults by patients in a general hospital occurred in a variety of locations. Although 41% of assaults occurred in the psychiatric units, they also occurred in emergency rooms (18%), medical units (13%), surgical units (80%) and even pediatric units (7%).

Carmel and Hunter (1989) found that the psychiatric nursing staff of a maximum security forensic hospital in California sustained 16 assault injuries per 100 employees per year. This investigation used the OSHA definition for occupational industry: an injury that results in death, lost work days, loss of consciousness, restriction of work or motion, termination of employment, transfer to another job or medical treatment other than first aid (Bureau of Labor Statistics, 1986). Work-related injuries reported on OSHA forms and reported to the Bureau of Labor Statistics (BLS) for 1989 occurred at a rate of 8.3 per 100 full-time workers in all industries combined. The highest rate, 14.2 per 100 full-time workers, was seen in the construction industry (Bureau of Labor Statistics, 1991). In comparison, data collected by Carmel and Hunter suggest that some psychiatric workers may be at a higher risk for injuries from all causes in the country's most hazardous industry (Lipscomb and Love, 1992).

Madden et al. (1976), Lanza (1983) and Poster and Ryan (1989) have reported that 46 to 100% of nurses, psychiatrists and other therapists in psychiatric facilities experienced at least one assault during their career. Research on the causes and methods of prevention of violence in psychiatric facilities was funded by the California Department of Industrial Relations after the death of a psychiatric hospital worker in 1989. This investigation is in progress at the forensic hospital at the present time.

Lavoie et al. (1988), investigated 127 large, university-based hospital emergency departments and reported that 43% (55) had at least one physical attack on a medical staff member per month. Of the reported acts of violence in the last 5 years, 7% (9) resulted in death. Emergency room personnel face a significant risk of injury from assaults by patients, but in addition, may be abused by relatives or other persons associated with the patient. Further, the violence that occurs in the emergency room is often shifted into the hospital when the patient is transferred to the receiving unit.

Bernstein (1981) reported that 26% of reported assaultive behaviors in a study of California psychotherapists occurred in the outpatient setting. The death of an outpatient psychiatric worker in 1989 in California at the hands of a homeless client underscores the risk that exists in this setting. Investigations by OSHA officials in two California counties identified a nearly complete lack of security measures in outpatient facilities, leaving workers unprotected and vulnerable to abuse and assaults.

Community service workers are at risk of hostile behavior from the public when they visit clients at hotels, apartments or homes in unfamiliar or dangerous locations, especially at night. Child welfare workers have reported that parents of children who are being taken to foster homes or other types of court action have become violent and assaulted workers with knives and fists. Sexual assaults with serious injury, other physical assaults and robberies have been reported by workers in the hospital and community. In addition, clients or their relatives and friends may direct their anger, which can be extreme or violent, at community workers. In Canada, in community settings, physical attacks by patients were reported by 1.1 to 14.1% of nurses surveyed by the Manitoba Association of Registered Nurses (Liss, 1993).

Few research investigations have focused on the incidence of violence to community workers, but reports have been received from many sources such as union workers or parking enforcement workers who have suffered abusive and at times violent behavior from hostile motorists. Hotel housekeepers are currently being studied after complaining of sexual abuse and threats in hotels in which they work. Such research is needed to identify the scope of violence in the medical field and the community as a whole.

A. Risk Factors

Risk factors may be viewed from the standpoint of 1) the environment, 2) work practices and 3) victim and perpetrator profile.

1. Environmental Factors

Health care and community service workers are at increased risk of assaults because of increased violence in our society. This increase in violence is thought to be a result of such factors as the easy availability of guns and weapons; the use of violence by many in the population as a means of solving problems; the increase in unemployment poverty and homelessness; the decrease in social services to the poor and mentally ill; the increase in gang-related activity and drug and alcohol use; violence depicted in television and movies and the increasing use of hospitals by police and criminal justice systems for acutely disturbed patients. These may be thought of as a partial listing, which may have a direct contribution to the safety and security of workers.

An important risk factor at hospital and psychiatric facilities is the carrying of weapons by patients and their family or friends. Wasserberger et al. (1989) reported that 25% of major trauma patients treated in the emergency room carried weapons. Attacks on emergency rooms in gang-related shootings have been documented in two Los Angeles hospitals (Long Beach Press Telegram, 1990). Goetz et al. (1991) found that 17.3% of psychiatric patients searched were carrying weapons.

Other risk factors include the early release from hospitals of the acute and chronically mentally ill, the right of patients to refuse psychotropic treatment, inability to involuntarily hospitalize mentally ill persons unless they pose an immediate threat to themselves or others and the use of hospitalization in lieu of incarceration of criminals. McNeil et al. (1991) found that police referrals were significantly more likely to have displayed violent behavior such as physical attacks and fear-inducing behavior during the two weeks before coming to the psychiatric emergency service and during the initial 24 hours of evaluation and treatment.

2. Work Practices

Many studies have implicated staffing patterns as contributors to violence. Both Jones (1985) and Fineberg et al. (1988) found that shortage of staff and the reduction of trained, regular staff increased the incidence of violence. Assaults were associated with meal times, visiting times and times of increased staff responsibilities. This suggests that staffing evaluations do not take into account the potential hazards associated with increased activity in the units or for times when transportation of clients is needed. Assaults were also noted at night when staffing is usually reduced. Frequency of exposure to and interaction with patients or clients are known factors that increase a health care or community worker's vulnerability. Work in high crime areas, at an isolated work station or working alone without systems for emergency assistance may increase the risk of assaults. In addition, typical work activities may arouse anger or fear in some patients and result in acts of violence. Long waits in emergency rooms and inability to obtain needed services are seen as contributors to the problem of violence. This was evidenced in the emergency department shooting in Los Angeles where three doctors were shot by an angry, dissatisfied and disturbed client.

3. Perpetrator and Victim Profile

It is difficult to predict when or which patients/clients will become violent since the majority of assaults are perpetrated by a minority of persons. More acute and untreated mentally ill persons are being admitted to and quickly released from psychiatric hospitals and are in need of intensive outpatient treatment and services. These services are often lacking due to funding cuts. Further, clearly only a small percentage of violence is perpetrated by the mentally ill. Gang members, distraught relatives, drug users, social deviants or threatened individuals are often aggressive or violent.

A history of violent behavior is one of the best indicators of future violence by an individual. This information, however, may not be available, especially for new patients or clients. Even if this information were available, workers not directly involved with the individual client would not have access to it. At times violence is not aimed at the actual care giver. Keep et al. (1992) reported on the gunshot death of a nurse and an emergency medical technician student who were targets of a disturbed family member of a patient who died in surgery the previous day.

Workers who make home visits or community work cannot control the conditions in the community and have little control over the individuals they may encounter in their work. Dillon (1992) reported the shooting death of four county workers in upstate New York and the beating death of a case worker who removed a 7-year-old child from a violent home. The victim of assault is often untrained and unprepared to evaluate escalating behavior and to know and practice methods of defusing hostility or protecting themselves from violence. Training, when provided, is often not required as part of the job and may be offered infrequently. However, using training as the sole safety program element, creates an impossible burden on the employee for safety and security for him or herself, co-workers or other clients. Personal protective measures may be needed and communication devices are often lacking.

B. Cost of Violence

Little has been done to study the cost to employers and employees of work--related injuries and illnesses, including assaults. A few studies have shown an increase in assaults over the past two decades. Adler et al. (1983) reported 42 work days lost over a two-year period due to violence to 28 workers, an increase from the previous two years in which 11 workers lost 62 work days. Carmel and Hunter (1989) reported that of 121 workers sustaining 134 injuries, 43% involved lost time from work with 13% of those injured missing more than 21 days from work. In this same investigation, an estimate of the costs of assault was that the 134 injuries from patient violence cost $766,000 and resulted in 4,291 days lost and 1,445 days of restricted duty. Lanza and Milner (1989) reported 78 assaults during a 4-month period. If this pattern were repeated for the remainder of the year, 312 assaults could be expected with a staggering cost per year from medical treatment and lost time. Additional costs may result from security or response team time, employee assistance program or other counseling services, facility repairs, training and support services for the unit involved, modified duty and reduction of effectiveness of work productivity in all staff due to a heightened awareness of the potential for violence.

True rates of violence at health care and community service facilities, however, must be assumed to be higher than documented rates. Episodes of violence are often unreported. If reported, records are not necessarily maintained. Nurses and other health care professionals are reluctant to report assaults or threatening behavior when the prevailing attitude of administrators and supervisors and sometimes other staff members, is that violence "comes with the territory" or "health professionals accept the risk when they enter the field." Administrators, peers and even the victims themselves, may initially assume that the violent act resulted from a failure to deal effectively or therapeutically with the client or patient and thus attribute the incident to professional incompetence. Lanza and Carifio (1991), in a study to determine causal attributions made to nurses who are victims of assault, found that women are blamed more than men and that if injured, "the nurse must have done something wrong."

In addition to the blame and potential for improper evaluation of the worker's skills, physical and emotional injury may have occurred. Poster and Ryan (1989a) report that cognitive emotional and physical sequelae may be present long after the victim has returned to work. Davidson and Jackson (1985), Lanza (1983 and 1985b) and Poster and Ryan (1989a) reported that assaulted workers experience feelings of self-doubt, depression, fear, post-traumatic stress syndrome, loss of sleep, irritability, disturbed relationships with family and peers, decreased ability to function effectively at the workplace, increased absenteeism and flight from the health care profession. The mental costs to the victim of violence should be recognized and even if physical injury did not occur, professional counseling services may be required to aid in an employee's recovery. The articles referenced all describe the need for and the conduct of counseling programs. Ryan and Poster (1989b) document the benefits of counseling for rapid recovery after assault. The costs to the employee are often unrecognized and thus are not included in any cost accounting of the problem.

White and Hatcher (1988) discuss costs to the organization and the victim of violence pointing to the increased costs due to the "2nd injury" phenomenon of perceived rejection of the victim by the agency, co-workers and even family, resulting in filing of lawsuits. These suits may cause substantial long-term costs to the agency.

C. Prevention

Although it is difficult to pinpoint specific causes and solutions for the increase in violence in the workplace and in particular health care settings, recognition of the problem is a beginning. Some solutions to the overall reduction of violence in this country may be found in actions such as eliminating violence in television programs, implementing effective programs of gun control, and reducing drug and alcohol abuse. All companies should investigate programs recently instituted by several convenience store chains of robbery deterrence strategies such as increased lighting, closed circuit T.V. monitors, visible money handling locations, if sales are involved, limiting access and egress and providing security staff.

Other methods of preventing assault may be in expanding the national data base with standardized reporting and information collection systems. It may also be necessary to fund and conduct research on post assault outcomes, the need for rehabilitation for returning to work, the length of employment after assault and on techniques of preventing injury and death from occupational violence.

In a San Francisco hospital, methods have been developed to attempt to deal with violence issues with the formation of two focus groups. One group, "the Violence Task Force," functions to advise the administration regarding modification of hospital policy toward reducing incidents of violence. The second group, "San Francisco Emergency Workers Critical Incident Stress Debriefing Team," counsels victims of physical, sexual or verbal assault. This group also provides needed support to staff who may be exposed to bloody and brutal scenes in their work environment.

White and Hatcher (1988) have outlined management and medical objectives and responses to violence-induced trauma as well as decision trees and checklists to aid in assessing and constructing a response plan. Although not necessarily incident preventing, a response plan should be incorporated into an overall plan of prevention.

Training employees in management of assaultive behavior or professional assault response has been shown by Carmel and Hunter (1990) to reduce the incidence of assaults to hospital staff. Infantino and Musingo (1985) and Blair and New (1991) also found that new and untrained staff were at highest risk for injury.

Keep and Gilbert (1992) report that legislation is being proposed in California to make violence to emergency personnel reportable to local police and criminal charges pressed if there is sufficient evidence. This action is also recommended by Morrison and Herzog (1992), especially in relation to emergency department staff. Other staff of facilities such as psychiatric units should be advised and policies established to assist in the decision of the appropriateness and effectiveness of such action.

Administrative controls and mechanical devices are being recommended and gradually implemented, but the problems appear to be escalating. Although long ignored by hospital and other administrators and professionals, the problem of workplace violence is being recognized. Increasing numbers of health care and community service workers, as well as OSHA professionals have come to the conclusion that injuries related to workplace violence should no longer be tolerated. In the past, little was done to protect workers from violence. Currently, as discussed, a variety of health care, community service facilities, unions and researchers are seeking solutions to the problem. Managers and administrators are being advised to make the provision of adequate measures to prevent violence a high priority. Some safety measures may seem expensive or difficult to implement but are needed to adequately protect the health and well being of health care and community service workers. It is also important to recognize that the belief that certain risks are "part of the job" contributes to the continuation of violence and possibly the shortage of trained health care and community service workers.

CAL/OSHA recognizes its obligation to develop standards and guidelines to provide safe workplaces for health care and community service workers. These workplaces should be free from health and safety hazards, including fear and the threat of assaults. The Injury and Illness Prevention Program as defined under the General Industry Safety Order, Section 3203, requires all employers to develop an Injury and Illness Prevention Program for hazards unique to their place of employment. This Injury and Illness Prevention Program should provide the framework for each employer's program of preventing assaults—one of the major hazards of work in health care and community service and perhaps in the community as a whole.

These CAL/OSHA guidelines are designed to assist managers and administrators in the development and implementation of programs to protect their workers. While not exhaustive, these guidelines include philosophical approaches as well as practical methods to prevent and control assaults. The potential for violence may always exist for health care and community service workers, whether at large medical centers, community-based drug treatment programs, mental health clinics or for workers making home visits in the community. Because of the potential for injury to workers, health care and community service organizations must comply with Title 8 of the CCR, Section 3203. This regulation requires an Injury and Illness Prevention Program which stipulates that responsible persons perform worksite analyses, identify sentinel events, and establish controls and training programs to reduce or eliminate hazards to worker health and safety. A copy of the

State of New Jersey OSHA Guidelines (Appendix C of Part 2) on measures and safeguards in dealing with violent or aggressive behavior in public sector health care facilities is provided as an example of one of the first state OSHA recommendations addressing this serious issue. We anticipate more states and federal OSHA will eventually follow suit.

Many health care providers, researchers, educators, unions and OSHA enforcement professionals contributed to the development of these guidelines. The cooperation and commitment of employers is necessary, however, to translate these guidelines into an effective program for the occupational health and safety of health care and community service workers.

II. PROGRAM DEVELOPMENT

The guidelines are divided into two major divisions: 1) general provisions and program development and 2) specific work setting requirements. General provisions and program development include provisions that must be adopted by all high risk industries to assess risk and to develop needed programs.

Within the specific work setting, guidelines will be subdivided into (a) engineering controls, (b) work practices, (c) personal protective measures and (d) individualized training measures by major worksite category, i.e., inpatient psychiatric hospitals and psychiatric units, hospital and emergency rooms, outpatient facilities and community workers.

A. General Program Essentials

1. Management Commitment and Employee Involvement

Commitment and involvement are essential elements in any safety and health program. Management provides the organizational resources and motivating forces necessary to deal effectively with safety and security hazards. Employee involvement, both individually and collectively, is achieved by encouraging participation in the worksite assessment, developing clear effective procedures and identifying existing and potential hazards. Employee knowledge and skills should be incorporated into any plan to abate and prevent safety and security hazards.

a. Commitment by Top Management

The implementation of an effective safety and security program includes a commitment by the employer to provide the visible involvement of administrators of hospitals, clinics and agencies, so that all employees, from managers to line workers, fully understand that management has a serious commitment to the program. An effective program should have a team approach with top management as the team leader and should include the following:

i. The demonstration of management's concern for employee emotional and physical safety and health by placing a high priority on eliminating safety and security hazards.

ii. A policy which places employee safety and health on the same level of importance as patient/client safety. The responsible implementation of this policy requires management to integrate issues of employee safety and security with restorative therapeutic services to assure that this protection is part of the daily hospital/clinic or agency activity.

iii. Employer commitment to security through the philosophical refusal to tolerate violence in the institution and to employees and the assurance that every effort will be made to prevent its occurrence.

iv. Employer commitment to assign and communicate the responsibility for various aspects of safety and security to supervisors, physicians, social workers, nursing staff and other employees involved so that they know what is expected of them. Also to ensure that recordkeeping is accomplished and utilized using good principles of epidemiology to aid in meeting program goals.

v. Employer commitment to provide adequate authority and resources to all responsible parties so that assigned responsibilities can be met.

vi. Employer commitment to ensure that each manager, supervisor, professional and employee responsible for the security and safety program in the workplace is accountable for carrying out those responsibilities.

vii. Employer develops and maintains a program of medical and emotional health care for employees who are assaulted or suffer abusive behavior.

viii. Development of a safety committee in keeping with requirements of GISO 3203 and which evaluates all reports and records of assaults and incidents of aggression. When this committee makes recommendations for correction, the employer reports back to the committee in a timely manner on actions taken on the recommendation.

b. Employee Involvement

An effective program includes a commitment by the employer to provide for, and encourage employee involvement in the safety and security program and in the decisions that affect worker safety and health as well as client well-being. Involvement may include the following:

i. An employee suggestion/complaint procedure which allows workers to bring their concerns to management and receive feedback without fear of reprisal or criticism of ability.

ii. Employees follow a procedure which requires prompt and accurate reporting of incidents with or without injury. If injury has occurred, prompt first aid or medical aid must be sought and treatment provided or offered.

iii. Employees participate in a safety and health committee that receives information and reports on security problems, makes facility inspections, analyzes reports and data and makes recommendations for corrections.

iv. Employees participate in case conference meetings, and present patient information and problems which may help employees to identify potentially violent patients and discuss safe methods of managing difficult clients. (Identification of potential perpetrators.)

v. Employees participate in security response teams that are trained and possess required professional assault response skills.

vi. Employees participate in training and refresher courses in professional assault response training such as PART®, see Section X, Appendix D, to learn techniques of recognizing escalating agitation, deflecting or controlling the undesirable behavior and, if necessary, of controlling assaultive behavior, protecting clients and other staff members.

vii. Participation in training as needed in non-hospital work settings, such as "dealing with the hostile client" or even the police department program of "personal safety" should be provided and, required to be attended by all involved employees.

2. Written Program

Effective implementation requires a written program for job safety, health and security that is endorsed and advocated by the highest level of management and professional practitioners or medical board. This program should outline the employer's goals and objectives. The written program should be suitable for the size, type and complexity of the facility and its operations and should permit these guidelines to be applied to the specific hazardous situation of each health care unit or operation.

The written program should be communicated to all personnel regardless of number of staff or work shift. The program should establish clear goals and objectives that are understood by all members of the organization. The communication needs to be extended to physicians, psychiatrists, etc. and all levels of staff including housekeeping, dietary and clerical.

3. Regular Program Review and Evaluation

Procedures and mechanisms should be developed to evaluate the implementation of the security program and to monitor progress. This evaluation and recordkeeping program should be reviewed regularly by top management and the medical management team. At least semi annual reviews are recommended to evaluate success in meeting goals and objectives. This will be discussed further as part of the recordkeeping and evaluation.

III. PROGRAM ELEMENTS

An effective occupational safety and health program of security and safety in medical care facilities and community service includes the following major program elements: (A) worksite analysis, (B) hazard prevention and control, (C) engineering controls, (D) administrative controls, (E) personal protective devices, (F) medical management and counseling, (G) education and training and (H) recordkeeping and evaluation.

A. Worksite Analysis

Worksite analysis identifies existing hazards and conditions, operations and situations that create or contribute to hazards, and areas where hazards may develop. This includes close scrutiny and tracking of injury/illness and incident records to identify patterns that may indicate causes of aggressive behavior and assaults.

The objectives of worksite analyses are to recognize, identify and to plan to correct security hazards. Analysis utilizes existing records and worksite evaluations including:

1. Record Review

a. Analyze medical, safety and insurance records, including the OSHA 200 log and information compiled for incidents or near incidents of assaultive behavior from clients or visitors. This process should involve health care providers to ensure confidentiality of records of patients and employees. This information should be used to identify incidence, severity and establish a baseline for identifying change.

b. Identify and analyze any apparent trends in injuries relating to particular departments, units, job titles, unit activities or work stations, activity or time of day. It may include identification of sentinel events such as threatening of providers of care or identification and classification of clients anticipated to be aggressive.

2. Identification of Security Hazards

Worksite analysis should use a systematic method to identify those areas needing in-depth scrutiny of security hazards. This analysis should do the following:

a. Identify those work positions in which staff is at risk of assaultive behavior.

b. Use a checklist for identifying high risk factors that includes components such as type of client, physical risk factors of the building, isolated locations/job activities, lighting problems, high risk activities or situations, problem clients, uncontrolled access and areas of previous security problems.

c. Identify low risk positions for light or relief duty or restricted activity work positions when injuries do occur.

d. Determine if risk factors have been reduced or eliminated to the extent feasible. Identify existing programs in place and analyze effectiveness of those programs, including engineering control measures and their effectiveness.

e. Apply analysis to all newly planned and modified facilities, or any public services program to ensure that hazards are reduced or eliminated before involving patients/clients or employees.

f. Conduct periodic surveys at least annually or whenever there are operation changes, to identify new or previously unnoticed risks and deficiencies and to assess the effects of changes in the building design, work processes, patient services and security practices. Evaluation and analysis of information gathered and incorporation of all this information into a plan of correction and ongoing surveillance should be the result of the worksite analysis.

B. Hazard Prevention and Control

Selected work settings have been utilized for discussion of methods of reducing hazards. Each of the selected work situations—psychiatric hospitals and psychiatric wards, hospitals and emergency rooms, outpatient facilities and community work settings—will be addressed with general engineering concepts, specific engineering and administrative controls, work practice controls and personal protective equipment as appropriate to control hazards. These methods are contained in B through F.

1. Engineering Administrative and Work Practice Controls for All Settings

a. General Building, Work Station and Area Designs.

Hospital, clinic, emergency room and nurse's station designs are appropriate when they provide secure, well-lighted protected areas which do not facilitate assaults or other uncontrolled activity.

i. Design of facilities should ensure uncrowded conditions for staff and clients. Rooms for privacy and protection, avoiding isolation are needed. For example, doors must be fitted with windows. Interview rooms for new patients or known assaultive patients should utilize a system which provides privacy but which may also permit other staff to see activity. In psychiatric units "time out" or seclusion rooms are needed. In emergency departments, rooms are needed in which agitated patients may be confined safely to protect themselves, other clients and staff.

ii. Patient care rooms and counseling rooms should be designed and furniture arranged to prevent entrapment of the staff and/or reduce anxiety in clients. Light switches in patient rooms should be located outside the room. Furniture may be fixed to the floor, soft or with rounded edges and colors restful and light.

iii. Nurse stations should be protected by enclosures which prevent patients from molesting, throwing objects, reaching into the station otherwise creating a hazard or nuisance to staff: such barriers should not restrict communication but should be protective.

iv. Lockable and secure bathroom facilities and other amenities must be provided for staff members separate from client restrooms.

v. Client access to staff counseling rooms and other facility areas must be controlled; that is, doors from client waiting rooms must be locked and all outside doors locked from the outside to prevent unauthorized entry, but permit exit in cases of emergency or fire.

vi. Metal bars or protective decorative grating on outside ground level windows should be installed (in accordance with fire department codes) to prevent unauthorized entry.

vii. Bright and effective lighting systems must be provided for all indoor building areas as well as grounds around the facility and especially in the parking areas.

viii. Curved mirrors should be installed at intersections of halls or in areas where an individual may conceal his or her presence.

ix. All permanent and temporary employees who work in secured areas should be provided with keys to gain access to work areas whenever on duty.

x. Metal detectors should be installed to screen patients and visitors in psychiatric facilities. Emergency rooms should have available hand-held metal detectors to use in identifying weapons.

b. Maintenance

i. Maintenance must be an integral part of any safety and security system. Prompt repair and replacement programs are needed to ensure the safety of staff and clients. Replacement of burned out lights, broken windows, etc. is essential to maintain the system in safe operating conditions.

ii. If an alarm system is to be effective, it must by used, tested and maintained according to strict policy. Any personal alarm devices should be carried and tested as required by the manufacturer and facility policy. Maintenance on personal and other alarm systems must take place monthly. Batteries and operation of the alarm devices must be checked by a security officer to ensure the function and safety of the system as prescribed by provisions of GISO 6184.

iii. Any mechanical device utilized for security and safety must be routinely tested for effectiveness and maintained on a scheduled basis.

C. Psychiatric Hospital/In-Patient Facilities

1. Engineering Control

Alarm systems are imperative for use in psychiatric units, hospitals, mental health clinics, emergency rooms or where drugs are stored. Whereas alarm systems are not necessarily preventive, they may reduce serious injury when a client is escalating in abusive behavior or threatening with or without a weapon.

a. Alarm systems which rely on the use of telephones, whistles or screams are ineffective and dangerous. A proper system consists of an electronic device which activates an alert to a dangerous situation in two ways, visually and audibly. Such a system identifies the location of the room or location of the worker by means of an alarm sound and a lighted indicator which visually identifies the location. In addition, the alarm should be sounded in a security area or other response team areas which will summon aid. This type of alarm system typically utilizes a pen-like device which is carried by the employee and can be triggered easily in an emergency situation. This system should be in accordance with provisions of California Title 8, GISO Section 6184, Emergency Alarm Systems (State of California, Department of Industrial Relations GISO). Back-up security personnel must be available to respond to the alarm.

b. "Panic buttons" are needed in medicine rooms, nurses stations, stairwells and activity rooms. Any such alarm system may incorporate a telephone paging system in order to direct others to the location of the disturbance but alarm systems must not depend on the use of a telephone to summon assistance.

c. Video screening of high risk areas or activities may be of value and permits one security guard to visualize a number of high risk areas, both inside and outside the building.

d. Metal detection systems such as hand-held devices or other systems to identify persons with hidden weapons should be considered. These systems are in use in courts, boards of supervisors, some Departments of Public Social Service, schools and emergency rooms. Although controversial, the fact remains that many people, including homeless and mentally ill persons do or are forced to carry weapons for defense while living on the streets. Some system of identifying persons who are carrying guns, knives, ice picks, screw drivers, etc. may be useful and should be considered. In psychiatric facilities, patients who have been on leave or pass should be screened upon return for concealed weapons.

2. Administrative Controls

A sound overall security program includes administrative controls that reduce hazards from inadequate staffing, insufficient security measures and poor work practices.

a. In order to enable staff members to identify and deal effectively with clients who behave in a violent manner, the administrator must insist on plans for patient treatment regimens and management of clients which include a gradual progression of measures given to staff to prevent violent behavior from escalating. These measures should not encourage inappropriate use of medication/restraints or isolation. However, the least restrictive yet appropriate and effective plan for preventing a client from injuring staff, other clients and self must be developed and be part of every unit and care plan. This enables a staff member to take primary prevention steps to stop escalating aggressive behavior. These procedures should cover verbal or physical threats or acting out of disturbed clients to help both the client and staff to feel a sense of control within the unit.

b. Security guards must be provided. These security guards should be trained in principles of human behavior and aggression. They should be assigned to areas where there may be psychologically stressed clients such as emergency rooms or psychiatric services.

c. In order to staff safely, a written acuity system should be established that evaluates the level of staff coverage vis-a-vis patient acuity and activity level. Staffing of units where aggressive behavior may be expected should be such that there is always an adequate, safe staff/patient ratio. The provision of reserve or emergency teams should be utilized to prevent staff members being left with inadequate support (regardless of staffing quotas) overwhelmed by circumstances of case load that would prevent adequate assessment of severity of illness. This also requires administrators to analyze and to identify times or areas where hostilities take place and provide a back-up team or staff at levels which are safe, such as in admission units, crisis or acute units or during the night hours or meal times or any other time or activity identified as high risk.

Provision of sufficient staff for interaction and clinical activity is important because patients/clients need access to medical assistance from staff. Possibility of violence often threatens staff when the structure of the patient/nurse relationship is weak. Therefore, sufficient staff members are essential to allow formation of therapeutic relationships and a safe environment.

d. It is necessary to establish on-call teams, reserve or emergency teams of staff who may provide services in hospitals such as, responding to emergencies, transportation or escort services, dining room assistance or many of the other activities which tend to reduce available staff where assigned.

e. All oncoming staff or employees should be provided with a census report which indicates precautions for every client. Methods must be developed and enforced to inform float staff, new staff members or oncoming staff at change of shifts of any potential assaultive behavior problems with clients. These methods of identification should include chart tags, log books, census reports and/or other information system within the facility. Other sources of information may include mandatory provision of probation reports of clients who may have had a history of violent behavior. However, the need for a program of "Universal Precautions for Violence" must be recognized and integrated in any patient care setting.

f. Staff members should be instructed to limit physical intervention in altercations between patients whenever possible unless there are adequate numbers of staff or emergency response teams, and security called. In the case where serious injury is to be prevented, emergency alarm systems should always be activated. Administrators need to give clear messages to clients that violence is not permitted. Legal charges may be pressed against clients who assault other clients or staff members. Administrators should provide information to staff who wish to press charges against assaulting clients.

g. Policies must be provided with regard to safety and security of staff when making rounds for patient checks, key and door opening policy, open vs. locked seclusion policies, evacuation policy in emergencies and for patients in restraints. Monitoring high risk patients at night and whenever behavior indicates escalating aggression, needs to be addressed in policy as well as medical management protocols.

h. Escort services by security should be arranged so that staff members do not have to walk alone in parking lots or other parking areas in the evening or late hours.

i. Visitors and maintenance persons or crews should be escorted and observed while in any locked facility. Often they have tools or possessions which could be inadvertently left and inappropriately used by clients.

j. Administrators need to work with local police to establish liaison and response mechanisms for police assistance when calls are made for help by a clinic or facility, and conversely to facilitate the hospital's provisions of assistance to local police in handling emergency cases.

k. Assaultive clients may need to be considered for placement in more acute units or hospitals where greater security may be provided. It is not wise to force staff members to confront a continuously threatening client, nor is it appropriate to allow aggressive behavior to go unchecked. Some programs may have the option of transferring clients to acute units, criminal units or to other more restrictive settings.

3. Work Practice Controls

a. Clothing should be worn which may prevent injury, such as low-heeled shoes, use of conservative earrings or jewelry and clothing which is not provocative.

b. Keys should be inconspicuous and worn in such a manner to avoid incidents yet be readily available when needed.

c. Personal alarm systems described under engineering controls must be utilized by staff members and tested as scheduled.

d. No employee should be permitted to work alone in a unit or facility unless backup is immediately available.

D. Clinics and Outpatient Facilities

1. Engineering Controls

a. An emergency personal alarm system is of the highest priority. An alarm system may be of two types: the personal alarm device as identified under hospitals and inpatient facilities or the type which is triggered at the desk of the counselor of medical staff. This desk system may be silent in the counseling room, but audible in a central assistance area and must clearly identify the room in which the problem is occurring. "Panic buttons" are needed in medicine rooms, bathrooms and other remote areas such as stairwells, nurses stations, activity rooms, etc.

Such systems may use a back-up paging or public address system on the telephone in order to direct others to the location for assistance but alarm systems must not depend on the use of telephone to summon assistance.

b. Maintenance is required for alarm systems as outlined in the Appendices, GISO, Section 6184 (Section X, Appendix B).

c. Reception areas should be designed so that receptionists and staff may be protected by safety glass and locked doors to the clinic treatment areas.

d. Furniture in crises treatment areas and quiet rooms should be kept to a minimum and be fixed to the floor. These rooms should have all equipment secured in locked cupboards.

e. First-aid kits shall be available as required in GISO Section 3400.

All requirements of the Bloodborne Pathogen Standard, GISO Section 5193, apply to clinics where blood exposure is possible.

2. Work Practice and Administrative Controls

a. Psychiatric clients/patients should be escorted to and from waiting rooms and not permitted to move about unsupervised in clinic areas. Access to clinic facilities other than waiting rooms should be strictly controlled with security provisions in effect.

b. Security guards trained in principles of human behavior and aggression should be provided during clinic hours. Guards should be provided where there may be psychologically stressed clients or persons who have taken hostile actions, such as in emergency facilities, hospitals where there are acute or dangerous patients or areas where drug or other criminal activity is commonplace.

c. Staff members should be given the greatest possible assistance in obtaining information to evaluate the history of, or potential for, violent behavior in patients. They should be required to treat and/or interview aggressive or agitated clients in open areas where other staff may observe interactions but still provide privacy and confidentiality.

d. Assistance and advice should be sought in case management conference with co-workers and supervisors to aid in identifying treatment of potentially violent clients. Whenever an agitated client or visitor is encountered, treatment or intervention should be provided when possible to defuse the situation. However, security or assistance should be requested to assist in avoiding violence.

e. No employee should be permitted to work or stay in a facility or isolated unit when they are the only staff member present in the facility, if the location is so isolated that they are unable to obtain assistance if needed, or in the evening or at night if the clinic is closed.

f. Employees must report all incidents of aggressive behavior such as pushing, threatening, etc. with or without injury, and logs maintained recording all incidents or near incidents.

g. Records, logs or flagging charts must be updated whenever information is obtained regarding assaultive behavior or previous criminal behavior.

h. Administrators should work with local police to establish liaison and response mechanisms for police assistance when calls are made for help by a clinic. Likewise, this will also facilitate the clinics provision of assistance to local police in handling emergency cases.

i. Referral systems and pathways to psychiatric facilities need to be developed to facilitate prompt and safe hospitalization of clients who demonstrate violent or suicidal behavior. These methods may include: direct phone link to the local police, exchange of training and communication with local psychiatric services and written guidelines outlining commitment procedures.

j. Clothing and apparel should be worn which will not contribute to injury such as low-heeled shoes, use of conservative earrings or jewelry and clothing which is not provocative.

k. Keys should be kept covered and worn in such a manner to avoid incidents, yet be available.

l. All protective devices and procedures should be required to be used by all staff.

E. Emergency Rooms and General Hospitals

1. Engineering Controls

a. Alarm systems or "panic buttons" should be installed at nurses stations, triage stations, registration areas, hallways and in nurse lounge areas. These alarm systems must be relayed to security police or locations where assistance is available 24 hours per day. A telephone link to the local police department should be established in addition to other systems.

b. Metal detection systems installed at emergency room entrances may be used to identify guns, knives or other weapons. Lockers can be used to store weapons and belongings or the weapons may be transferred to the local police department for processing if the weapons are not registered. Hand-held metal detection devices are needed to identify concealed weapons if there is no larger system. Signs posted at the entrance will notify patients and visitors that screening will be performed.

c. Seclusion or security rooms are required for containing confused or aggressive clients. Although privacy may be needed both for the agitated patient and other patients, security and the ability to monitor the patient and staff is also required in any secluded or quiet room.

d. Bullet-resistant glass should be used to provide protection for triage, admitting or other reception areas where employees may greet or interact with the public.

e. Strictly enforced limited access to emergency treatment areas are needed to eliminate unwanted or dangerous persons in the emergency room. Doors may be locked or key-coded.

f. Closed circuit T.V. monitors may be used to survey concealed areas or areas where problems may occur.

2. Work Practices and Administrative Controls

a. Security guards trained in principles of human behavior and aggression must be provided in all emergency rooms. Death and serious injury have been documented in emergency areas in hospitals, but the presence of security persons often reduces the threatening or aggressive behavior demonstrated by patients, relatives, friends, or those seeking drugs. Armed guards must be considered in any risk assessment in high volume emergency rooms.

b. No staff person should be assigned alone in an emergency area or walk-in clinic.

c. After dark, all unnecessary doors are locked, access into the hospital is limited and patrolled by security.

d. A regularly updated policy should be in place directing hostile patient management, use of restraints or other methods of management. This policy should be detailed and provide guidelines for progressively restrictive action as the situation calls for.

e. Any verbally threatening, aggressive or assaultive incident must be reported and logged.

f. Name tags need to be worn at all times in the hospital and emergency room. Hospital policy must demand that persons, including staff, who enter into the treatment area of the emergency room have or seek permission to enter the area to reduce the volume of unauthorized individuals.

g. When transferring a hostile or agitated patient (or one who may have relatives, friends or enemies who pose a security problem) to a unit within the hospital, security is required during transport and transfer to the unit. This security presence may be required until the patient is stabilized or controlled to protect staff who are providing care.

h. Emergency or hospital staff who have been assaulted should be permitted and/or assisted to request police assistance or file charges of assault against any patient or relative who injures, just as a private citizen has the right to do so. Being in the helping professions does not reduce the right of pressing charges or damages.

3. General Hospitals

a. Information must be clearly transmitted to the receiving unit of security problems with the patient. Charts must be flagged clearly noting and identifying the security risks involved with this patient.

b. If patients with any disorder or illness have a known history of violent acts, it is encumbent upon the administration to demand health care providers or physicians to disclose that information to hospital staff at the onset of hospitalization.

c. Whenever patients display aggressive or hostile behavior to hospital staff members, it must be made part of the care plan that supervisors and managers are notified and protective measures and action are initiated.

d. Prompt medical or emotional evaluation treatment must be made available to any staff who has been subjected to abusive behavior from a client/patient, whether in emergency rooms, psychiatric units or general hospital settings.

e. Visitors should sign in and have an issued pass particularly in newborn nursery, pediatric departments or any other risk departments.
   Any patient who may be deemed at risk should be placed on a "restricted visitor list." Restricted visitor lists must be maintained by security, nurses station and visitor sign-in areas.

f. Social service/worker staff should be utilized to defuse situation. In-house social workers are an important part of the hospital staff as are employee heath staff.

F. Home/Field Operations - Community Service Workers

1. Engineering Controls

a. In order to provide some measure of safety and to keep the employee in contact with headquarters or a source of assistance, cellular car phones should be installed/provided for official use when staff are assigned to duties which take them into private homes and the community. The workers may include (to name a few) parking enforcers, union business agents, psychiatric evaluators, public social service workers, childrens' service workers, visiting nurses and home health aides.

b. Hand-held alarm or noise devices or other effective alarm devices are highly recommended to be provided for all field personnel.

c. Beepers or alarm systems which alert a central office of problems should be investigated and provided.

d. Other protective devices should be investigated and provided such as pepper spray.

2. Work Practice and Administrative Controls

a. Employees are to be instructed not to enter any location where they feel threatened or unsafe. This decision must be the judgment of the employee. Procedures should be developed to assist the employee to evaluate the relative hazard in a given situation. In hazardous cases, the managers must facilitate and establish a "buddy system." This "buddy system" should be required whenever an employee feels insecure regarding the time of activity, the location of work, the nature of the clients health problem and history of aggressive or assaultive behavior or potential for aggressive acts.

b. Employers must provide for the field staff a program of personal safety education. This program should be at the minimum, one provided by local police departments or other agencies which include training on awareness, avoidance, and action to take to prevent mugging, robbery, rapes and other assaults.

c. Procedures should be established to assist employees to reduce the likelihood of assaults and robbery from those seeking drugs or money, as well as procedures to follow in the case of threatening behavior and provision for a fail-safe backup in administration offices.

d. A fail-safe back-up system is provided in the administrative office at all times of operation for employees in the field who may need assistance.

e. All incidents of threats or other aggression must be reported and logged. Records must be maintained and utilized to prevent future security and safety problems.

f. Police assistance and escorts should be required in dangerous or hostile situations or at night. Procedures for evaluating and arranging for such police accompaniment must be developed and training provided.

IV. MEDICAL MANAGEMENT

A medical program which provides knowledgeable medical and emotional treatment should be established. This program shall assure that victimized employees are provided with the same concern that is often shown to the abusive client. Violence is a major safety hazard in psychiatric and acute care facilities, emergency rooms, homeless shelters and other health care settings and workplaces. Medical and emotional evaluation and treatment are frequently needed but often difficult to obtain.

The consequences to employees who are abused by clients may include death and severe and life-threatening injuries, in addition to short- and long-term psychological trauma, post-traumatic stress, anger, anxiety, irritability, depression, shock, disbelief, self-blame, fear of returning to work, disturbed sleep patterns, headache and change in relationships with co-workers and family. All have been reported by health care workers after assaults, particularly if the attack has come without warning. They may also fear criticism by managers, increase use of alcohol and medication to cope with stress, suffer from feelings of professional incompetence, physical illness, powerlessness, increase in absenteeism and experience performance difficulties.

Administrators and supervisors have often ignored the needs of the physically or psychologically abused or assaulted staff, requiring them to continue working, obtain medical care from private medical doctors or blame the individual for irresponsible behavior. Injured staff must have immediate physical evaluations, be removed from the unit and treated for acute injuries. Referral should be made for appropriate evaluation, treatment, counseling and assistance at the time of the incident and for any required follow-up treatment.

A. Medical Services

1. This should include provision of prompt medical evaluation and treatment whenever an assault takes place regardless of severity. A system of immediate treatment is required regardless of time of day or night. Injured employees should be removed from the unit until order has been restored. Transportation of the injured to medical care must be provided if it is not available on-site or in an employee health service. Follow-up treatment provided at no cost to employees must also be provided.

B. Counseling Services

1. A trauma-crisis counseling or critical incident debriefing program must be established and provided on an ongoing basis whenever staff are victims of assaults. This "counseling program" may be developed and provided by in-house staff as part of an employee health service, by a trained psychologist, psychiatrist, or other clinical staff member such as a clinical nurse specialist, a social worker or referral may be made to an outside specialist. In addition, peer counseling or support groups may be provided. Any counseling provided should be by well-trained psychosocial counselors whether through EAP programs, in-house programs or by other professionals away from the facility who must understand the issues of assault and its consequences.

2. Reassignment of staff should be considered when assaults have taken place. At times it is very difficult for staff to return to the same unit to face the assailant. Assailants often repeat threats and aggressive behavior and actions need to be taken to prevent this from occurring. Staff development programs should be provided to teach staff and supervisors to be more sensitive to the feelings and trauma experienced by victims of assaults. Some professionals advocate joint counseling sessions including the assaultive client and staff member to attempt to identify the motive

when it occurs in inpatient facilities and to defuse situations which may lead to continued problems.

3. Unit staff should also receive counseling to prevent "blaming the victim syndrome" and to assist them with any stress problems they may be experiencing as a result of the assault. Violence often leaves staff fearful and concerned. They need to have the opportunity to discuss these fears and to know that administration is concerned and will take measures to correct deficiencies. This may be called a defusing or debriefing secession and unit staff members may need this activity immediately after an incident to enable them to continue working. First-aid kits or materials must be provided on each unit or facility.

4. The replacement and transportation of the injured staff member must be provided for at the earliest time. Do not leave a unit short staffed in the event of an assault. The development of an employee health service, staffed by a trained occupational health specialist, may be an important addition to the hospital team. Such employee health staff can provide treatment, arrange for counseling, refer to a specialist and should have procedures in place for all shifts. Employee health nurses should be trained in post-traumatic counseling and may be utilized for group counseling programs or other assistance programs.

5. Legal advice regarding pressing charges should be available, as well as information regarding workers' compensation benefits, and other employee rights must be provided regardless of apparent injury. If assignment to light duty is needed or disability is incurred, these services are to be provided without hesitation. Reporting to the appropriate local law enforcement agency and assistance in making this report is to be provided. Employees may not be discouraged or coerced when making reports or workers' compensation claims.

6. All assaults must be investigated, reports made and needed corrective action determined. However, methods of investigation must be such that the individual does not perceive blame or criticism for assaultive actions taken by clients. The circumstances of the incident or other information which will help to prevent further problems, needs to be identified, but not to blame the worker for incompetence and compound the psychological injury which is most commonly experienced.

V. RECORDKEEPING

Within the major program elements, recordkeeping is the heart of the program, providing information for analysis, evaluation of methods of control, severity determinations, identifying training needs and overall program evaluations.

Records shall be kept of the following:

1. OSHA 200 log. OSHA regulations require entry on the Injury and Illness Log 200, of any injury which requires more than first aid, is a lost time injury, requires modified duty or causes loss of consciousness. Assaults should be entered on the log. Doctors' reports of work injury and supervisors' reports shall be kept of each recorded assault.

2. Incidents of abuse, verbal attacks or aggressive behavior which may be threatening to the worker but not resulting in injury, such as pushing, shouting or an act of aggression toward other clients requiring action by staff should be recorded. This record may be an assaultive incident report or documented in some manner which can be evaluated on a monthly basis by department safety committee.

3. A system of recording and communicating should be developed so that all staff who may provide care for an escalating or potentially aggressive, abusive violent client will be aware of the status of the client and of any problems experienced in the past. This information regarding history of past violence should be noted on the patient's chart, communicated in shift change report and noted in an incident log.

4. An information gathering system should be in place which will enable incorporation of past history of violent behavior, incarceration, probation reports or any other information which will assist health care staff to assess violence status. Employees are to be encouraged to seek and obtain information regarding history of violence whenever possible.

5. Emergency room staff should be encouraged to obtain a record from police and relatives, information regarding drug abuse, criminal activity or other information to adequately assist in assessing a patient. This would enable them to appropriately house, treat and refer potentially violent cases. They should document the frequency of admission of violent clients or hostile encounters with relatives and friends.

6. Records need to be kept concerning assaults, including the type of activity, i.e., unprovoked sudden attack, patient-to-patient altercation and management of assaultive behavior actions. Information needed includes who was assaulted and circumstances of the incident without focusing on any alleged wrongdoing of staff persons. These records also need to include a description of the environment, location or any contributing factors, corrective measures identified, including building design, or other measures needed. Determination must be made of the nature of the injuries sustained: severe, minor or the cause of long-term disability, and the potential or actual cost to the facility and employee. Records of any

lost time or other factors which may result from the incident should be maintained.

7. Minutes of the safety meetings and inspections shall be kept in accordance with requirements of Title 8, Section 3203. Corrective actions recommended as a result of reviewing reports or investigating accidents or inspections need to be documented with the administration's response and completion dates of those actions should be included in the minutes and records.

8. Records of training program contents and sign-in sheets of all attendees should be kept. Attendance records at all "PART" or "MAB" training should be retained. Qualifications of trainers shall be maintained along with records of training.

VI. TRAINING AND EDUCATION

A. General

A major program element in an effective safety and security program is training and education. The purpose of training and education is to ensure that employees are sufficiently informed about the safety and security hazards to which they may be exposed and thus, are able to participate actively in their own and co-workers protection. All employees should be periodically trained in the employer's safety and security program.

Training and education are critical components of a safety and security program for employees who are potential victims of assaults. Training allows managers, supervisors and employees to understand security and other hazards associated with a job or location within the facility, the prevention and control of these hazards and the medical and psychological consequences of assault.

1. A training program should include the following individuals:

   a. All affected employees including doctors, dentists, nurses, teachers, counselors, psychiatric technicians, social workers, dietary and housekeeping, in short, all health care and community service staff and all other staff members who may encounter or be subject to abuse or assaults from clients/patients.
   b. Engineers, security officers, maintenance personnel.
   c. Supervisors and managers.
   d. Health care providers and counselors for employees and employee health personnel.

2. The program should be designed and implemented by qualified persons. Appropriate special training should be provided for personnel responsible for administering the training program.

3. Several types of programs are available and have been utilized, such as Management of Assaultive Behavior (MAB), Professional Assault Response Training (PART), Police Department Assault Avoidance Programs or Personal Safety training. A combination of such training may be incorporated depending on the severity of the risk and assessed risk. These management programs must be provided and attendance required at least yearly. Updates may be provided monthly/quarterly.

4. The program should be presented in the language and at a level of understanding appropriate for the individuals being trained. It should provide an overview of the potential risk of illness and injuries from assault and the cause and early recognition of escalating behavior or recognition of situations which may lead to assaults. The means of preventing or defusing volatile situations, safe methods of restraint or escape or use of other corrective measures or safety devices which may be necessary to reduce injury and control behavior are critical areas of training. Methods of self-protection and protection of co-workers, the proper treatment of staff and patient procedures, recordkeeping and employee rights need to be emphasized.

5. The training program should also include a means for adequately evaluating its effectiveness. The adequacy of the frequency of training should be reviewed. The whole program evaluation may be achieved by using employee interviews, testing and observing and/or reviewing reports of behavior of individuals in situations that are reported to be threatening in nature.

6. Employees who are potentially exposed to safety and security hazards should be given formal instruction on the hazards associated with the unit or job and facility. This includes information on the types of injuries or problems identified in the facility, the policy and procedures contained in the overall safety program of the facility, those hazards unique to the unit or program and the methods used by the facility to control the specific hazards. The information should discuss the risk factors that cause or contribute to assaults, etiology of violence and general characteristics of violent people, methods of controlling aberrant behavior, methods of protection and reporting procedures and methods to obtain corrective action.

   Training for affected employees should consist of both general and specific job training. "Specific job training" is contained in the following section or may be found in administrative controls in the specific work location section.

B. Job Specific Training

New employees and reassigned workers or registry staff should receive an initial orientation and hands-on-training prior to being placed in a treatment unit or job. Each new employee should receive a demonstration of alarm systems and protective devices and the required maintenance schedules and procedures. The training should also contain the use of administrative or work practice controls to reduce injury.

1. The initial training program should include:

    a. Care, use and maintenance of alarm tools and other protection devices.
    b. Location and operation of alarm systems.
    c. MAB, PART or other training.
    d. Communication systems and treatment plans.
    e. Policies and procedures for reporting incidents and obtaining medical care and counseling.
    f. Injury and Illness Prevention Program (8 CCR 3203).
    g. Hazard Communication Program (8 CCR 5194).
    h. Bloodborne Pathogen Program if applicable (8 CCR 5193).
    i. Rights of employees, treatment of injury and counseling programs.

2. On-the-job training should emphasize development and use of safe and efficient methods of de-escalating aggressive behavior, self-protection techniques, methods of communicating information which will help other staff to protect themselves and discussions of rights of employees vis-a-vis patient rights.

3. Specific measures at each location, such as protective equipment, location and use of alarm systems, determination of when to use the buddy system and so on as needed for safety, must be part of the specific training.

4. Training unit co-workers from the same unit and shift may facilitate team work in the work setting.

C. Training for Supervisors and Managers Maintenance and Security Personnel

1. Supervisors and managers are responsible for ensuring that employees are not placed in assignments that compromise safety and that employees feel comfortable in reporting incidents. They must be trained in methods and procedures which will reduce the security hazards and train employees to behave compassionately with co-workers when an incident does occur. They need to ensure that employees follow safe work practices and receive appropriate training to enable them to do this. Supervisors and

managers, therefore, should undergo training as comparable to that of the employee and such additional training as will enable them to recognize a potentially hazardous situation, make changes in the physical plant, patient care treatment program, staffing policy and procedures or other such situations which are contributing to hazardous conditions. They should be able to reinforce the employer's program of safety and security, assist security guards when needed and train employees as the need arises.

2. Training for engineers and maintenance should consist of an explanation or a discussion of the general hazards of violence, the prevention and correction of security problems and personal protection devices and techniques. They need to be acutely aware of how to avoid creating hazards in the process of their work.

3. Security personnel need to be recruited and trained whenever possible for the specific job and facility. Security companies usually provide general training on guard or security issues. However, specific training by the hospital or clinic should include psychological components of handling aggressive and abusive clients, types of disorders and the psychology of handling aggression and defusing hostile situations. If weapons are utilized by security staff, special training and procedures need to be developed to prevent inappropriate use of weapons and the creation of additional hazards.

VII. EVALUATION OF THE PROGRAM

Procedures and mechanisms should be developed to evaluate the implementation of the safety and security programs and to monitor progress and accomplishments. Top administrators and medical directors should review the program regularly. Semi-annual reviews are recommended to evaluate success in meeting goals and objectives. Evaluation techniques include some of the following:

A. Establishment of a uniform reporting system and regular review of reports.

B. Review of reports and minutes of Safety and Security Committee.

C. Analyses of trends and rates in illness/injury or incident reports.

D. Survey employees.

E. Before and after surveys/evaluations of job or worksite changes or new systems.

F. Up to date records of job improvements or programs implemented.

G. Evaluation of employee experiences with hostile situations and results of medical treatment programs provided. Follow up should be repeated several weeks and several months after an incident.

Results of management's review of the program should be a written progress report and program update which should be shared with all responsible parties and communicated to employees. New or revised goals arising from the review identifying jobs, activities, procedures and departments should be shared with all employees. Any deficiencies should be identified and corrective action taken. Safety of employees should not be given a lesser priority than client safety as they are often dependent on one another. If it is unsafe for employees, the same problem will be the source of risk to other clients or patients.

Managers, administrators, supervisors, medical and nursing directors should review the program frequently to reevaluate goals and objectives and discuss changes. Regular meetings with all involved including the Safety Committee, union representatives and employee groups at risk should be held to discuss changes in the program.

If we are to provide a safe work environment, it must be evident from administrators, supervisors and peer groups that hazards from violence will be controlled. Employees in psychiatric facilities, drug treatment programs, emergency rooms, convalescent homes, community clinics or community settings are to be provided with a safe and secure work environment and injury from assault is not to be accepted or tolerated and is no longer "part of the job."

VIII. REFERENCES AND ADDITIONAL READINGS

Adler WN, Kreeger C, Ziegler P: Patient violence in a psychiatric hospital, in Lion JR, Reid WH (Eds.) *Assaults within Psychiatric Facilities,* Orlando, FL: Grune & Stratton, Inc. pp. 81–90, 1983.

Bell C: Female homicides in United States workplaces, 1980–1985. *American Journal of Public Health,* 81(6):729–732, 1991.

Blair T, New SA: Assaultive behavior. *Journal of Psychosocial Nursing,* 29(11):25–29, 1991.

Bernstein HA: Survey of threats and assaults directed toward psychotherapists. *American Journal of Psychotherapy,* 35(4):542–549, 1981.

California Department of Industrial Relations, California Code of Regulations, Title 8, General Industry Safety Orders. Sections 3203, 6184 and 3400.

Carmel H, Hunter M: Staff injuries from inpatient violence. *Hospitals and Community Psychiatry,* 40(1):41–46, 1989.

Carmel H, Hunter M: Compliance with training in managing assaultive behavior and injuries from in-patient violence. *Hospital & Community Psychiatry,* 41(5):558–560, 1990.

Centers for Disease Control (CDC). Occupational Homicides Among Women - United States, 1980–1985. *MMWR* 39, 543–544, 551–552, 1990.

Cohen S, Kamarck T, Mermelstein R: A global measure of perceived stress. *Journal of Health and Social Behavior,* 24, 385–396, December, 1983.

Conn LM, Lion JR: Assaults in a university hospital. *Assaults Within Psychiatric Facilities,* Philadelphia, PA: W.B. Saunders & Co., pp. 61–69, 1983.

Craig TJ: An epidemiological study of problems associated with violence among psychiatric inpatients. *American Journal of Psychiatry,* 139(10):1262–1266, 1982.

Cronin M: New law aims to reduce kidnappings. *Nurse Week,* 5(3):1 and 24, 1991.

Davidson P, Jackson C: The nurse as a survivor: Delayed post-traumatic stress reaction and cumulative trauma in nursing. *International Journal of Nursing Studies,* 22(1):1–13, 1985.

Dillon S: Social workers: Targets in a violent society. *New York Times,* 11/18/92, pp. A1 and A18, 1992.

Edelman SE: Managing the violent patient in a community mental center. *Hospital & Community Psychiatry,* 29(7):460–462, 1978.

Eichelman E: A behavioral emergency plan. *Hospital & Community Psychiatry,* 35(10):1678, 1984.

Engle F, Marsh S: Helping the employee victim of violence in hospitals. *Hospital & Community Psychiatry,* 37(2):159–162, 1986.

Fineberg NA, James DV, Shah AK: Agency nurses and violence in psychiatric ward. *The Lancet,* 1, 474, 1988.

Goetz RR, Bloom JD, Chenell SL, Moorhead JC: Weapons possessed by patients in a university emergency department. *Annals of Emergency Medicine,* 20(1):8–10, 1981.

Gosnold DK: The violent patient in the accident and emergency department. *Royal Society of Health Journal,* 98(4):189–190, 1978.

Haffke EA, Reid WH: Violence against mental health personnel in Nebraska. In Lion JR, Reid WH (Eds.), *Assaults within Psychiatric Facilities* Orlando, FL: Grune and Stratton, Inc., pp 91–102, 1983.

Hatti S, Dubin WR, Weiss KJ: A study of circumstances surrounding patient assaults on psychiatrists. *Hospitals & Community Psychiatry,* 33(8):660–661, 1982.

Hodgkinson P, Hillis T, Russell D: Assaults on staff in psychiatric hospitals. *Nursing Times,* 80, 44–46, 1984.

Infantino AJ, Musingo S: Assaults and injuries among staff with and without training in aggression control techniques. *Hospital & Community Psychiatry,* 36, 1312–1314, 1983.

Ionno JA: A prospective study of assaultive behavior in female psychiatric inpatients. In J.R. Lion and W.H. Reid (Eds.). *Assaults within Psychiatric Facilities,* Orlando, FL: Grune & Stratton, Inc., pp. 71–80, 1983.

Jenkins LE, Layne L, Kesner S: Homicides in the Workplace. *The Journal of the American Association of Occupational Health Nurses,* 40(5):215–218, 1992.

Jones MK: Patient violence report of 200 incidents. *Journal of Psychosocial Nursing and Mental Health Services,* 23(6):12–17, 1985.

Keep N, Gilbert P, et al.: California Emergency Nurses Association's informal survey of violence in California emergency departments. *Journal of Emergency Nursing,* 18(5):433–442, 1992.

Kraus JF: Homicide while at work: Persons, industries and occupations at high risk. *American Journal of Public Health,* 77, 1285–1289, 1987.

Kurlowitcz L: Violence in the Emergency Department. *American Journal of Nursing,* 90(9):34–37, 1990.

Kuzmits FE: When employees kill other employees: The case of Joseph T. Wesbecker. *Journal of Occupational Medicine,* 32(10):1014–1020, 1990.

La Brash L, Cain J: A near-fatal assault on a psychiatric unit. *Hospital & Community Psychiatry,* 35(2):168–169, 1984.

Lanza ML: The reactions of nursing staff to physical assault by a patient. *Hospital and Community Psychiatry,* 34(1):44–47, 1983.

Lanza ML: Factors affecting blame placement for patient assault upon nurses. *Issues in Mental Health Nursing,* 6(1–2):143–161, 1984a.

Lanza ML: A follow-up study of nurses' reactions to physical assault. *Hospital & Community Psychiatry,* 35(5):492–494, 1984b.

Lanza ML: Victim assault support team for staff. *Hospital & Community Psychiatry,* 35(5):414–417, 1984c.

Lanza ML: Counseling services for staff victims of patient assault. *Administration in Mental Health,* 12(3):205–207, 1985a.

Lanza ML: How nurses react to patient assault. *Journal of Psychosocial Nursing,* 23(6):6–11, 1985b.

Lanza ML, Carifio J: Blaming the victim: Complex (non-linear) patterns of causal attribution by nurses in response to vignettes of a patient assaulting a nurse. *Journal of Emergency Nursing,* 17(5):299–309, 1991.

Lanza ML, Miller J: The dollar cost of patient assaults. *Hospital and Community Psychiatry,* 40(12):1227–1229, 1989.

Lavoie F, Carter GL, Denzel DF, Berg RL: Emergency department violence in United States teaching hospitals. *Annals of Emergency Medicine,* 17(11):1227–1233, 1988.

Levin PF, Hewitt J, Misner S: Female workplace homicides. *The Journal of the American Association of Occupational Health Nurses,* 40(8):229–236, 1992.

Levy P, Hartocollis P: Nursing aides and patient violence. *American Journal of Psychiatry,* 133(4):429–431, 1976.

Lion JR, Pasternak SA: Countertransference reactions to violent patients. *American Journal of Psychiatry,* 130(2):207–210, 1973.

Lion JR, Reid WH (Eds.): *Assaults within Psychiatric Facilities.* Orlando, FL: Grune & Stratton, Inc., 1983.

Lion JR, Snyder W, Merrill GL: Underreporting of assaults on staff in a state hospital. *Hospital & Community Psychiatry,* 32(7):497–498, 1981.

Lipscomb JA, Love C: Violence toward health care workers. *The Journal of the American Association of Occupational Health Nurses,* 40(5):219–228, 1992.

Liss GM: Examination of workers' compensation claims among nurses in Ontario for injuries due to violence. *Unpublished report, Health & Safety Studies Unit -Ministry of Labor,* 1993.

Long Beach (Calif.) Press Telegram (1990), April 15, 1.

Lusk SL: Violence experienced by nurses aides in nursing homes. *The Journal of the American Association of Occupational Health Nurses,* 40(5):237–241, 1992.

Madden DJ, Lion JR, Penna MW: Assaults on psychiatrists by patients. *American Journal of Psychiatry,* 133(4):422–425, 1976.

Mantell M: The crises response team reports on Edmond, Oklahoma massacre. *Nova Newsletter 11,* 1987.

McNeil DE, et al.: Characteristics of persons referred by police to psychiatric emergency room. *Hospital & Community Psychiatry,* 42(4):425–427, 1991.

Meddis SV: 7 cities lead violence epidemic. *USA Today,* April 29, 1991.

Monahan J, Shah SA: Dangerousness and commitment of the mentally disordered in the United States. *Schizophrenia Bulletin,* 15(4):541–553, 1989.

Morrison EF, Herzog EA: What therapeutic and protective measures, as well as legal actions, can staff take when they are attached by patients. *Journal of Psychosocial Nursing,* 30(7):41–44, 1992.

Navis ES: Controlling violent patients before they control you. *Nursing,* 87, 17, 52–54, 1987.

Ochitill HN: Violence in a general hospital. In Lion JR, Reid WH (Eds.), *Assaults within Psychiatric Facilities,* Orlando, FL: Grune & Stratton, Inc., pp. 103–118, 1983.

Phelan LA, Mills MJ, Ryan JA: Prosecuting psychiatric patients for assaults. *Hospital & Community Psychiatry,* 36(6):581–582, 1985.

Poster EC, Ryan JA: Nurses' attitudes toward physical assaults by patients. *Archives of Psychiatric Nursing,* 3(6):315–322, 1989.

Rossi AM, Jacobs M, Monteleone M, Olson R, Surber RW, Winkler E, Wommack A: Violent or fear-inducing behavior associated with hospital admission. *Hospital & Community Psychiatry,* 36(6):643–647, 1985.

Ruben I, Wolkon G, Yamamoto J: Physical attacks on psychiatric residents by patients. *Journal of Nervous and Mental Disease,* 168(4):243–245, 1980.

Ryan JA, Poster EC: The assaulted nurse: Short-term and long-term responses. *Archives of Psychiatric Nursing,* 3(6):323–331, 1989a.

Ryan JA, Poster EC: Supporting your staff after a patient assault. *Nursing,* 89(12):32k, 32n, 32p., 1989b.

Ryan JA, Poster EC: When a patient hits you. *Canadian Nurse,* 87(8):23–25, 1991.

Schwartz CJ, Greenfield GP: Charging a patient with assault of a nurse on psychiatric unit. *Canadian Psychiatric Association Journal,* 23(4):197–200, 1978.

Scott JR, Whitehead JJ: An administrative approach to the problem of violence. *Journal of Mental Health Administration,* 8(2):36–40, 1981.

Sosowsky L: Explaining the increased arrest rate among mental patients: A cautionary note. *American Journal of Psychiatry,* 137(12):1602–1605, 1980.

State of California/Internal Memorandum Employee lost workday injuries from client violence, 1973–1980, 1980.

Tardiff K: A survey of assault by chronic patients in a state hospital system. In Lion JR, Ried WH (Eds.), *Assaults within Psychiatric Facilities* Orlando, FL: Grune & Stratton, Inc., pp 3–20, 1983.

Tardiff K, Koenigsberg HW: Assaultive behavior among psychiatric outpatients. *American Journal of Psychiatry,* 142(8):960–963, 1985.

Tardiff K, Sweillam A: Assault, suicide and mental illness. *Archives of General Psychiatry,* 37(2):164–169, 19890.

Tardiff K, Sweillam A: Assaultive behavior among chronic inpatients. *American Journal of Psychiatry,* 139(2):212–215, 1982.

Teplin L: The prevalence of severe mental disorder among male urban jail detainees: Comparison with the epidemiologic catchment area program. *American Journal of Public Health,* 80(6):663–669, 1990.

U.S. Department of Labor, Bureau of Labor Statistics, *Occupational Injuries and Illnesses in the United States by Industry,* 1989. Bulletin 2379, 1991.

U.S. Department of Labor, Bureau of Labor Statistics, *A Brief Guide to Recordkeeping Requirements for Occupational Injuries and Illness,* 29 CFR 1904, 1986.

Wasserberger J, Ordog GJ, Harden E, Kolodny M, Allen K: Violence in the Emergency Department. *Topics in Emergency Medicine,* 14(2):71–78, 1992.

Wasserberger J, Ordog GJ, Kolodny M, Allen K: Violence in a Community Emergency Room. *Archives of Emergency Medicine,* 6, 266–269, 1989.

White SG, Hatcher C: Violence and trauma response. Larsen RC, Felton JS (Eds.), Psychiatric injury in the workplace. *Occupational Medicine: State of the Art Reviews,* Hanley & Belfus, Inc., Philadelphia, 3(4):677–694, 1988.

Whitman RM, Armao BB, Dent OB: Assault on the therapist. *American Journal of Psychiatry,* 133(4):426–429, 1976.

Wilkinson T: Drifter judged sane in killing of mental health therapist, *Los Angeles Times,* December 11, 1990, B1-B4.

Winterbottom S: Coping with the violent patient in accident and emergency. *Journal of Medical Ethics,* 5(3):124–127, 1979.

Yesavage JA, Werner PD, Becker J, et al.: Inpatient evaluation of aggression in psychiatric patients. *Journal of Nervous and Mental Disease,* 169(5):299–302, 1981.

Zitrin A, Herdesty AS, Burdock EL, Drossman AK: Crime and violence among mental patients. *American Journal of Psychiatry,* 133(2):142–149, 1976.

IX. GLOSSARY

Abusive behavior: Actions which result in injury such as slapping, pinching, pulling hair or other actions such as pulling clothing, spitting, threats or other fear producing actions such as racial slurs, posturing, damage to property, throwing food or objects.

Assault: Any aggressive act of hitting, kicking, pushing, biting, scratching, sexual attack or any other such physical or verbal attacks directed to the worker by a patient/client, relative or associated individual which arises during or as a result of the performance of duties and which results in death, physical injury or mental harm.

Assaultive incident: An aggressive act or threat by a patient/client, relative or associated individual which may cause physical or mental injury, even of a minor nature, requiring first aid or reporting.

Community worker: All employed workers who provide service to the community in private homes, places of business or other locations which may present an unsafe or hostile environment. Examples of such workers include, but are not limited to parking enforcement officers, psychiatric social workers, home health workers, union representatives, visiting or public health nurses, social service workers and home health aids. The location of the workplace may be mobile or fixed.

Inpatient facility: A hospital, convalescent hospital, nursing home, board and care facility, homeless shelter, developmentally disabled facility, correction facility or any facility which provides 24-hour staffing and health care, supervision and protection.

Injury: Physical or emotional harm to an individual resulting in broken bones, lacerations, bruises and contusions, scratches, bites, breaks in the skin, strains and sprains, or other pain and discomfort immediate or delayed, caused by an interaction with a patient/client or in the performance of the job.

Management of Assaultive Behavior (MAB): A training program which trains staff to prevent assaultive incidents and to implement emergency measures when prevention fails.

Mental harm: Anxiety, fear, depression, inability to perform job functions, post-traumatic stress syndrome, inability to sleep or other manifestations of emotional reactions to an assault or abusive incident.

Outpatient facility: Any health care facility or clinic, emergency room, community mental health clinic, drug treatment clinic or other facility which provides drop-in or other "as needed care" or service to the community in fixed locations.

Professional Assault Response Training (PART): A training program designed to provide a systematic approach to recognition and control of escalating aggressive and assaultive behavior in a patient/client or of other hostile situations.

Psychiatric inpatient facility: Public or private psychiatric inpatient treatment facilities.

Threat: A serious declaration of intent to harm at the time or in the future.

Threat or verbal attack: Any words, racial slurs, gestures or display of weapons which are perceived by the worker as a clear and real threat to their safety and which may cause fear, anxiety or inability to perform job functions.

APPENDIX A

General Industry Safety Orders (Section 3203) § 3203. Injury and Illness Prevention Program

(a) Effective July 1, 1991, every employer shall establish, implement and maintain an effective Injury and Illness Prevention Program (Program). The Program shall be in writing and, shall, at a minimum:

(1) Identify the person or persons with authority and responsibility for implementing the Program.

(2) Include a system for ensuring that employees comply with safe and healthy work practices. Substantial compliance with this provision includes recognition of employees who follow safe and healthful work practices, training and retraining programs, disciplinary actions, or any other such means that ensures employee compliance with safe and healthful work practices.

(3) Include a system for communicating with employees in a form readily understandable by all affected employees on matters relating to occupational safety and health, including provision designed to encourage employees to inform the employer of hazards at the worksite without fear of reprisal. Substantial compliance with this provision includes meetings, training programs, posting, written communications, a system of anonymous notification by employees about hazards, labor/management safety and health committees, or any other means that ensure communication with employees.

EXCEPTION: Employers having fewer than 10 employees shall be permitted to communicate to and instruct employees orally in general safe work practices with specific instructions with respect to hazards unique to the employees' job assignments as compliance with subsection (a)(3).

(4) Include procedures for identifying the evaluating work place hazards including scheduled periodic inspections to identify unsafe conditions and work practices. Inspections shall be made to identify and evaluate hazards.

(A) When the Program is first established.

EXCEPTION: Those employers having in place on July 1, 1991, a written Injury and Illness Prevention Program complying with previously existing section 3203.

(B) Whenever new substances, processes, procedures, or equipment are introduced to the workplace that represent a new occupational safety and health hazard; and

(C) Whenever the employer is made aware of a new or previously unrecognized hazard.

(5) Include a procedure to investigate occupational injury or occupational illness.

(6) Include methods and/or procedures for correcting unsafe or unhealthy conditions, work practices and work procedures in a timely manner based on the severity of the hazard:

(A) When observed or discovered; and

(B) When an imminent hazard exists which cannot be immediately abated without endangering employee(s) and/or property, remove all exposed personnel from the area except those necessary to correct the existing condition. Employees necessary to correct the hazardous condition shall be provided the necessary safeguards.

(7) Provide training and instruction:

(A) When the program is first established,

EXCEPTION: Employers having in place on July 1, 1991, a written Injury and Illness Prevention Program complying with the previously existing Accident Prevention Program in section 3203.

(B) To all new employees;

(C) To all employees given new job assignments for which training has not previously been received;

(D) Whenever new substances, processed procedures or equipment are introduced to the workplace and represent a new hazard.

(E) Whenever the employer is made aware of a new or previously unrecognized hazard; and,

(F) For supervisors to familiarize them with the safety and health hazards to which employees under their immediate direction and control may be exposed

(b) Records of the steps taken to implement and maintain the Program shall include:

(1) Records of scheduled and periodic inspections required by subsection (a)(4) to identify unsafe conditions and work practices, including person(s) conducting the inspection, the unsafe conditions and work practices that have been identified and action taken to correct the identified unsafe conditions and work practices. These records shall be maintained for three (3) years; and

EXCEPTION: Employers with fewer than 10 employees may elect to maintain its inspection records only until the hazard is corrected.

(2) Documentation of safety and health training required by subsection (a)(7) for each employee, including employee name or other identified training dates, type(s) of training, and training providers. This documentation shall be maintained for three (3) years.

EXCEPTION NO. 1: Employers with fewer than 10 employees can substantially comply with the documentation provision by maintaining a log of instruction provided to the employee with respect to the hazards unique to the employees work assignment when he is first hired or assigned new duties.

EXCEPTION NO. 2: Training records of employees who have worked for less than one (1) year for the employer need not be retained beyond the term of employment if there are provided to he employee upon termination of employment.

(c) Employers who elect to use a labor/management safety and healthy committee to comply with the communication requirements of subsection (a)(3) of this section shall be presumed to be in substantial compliance with subsection (a)(3) if the committee:

(1) Meets regularly, but not less than quarterly;

(2) Prepares and makes available to the affected employees, written records of the safety and health issues discussed at the committee meetings and, maintained for review by the Division upon request;

(3) Reviews results of the periodic, scheduled worksite inspections;

(4) Reviews investigations of occupational accidents and causes of incidents resulting in occupational injury, occupational illness, or exposure to hazardous substances and, where appropriate, submits suggestions to management for the prevention of future incidents.

(5) Reviews investigations of alleged hazardous conditions brought to the attention of any committee member. When determined necessary by the committee, the committee may conduct its own inspection and investigation to assist in remedial solutions:

(6) Submits recommendations to assist in the evaluation of employee safety suggestions; and

(7) Upon request from the Division, verifies abatement action taken by the employer to abate citations issued by the Division.

NOTE: Authority cited Sections 142.3 and 6401.7, Labor Code Reference. Sections 142.3 and 6401.7 Labor Code.

HISTORY

1. New section filed 4-1-77, effective thirtieth day thereafter (Register 77, No. 14). For former history, see Register 74, No. 43.
2. Editorial correction of subsection (a)(1)(Register 77, No. 41).
3. Amendment of subsection (a)(2) filed 4-12-83, effective thirtieth day thereafter (Register 83, No. 16).
4. Amendment filed 1-16-91, operative 2-15-91 (Register 91, No. 8).
5. Editorial correction of subsections (a), (a)(2), (a)(4)(A) and (a)(7)(Register 91, No. 31).

Appendix B

General Industry Safety Order (Section 6184)
Article 165. Employee Alarm Systems

(a) Scope and Application.

(1) This section applies to all emergency employee alarms. This section does not apply to those discharge or supervisory alarms required on various fixed extinguishing systems or to supervisory alarms on fire suppression, alarm or detection systems unless they are intended to be employee alarm systems.

(2) The requirements in this section that pertain to maintenance, testing and inspection shall apply to all local fire alarm signaling systems used for alerting employees regardless of the other functions of the system.

(3) All pre-discharge employee alarms shall meet the requirements of subsection (b)(1) through (b)(4), (c) and (d)(1) of this section.

(4) The employee alarm shall be distinctive and recognizable as a signal to evacuate the work area or to perform actions designated under the emergency action plan.

(5) All employees shall be made aware of means and methods of reporting emergencies. These methods may be but not limited to manual pull box alarms, public address systems, radio or telephones. When telephones are used as a means of reporting an emergency, telephone numbers shall be conspicuously posted nearby.

(6) The employer shall establish procedures for sounding emergency alarms in the workplace. For those employers with 10 or fewer employees in a particular workplace, direct voice communication is an acceptable procedure for sounding the alarm provided all employees can hear the alarm. Such workplaces need not have a back-up system.

(b) General Requirements.

(1) Where local fire alarm signaling systems are required by these orders, they shall meet the design requirements of the National Fire Protection Association's "Standard for the Installation, Maintenance, and Use of Local Protective Signaling Systems for Watchman, Fire Alarm and Supervisory Service," NFPA No. 72A—1975 and the requirements of this section.

(2) The employee alarm system shall provide warning for necessary emergency action as called for in the emergency action plan, or for reaction time for safe escape of employees from the workplace or the immediate work area, or both.

(3) The employee alarm shall be capable of being perceived above ambient noise or light levels by all employees in the affected portions of the workplace. Tactile devices may be used to alert those employees who would not otherwise be able to recognize the audible or visual alarm.

(c) Installation and Restoration.

(1) The employer shall assure that all devices, components, combinations of devices or systems constructed and installed to comply with this standard shall be approved. Steam whistles, air horns, strobe lights or similar lighting devices, or tactile devices meeting the requirements of this section are considered to meet this requirement for approval.

(2) All employee alarm systems shall be restored to normal operating condition as promptly as possible after each test or alarm.

(d) Maintenance and Testing.

(1) All employee alarm systems shall be maintained in operating condition except when undergoing repairs or maintenance.

(2) A test of the reliability and adequacy of non-supervised employee alarm systems shall be made every two months. A different actuation device shall be used in each test of a multi-actuation device system so that no individual device is used for two consecutive tests.

(3) The employer shall maintain or replace power supplies as often as is necessary to assure a fully operational condition. Back-up means of alarm, such as employee runners or telephones, shall be provided when systems are out of service.

(4) Employee alarm circuitry installed after July 1, 1981, shall be supervised and provide positive notification to assigned personnel whenever a deficiency exists in the system. All supervised employee alarm systems shall be tested at least annually for reliability and adequacy.

(5) Servicing, maintenance and testing of employee alarms shall be performed by persons trained in the designed operation and functions necessary for reliable and safe operations of the system.

(e) Manual Operation.

(1) Manually operated actuation devices for use in conjunction with employee alarms shall be unobstructed, conspicuous and readily accessible.

NOTE: Authority and reference cited: Section 142.5, Labor Code.

HISTORY:

1. New Article 165 (Section 6184) filed 9-8-81; effective thirtieth day thereafter (Register 81, No. 37).
2. Editorial correction of subsections (b)(1) and (e)(1) filed 11-9-81; effective thirtieth day thereafter (Register 81, No. 45).
3. Editorial correction of subsections (b) and (d) filed 6-30-82 (Register 82, No. 27).
4. Change without regulatory effect deleting Title 24 reference (Register 87, No. 49).
5. Editorial correction of subsection (e)(1) deleting obsolete Title 24 reference (Register 88, No. 9).

Appendix C

The New Jersey Department of Labor Bulletin

Guidelines on Measures and Safeguards in Dealing with Violent or Aggressive Behavior in Public Sector Health Care Facilities

INTRODUCTION

The State of New Jersey and its local government agencies are concerned with the management of clients with violent or aggressive behavior toward employees. The level of awareness and sensitivity to effective and available safeguards must be continuously emphasized.

These guidelines are being issued by the New Jersey Department of Labor (NJDOL) under the Public Employees Occupational Safety and Health (PEOSH) Program to assist public employers in health care facilities in adopting measures and procedures which will protect the safety of their employees. This is consistent with the legislative intent of the New Jersey Public Employees Occupational Safety and Health Act, N.J.S.A. 34:6A-25 et seq.

AUTHORITY

Pursuant to N.J.S.A. 34:6A-31, and the Safety and Health Standards for Public Employees, N.J.A.C. 12:100, Commissioner of Labor is authorized to promulgate safety and health standards for employees working within public facilities in the State of New Jersey.

Scope

Health care facility workers employed in jobs with patients/clients assessed as a safety risk because of violent or aggressive behavior.

1. Safety measures and procedures shall include:

    a. A system for patient/client assessment with respect to client/patient behaviors.
    b. A system for communicating such information to employees assigned to work with the assessed patient/client.
    c. A system for summoning assistance in a cottage or ward if a patient/client behaves in a violent or aggressive manner.
    d. Provision or creation or an area where violent or aggressive patients/clients can be contained if necessary to ensure the safety of others.
    e. A procedure for patient/client restraint to be used when necessary to ensure the safety of others.

2. Instruction and training in the management of violent and aggressive patients/clients shall be provided to all patient/client care providers during orientation and a review program shall be offered periodically.
3. First-aid supplies and personnel trained in first aid shall be available in all cottages and wards.
4. A system of supportive intervention shall be made available to any employee involved in an incident with a violent or aggressive patient/client.
5. Safe staffing levels as determined by facility or Division policy or mandated by regulation shall be maintained.
6. In instances of violent or aggressive patient/client activity, systems shall be in place to initiate appropriate interventions. Concerns about the possibility of a violent patient/client incident shall be reported to an immediate supervisor who shall immediately investigate and take whatever action is necessary to manage the situation in a manner that prevents or minimizes harm.

PEOSH REPORTING REQUIREMENTS

In compliance with the New Jersey Public Employees Occupational Safety and Health (PEOSH) Act, N.J.S.A. 34:6A-25 et seq., all public employers shall maintain records and file reports on occupational injuries and illness occurring to their workers. Such documentation shall be maintained on the Public Employees Occupational Safety and Health Program Log and Summary of Occupational Injuries and Illness (NJOSH No. 200).

Appendix D

Sources of Assistance in Training and Program Development

Note: These programs and groups have not been evaluated, not recommended but are provided as an available source.

1. Part - "Professional Assault Response Training"
   Paul A. Smith, PhD
   Professional Growth Facilitators
   Post Office Box 5981
   San Clemente, CA 92674-5981
   Telephone & Fax #(714)498-3529

   Part ® — Basic Course
   Part ® — Advance Consultation
   Part ® — Training for Trainers
   Part ® — Trainer Re-Certification

2. Non-violent Crises Intervention
   A two-tape video series
   National Crises Prevention Institute
   3 315-K N. 124th Street
   Brookfield, WI 53005
   1(800)558-8976

3. Service Employees International Union (SEIU)
   Maggie Robbins & John Mehring
   Western Health & Safety Coordinator
   3055 Wilshire Boulevard, Suite 1050
   Los Angeles, CA 90010

PART 3

Assault on the Job: We Can Do Something About It!

This material is by the Service Employees International Union, representing 1 million workers employed in the public and private sector. The SEIU is also the nations largest health care workers union representing over 450,000 health care workers.

On any given day, in any given place, assaults are a reality of our working lives.

Joe, a mental health aide, must often work alone due to cutbacks in staffing. In the middle of the night he is attacked by a patient, suffering cuts, bruises, and a back injury. This is his second assault in a year.

As a receptionist in a social service agency, Donna is on the front line, dealing with clients who are frustrated and impatient with the system. Yet she has little power to do anything to help them. The abuse, hostility, and threats of violence take their toll on Donna's health. She fears the day when someone will attack her with a weapon.

The workers are repairing gas lines when they get caught in the cross fire of a gang shootout.

Maria works in a nursing home. One of her patients continually hassles her with insults and sexual advances. When her attempts to transfer fail, she begins to suffer frequent headaches and has trouble sleeping.

WHAT IS ASSAULT ON THE JOB?

As the above cases show, assault is not just physical violence. It also includes near misses, verbal abuse, unwanted sexual advances, or the threat of any of these. Even if a worker is never physically injured, the stress from the fear of assault may lead to serious health problems.

Assault on the Job can Also Mean Death

The National Institute for Occupational Safety and Health (NIOSH) estimates that during the early 1980s, 13% of the 7,000 annual work-related deaths were no

1-56670-083-3/94/$0.00+$.50

accidents—they were preventable homicides. According to the Centers for Disease Control and Prevention, murder is the leading cause of death for women in the workplace.

How Big a Problem is it?

State workers' compensation agencies don't keep figures on job-related assaults so it's not known how many work-related injuries are caused by on-the-job violence. And workers often do not report assaults. Because so many attacks go unreported, employers, the government and often workers as well, don't always recognize violence as a workplace health and safety issue. Still, common sense should tell us that society's problems with violence doesn't stop outside the walls of our workplace.

Why is Workplace Assault Usually not Reported?

The reasons are many:

- "Part of the job" syndrome: In certain jobs workers are expected to put up with attacks, threats and verbal abuse.
- A violent society is to blame: The assault is considered a consequence of living in a violent society, rather than working in an unsafe workplace.
- Fear of blame or reprisal: Workers may be afraid that they'll be held responsible for any violent act that involved a patient or client.
- Lack of management support: Workers are often discouraged from reporting problems.
- No serious injuries: When physical injuries are minor, or when a worker doesn't miss a day of work, the injury is not reported to the worker's compensation board.
- Not worth the effort: If workers think nothing will be done, they feel there is no reason to file a report.

Who is Affected?

Workers who are most likely to be assaulted are:

- Those who work with the public;
- Those who must work alone;
- Those who handle money;
- Those who come in contact with patients or clients who may be violent;
- New employees.

Many SEIU members are affected, especially those who work in hospitals, nursing homes, mental health institutions and facilities for the developmentally disabled, shelters, prisons, social service departments and social/family service agencies, or educational institutions, as well as visiting nurses, home-care providers and utility workers.

Why are Workers Assaulted?

Each incident of violence has its own set of causes. Working with clients or patients who may be frustrated, anxious, impatient, angry, in shock, mentally disturbed or under the influence of drugs or alcohol inevitably carries with it the potential for violence. These people may lash out against whoever is closest to them—often an employee. Assaults are often unprovoked but also occur when workers are performing their duties in restraint and seclusion procedures.

Some specific factors which commonly play a role are:

- Understaffing, which forces people to work alone or without enough staff to provide good coverage, thus allowing tensions to rise among patients or clients;
- Deinstitutionalization, which leaves institutions with patients and clients who need greater attention and care and, in the absence of proper staffing and safeguards, can become dangerous;
- Lack of training for workers in recognizing and defusing potentially violent situations;
- Failure to alert workers to which patients or clients have a history of violence;
- Failure to design safe workplaces and emergency procedures;
- Failure to identify hazardous conditions and develop proper controls, policies and education programs.

Many of these factors result from budget cuts that are all too common in health care institutions and government agencies these days. These cutbacks make it hard for workers to give the care and service patients and clients deserve. Everyone suffers as a result. Increased violence is one very real effect of the budget axe.

IS SEXUAL HARASSMENT A FORM OF ASSAULT ON THE JOB?

Sexual harassment is unwanted, repeated sexual attention at work. It may be expressed in the following ways:

- Unwelcome touching or patting;
- Suggestive remarks or other verbal abuse;
- Staring or leering;
- Requests for sexual favors;
- Compromising invitations;
- Physical assault;
- Offensive work environment (pinups/pornography)

Sexual harassment *is* another form of assault. Even if a worker is never physically injured, the stress of a repeated verbal abuse or fear of impending violence can result in serious health problems. Those who have experienced severe harassment cite a long list of physical symptoms including headaches, backaches, nausea, stomach ailments, fatigue and sleep and eating disorders.

Sexual harassment is a form of sex discrimination and it is illegal under Title VII of the Civil Rights Act of 1964.

Specifically, sexual harassment is illegal if:

- Your job or promotion depends on your saying yes;
- The harassment creates an intimidating, hostile or offensive workplace.

A Widespread Problem

Sexual harassment happens in every kind of work environment, at all levels. It is carried out by superiors and subordinates, co-workers and clients. Every race, gender or age group can become a target. Men are also sexually harassed, though on a much smaller scale.

The victim does not have to be of the opposite sex. In fact, according to Equal Employment Opportunity Commission (EEOC) guidelines, the victim does not even have to be the person harassed but could be anyone affected by the offensive conduct.

> Nancy and Susan work together. Susan is being harassed by a male co-worker who uses foul language and tells off-color jokes to embarrass Susan. He also leaves pornographic pictures on Susan's desk. Nancy overhears these remarks and sometimes sees the pictures. In this instance, Susan is not the only person who is being sexually harassed!

- A survey of federal employees in 1980, showed that 42% of females and 15% of males said they'd been harassed on the job. A follow-up survey in 1987 yielded nearly identical results.
- Since 1980, more than 38,500 charges of sexual harassment have been filed with the federal government, but this figure is just the tip of the iceberg because many workers never file a complaint.

What is Your Employer's Responsibility?

A significant number of workplaces—especially small and medium-sized firms—have no policy regarding sexual harassment, but they should. While there is no one model for a good sexual harassment policy, all policies should send a clear message: *sexual harassment will not be tolerated.*

Elements of a Good Sexual Harassment Policy

- Union representatives are involved in policy development.
- Employees are involved at all stages of policy development.
- The policy is in writing and widely distributed through an employee handbook and orientation materials.
- The policy is publicized to both staff and clients.
- Top management is seen actively supporting the policy.
- Training is ongoing and occurs on work time.
- Procedures for how complaints are to be reported, recorded and investigated are clear and concise.
- Managers and supervisors are clearly instructed to begin investigations within seven days after a formal complaint is made.
- Discipline and counseling procedures are clearly stated. Discipline can range from verbal and written warnings to formal reprimands, suspension, transfer, probation, demotion or dismissal. Counseling and/or sensitivity training may be appropriate.
- Confidentiality of an incident is maintained throughout the process.

Sexual Harassment: Fighting Back!

Dealing with sexual harassment may be difficult, but ignoring sexual harassment does not make it go away. Use the following guidelines to establish a strong case and fight back!

1. *Say no clearly.* State frankly that you find the harasser's behavior offensive. Firmly refuse all invitations. If harassment persists, write a memo asking the harasser to stop; keep a copy.

2. *Document the harassment.* Detail what, when and where it happened, and include your response. This information is vital when a pattern of offensive conduct must be proven.

3. *Get emotional support from friends and family.* Don't try to fight this alone.

4. *Keep records of your job performance.* Your harasser may question your job performance in order to justify his/her behavior.

5. *Look for witnesses and other victims.* Two accusations are harder to ignore.

6. *Grieve it.* Get your steward and union involved right away.

7. *File a complaint.* Contact your state anti-discrimination agency or the federal Equal Employment Opportunity Commission (EEOC) if you decide to pursue a legal solution. You don't need an attorney to file a claim, but it may help to speak with a lawyer who specializes in employment discrimination.

8. *Act promptly.* There are state and federal time limits on how long after an act of harassment a complaint can be filed.

9. *Don't be intimidated.* It is unlawful for employers to retaliate after a complaint is filed. File another complaint based on the retaliation.

10. *Negotiate* a strong clause in your union contract that protects workers from sexual harassment.

11. *Educate and Agitate.* Organize discussions on sexual harassment. Find out if others are experiencing the problem. Use posters, buttons and flyers to send a strong message to management that workers will not accept such hostile working conditions.

Sexual Harassment Can be Stopped

Break the silence. Use all means to demonstrate that sexual harassment will not be tolerated in the workplace!

POST-TRAUMATIC STRESS DISORDER (PTSD)

Assault victims have something in common with combat veterans and victims of terrorism, natural and man-made disasters, street crime, rape and incest—an increased risk of post-traumatic stress disorder (PTSD).

What is PTSD?

PTSD is the way a person reacts to emotional stress or physical injury, assault or other forms of extreme stress outside everyday experience. It includes physical pain from the assault, as well as anger, anxiety, depression, fatigue and preoccupation with the event. Other common symptoms are depression, flashbacks and nightmares. PTSD can also do serious damage to family relations and social life.

Should PTSD be Treated?

Yes. Voluntary individual counseling is the best form of treatment for an assaulted worker. Often, however, assault victims fail to seek help and blame themselves for the incident. They may also see a psychiatric referral for treatment as an indication of mental illness, rather than as simply continuing treatment for their injuries.

What are the Employers' Obligations?

Employers are responsible for providing employees a safe healthy work environment free from recognized hazards. The location of the assault, which is likely to be the everyday worksite, can provoke anxiety and flashbacks (sometimes for weeks or months) and make it difficult for an employee to return to work. Therefore, injured workers need far more than a few minutes of counseling. They need ongoing support. Most assaulted workers depend on the informal support of family and friends which may not be enough.

Workers will also feel less victimized if they feel they have some control of their safety on the job. A staff's ability to make changes in policies and procedures that lead to a safer workplace will help foster a sense of empowerment and control.

What Can the Union Do?

A system for providing ongoing support, counseling and assistance for assaulted members is a necessary part of any policy dealing with on-the-job assault. As trade unionists, we must educate our bosses and our members to the reality of post-traumatic stress. We must push to have post-traumatic stress disorder recognized as a consequence of violence in the workplace.

SAMPLE CONTRACT LANGUAGE TO PROTECT WORKERS FROM ON-THE-JOB ASSAULT

One of the most effective ways to provide a safe and healthy workplace is to negotiate specific contract language. Here are some sample clauses relating to on-the-job assault.

General Clause

The employer is responsible for taking all necessary steps to protect employees from assault on the job.

Employer's Policy for Dealing with Assault on the Job

The Employer shall develop written policies and procedures to deal with on-the-job assault. Such policies must address the prevention of assault on-the-job, the management of situations of assault and the provision of legal counsel and post-traumatic support to employees who have been assaulted on the job by clients or the public.

This policy shall be part of the Employer's comprehensive health and safety policy. A written copy of the policy shall be given to every employee. The Employer must also establish a procedure for the documentation of all incidents, and shall take immediate and appropriate action, as outlined in the written policy, to deal with each incident.

Policy/Plan for Health Care/Social Service Facilities

The Employer shall conduct an ongoing security and safety assessment, and develop a security plan with measures to protect personnel, patients, and visitors from aggressive or violent behavior. The security and safety assessment shall examine trends of aggressive or violent behavior. A security plan shall include, but not be limited to, security considerations relating to all of the following:

- physical layout;
- staffing;
- security personnel availability;
- policy and training related to appropriate responses to aggressive or violent acts.

Joint Labor/Management Health and Safety Committee

The purpose of the committee is to identify and investigate health and safety hazards and make recommendations on preventive measures.

The committee shall be composed of an equal number of representatives from the Union and the Employer, and the Union shall have the sole power to appoint its representatives to the committee. The committee shall make recommendations on policies to prevent on-the-job assault, on the management of violent situations and on how to provide support to workers who have experienced or face on-the-job assault.

All incidents of assault will be brought to the attention of the Health and Safety Committee. The parties agree that the Health and Safety Committee shall:

- assist in the development of policies and workplace design changes that will reduce the risk of assault on the job;
- regularly review all reports of incidents of assault;
- assist in the development and implementation of training programs that will reduce the risk of on-the-job assault.

Staffing Levels

The employer agrees to provide an adequate level of trained staff. Employees will not be required to work alone in potentially violent situations.

Employees will be notified as to potentially violent or aggressive patients, residents or clients, and will work/travel in pairs when required to work under such circumstances.

Workplace Design

Where appropriate, the Employer shall institute additional security measures including, but not limited to:

- installation of metal detectors;
- installation of surveillance cameras;
- limiting public access to the facility and specific departments or units;
- installation of bullet-proof glass;
- installation of emergency "panic" buttons to alert security personnel.

Two-Way Radios, Alarms and Paging Systems

The Employer shall provide two-way radios, alarms and/or paging systems, or other electronic warning devices or means of summoning immediate aid where employees ascertain a need. All equipment shall be maintained and periodically tested, and employees will receive training in the operation of the equipment.

Training

The Employer shall provide training to all employees at risk of assault on how to defuse potentially violent situations and verbal confrontation. Employees shall also be trained in self-protection. Training should include, but not be limited to: discussion of how to recognize warning signs and possible triggers to violence; how to resist attack and avoid escalation of the situation; how to control and defuse aggressive situations; and a full review of the Employer's written policy for dealing with assault on the job.

All employees at risk of assault by patients, clients, or the public shall receive security education and training relating to the following topics:

- general safety measures;
- personal safety measures;
- the assault cycle;
- aggressive and violence-predicting behavior;
- obtaining patient history from a patient with aggressive or violent behavior;
- characteristics of aggressive and violent patients, and victims;
- verbal and physical maneuvers to diffuse and avoid violent behavior;
- strategies to avoid physical harm;
- restraining techniques;
- appropriate use of medications as chemical restraints;

- critical incident stress briefing;
- available employee assistance programs.

Post-Traumatic Stress/Referral Services

The Employer shall, in the event of an incident of assault, provide counseling and support for the affected employee(s). Employees are to be compensated for lost days of work, counseling sessions, hospitalization and other relevant expenses.

The Employer shall offer referral information and assistance to any employee who is assaulted by a patient, visitor or member of the public. Such information shall include, but not be limited to, the employee's legal right to press charges in a court of law.

Prosecuting Offenders

The Employer shall assist the assaulted employee in any legal actions that she/he undertakes against the offender. If the employee decides not to press charges, the Employer must provide a written explanation to the union and to the employee of its decision.

No-Retaliation Clause

No employee shall be discharged, penalized or disciplined for his/her victimization in an incident of assault. The Employer agrees not to retaliate or discriminate against that employee.

Union Non-Liability

The Employer has the sole responsibility to provide a safe workplace and to correct health and safety hazards, and that nothing in this Agreement shall imply that the Union has undertaken or assumed any portion of that responsibility.

Other Union Rights

The Employer agrees that the Union has the right to bring into the workplace any union staff or other union representatives to assist investigating health and safety conditions.

Nothing in this article shall be deemed to wave any statutory rights that the Union may have.

RESOURCE LIST

California Occupational Safety and Health Administration (CALOSHA) had *Guidelines for Security and Safety of Health Care and Community Service Workers*

CAL-OSHA
Department of Industrial Relations
455 Golden Gate Avenue, Suite 5202
San Francisco, CA 94102
(415)703-4341

New Jersey Department of Labor/Public Employee Occupational Safety and Health Administration has *Guidelines on Measures and Safeguards in Dealing with Violent or Aggressive Behavior in Public Sector Health Care Facilities*

NJDOL/Office of Public Employees Safety
CN 386
Trenton, NJ 08625
(609)633-3796

Organizations

Service Employees International Union, Health & Safety Staff

International Office
1313 L Street, NW
Washington, DC 20005
(202)898-3200

Eastern Regional Office
145 Tremont Street, Suite 202
Boston, MA 02111
(617)482-4471

New York Regional Office
330 West 42nd Street, Suite 1905
New York, NY 10036
(212)947-1944

New England Office
14 Quentin Street
Waterbury, CT 06706
(203)574-7966

West Coast Office
3055 Wilshire Blvd., Suite 1050
Los Angeles, CA 90010

Western Region Field Office
150 Denny Way
P.O. Box 19360
Seattle, WA 98109
(206)448-7348

Central States Regional Office
228 S. Wabash, Suite 300
Chicago, IL 60604
(312)427-7637

Michigan State Council Office
419 S. Washington Street
Lansing, MI 48933
(517)372-0903

SEIU Canadian Office
75 The Donway West, Suite 1410
Don Mills, ONT M3C2E9
Canada
(416)447-2311

9to5 National Association of Working Women
1224 Huron Road
Cleveland, OH 44115
(216)566-9308

National Institute for Occupational Safety and Health
(NIOSH is the research arm of the Occupational Safety and Health Administration, and will conduct a health hazard evaluation at your workplace on request.)
1600 Clifton Road, NE
Atlanta, GA 30333
(800)356-4673

Occupational Safety and Health Administration (OSHA)
(OSHA develops and enforces workplace health and safety standards.)

OSHA Offices

National Office
200 Constitution Avenue, NW
Washington, DC 20210
(202)523-8091

Region I (Connecticut, Maine, Massachusetts, New Hampshire, Rhode Island, Vermont)
133 Portland Street, 1st Floor
Boston, MA 02114
(617)565-7164

Region II (New Jersey, New York, Puerto Rico, Virgin Islands)
201 Varick Street, Room 670
New York, NY 10014
(212)337-2378

Region III (Delaware, District of Columbia, Maryland, Pennsylvania, Virginia, West Virginia)
Gateway Building, Suite 2100
3535 Market Street
Philadelphia, PA 19104
(212)596-1201

Region IV (Alabama, Florida, Georgia, Kentucky, Mississippi, North Carolina, South Carolina, Tennessee)
Suite 587
1375 Peachtree Street, NE
Atlanta, GA 30367
(404)347-3573

Region V (Illinois, Indiana, Michigan, Minnesota, Ohio, Wisconsin)
32nd Floor, Room 3244
230 S. Dearborn Street
Chicago, IL 60604
(312)353-2220

Region VI (Arkansas, Louisiana, New Mexico, Oklahoma, Texas)
525 Griffin Street, Room 602
Dallas, TX 75202
(212)767-4731

Region VII (Iowa, Kansas, Missouri, Nebraska)
911 Walnut Street, Room 406
Kansas City, MO 64106
(816)844-3061

Region VIII (Colorado, Montana, North Dakota, South Dakota, Utah, Wyoming)
Federal Building, Room 1576
1961 Stout Street
Denver, CO 80294
(303)844-3061

Region IX (Arizona, California, Hawaii, Nevada, American Samoa, Guam, Trust Territory of the Pacific Islands)
71 Stevenson Street, 4th Floor
San Francisco, CA 94105
(415)744-6670

Region X (Alaska, Idaho, Oregon, Washington)
Federal Office Building
Room 6003
909 1st Avenue
Seattle, WA 98174
(206)442-5930

OSHA Federally-Approved State Plan Offices

Alaska
Alaska Department of Labor
P.O. Box 1149
Juneau, AK 99802
(907)465-2700

Arizona
Division of Occupational Safety and Health
Industrial Commission of Arizona
800 West Washington
Phoenix, AZ 85007
(602)255-5795

California
Department of Industrial Relations
525 Golden Gate Avenue
San Francisco, CA 94102
(415)557-3356

Connecticut
(Public Employees Only)
Connecticut Department of Labor
200 Folly Brook Boulevard
Wetherfield, CT 06109
(203)566-5123

Delaware
Department of Labor
820 North French Street
Wilmington, DE 19801

Hawaii
Department of Labor and Industrial Relations
830 Punchbowl Street
Honolulu, HI 93813
(808)548-3150

Indiana
Division of Labor
1013 State Office Building
100 North Senate Avenue
Indianapolis, IN 46204
(317)232-2665

Iowa
Division of Labor Services
1000 East Grand Avenue
Des Moines, IA 50319

Kentucky
Kentucky Labor Cabinet
U.I. Highway 127 South
Frankfort, KY 40601
(502)564-3070

Maryland
Division of Labor and Industry
Department of Licensing and Regulations
502 St. Paul Place
Baltimore, MD 21202
(301)333-4176

Michigan
Department of Public Health
3423 North Logan Street
P.O. Box 30195
Lansing, MI 48909
(517)335-8022

Michigan Department of Labor
309 N. Washington
P.O. Box 30015
Lansing, MI 48909
(517)373-9600

Minnesota
Department of Labor and Industry
443 Lafayette Road
St. Paul, MN 55101
(612)296-2342

Nevada
Department of Industrial Relations
Division of Occupational Safety and Health
Capitol Complex
1370 South Curry Street
Carson City, NV 89710
(702)885-5240

New Mexico
Environmental Improvement Division
Health and Environment Department
1190 St. Francis Drive, N2200
Santa Fe, NM 87503
(505)827-2850

New York
(Public Employees Only)
New York Department of Labor
Division of Safety and Health State Campus
Bldg. 12, Suite 159
Albany, NY 12240
(518)457-5508

North Carolina
Department of Labor
4 West Edenton Street
Raleigh, NC 27603
(919)733-7166

Oregon
Department of Insurance and Finance
21 Labor and Industries Building
Salem, OR 97310
(503)378-3304

Puerto Rico
Department of Labor and Human Resources
Prudencio Rivera Martinez Building
505 Munoz Rivera Avenue
Hata Rey, Puerto Rico 00918
(809)754-2119/2122

South Carolina
Department of Labor
3600 Forest Drive
P.O. Box 11329
Columbia, SC 29211
(803)734-9594

Tennessee
Department of Labor
501 Union Building
Suite "A," 2nd Floor
Nashville, TN 37219
(615)741-2582

Utah
Utah Occupational Safety and Health
160 East 300 South
P.O. Box 5800
Salt Lake City, UT 84110-5800
(801)530-6900

Vermont
Department of Labor and Industry
120 State Street
Montpelier, VT 05602
(802)828-2765

Virgin Islands
Department of Labor
P.O. Box 890, Christiansted
St. Croix, Virgin Islands 00820
(809)773-1994

Virginia
Department of Labor and Industry
P.O. Box 12064
Richmond, VA 23241-0064
(804)786-2376

Washington
Department of Labor and Industry
General Administration Building
Room 344-AX31
Olympia, WA 98504
(206)753-6307

Wyoming
Occupational Health and Safety Department
604 East 25th Street
Cheyenne, WY 82002
(307)777-7786/7787

Committees on Occupational Safety and Health (COSH Groups)

Alaska
Alaska Health Project
420 W. 7th Avenue, Suite 101
Anchorage, AK 99501
(907)276-2864

California
BACOSH (San Francisco Bay Area COSH)
c/o Mr. Glenn Shor
Labor Occupational Health Program
Institute of Industrial Relations
2521 Channing Way
Berkeley, CA 94720
(415)642-5507

LACOSH (Los Angeles COSH)
2501 South Hill Street
Los Angeles, CA 90007
(213)749-6161

Sacramento COSH
c/o Fire Fighters Local 522
3101 Stockton Boulevard
Sacramento, CA 95820
(916)444-8134 or (916)924-8060

SCCOSH (Santa Clara Center for Occupational Safety and Health)
Occupational Safety and Health
760 N. 1st Street
San Jose, CA 95112
(408)998-4050

Connecticut
ConnectiCOSH
P.O. Box 3117
Hartford, CT 06103
(203)549-1877

District of Columbia
Alice Hamilton Center for Occupational Safety and Health
410 Seventh Street, SE
Washington, DC 20003
(202)543-0005

Illinois
CACOSH (Chicago COSH)
37 South Ashland
Chicago, IL 60607
(312)666-1611

Maine
Maine Labor Group on Health, Inc.
Box V
Augusta, ME 04332
(207)622-7823

Massachusetts
MassCOSH
555 Amory Street
Boston, MA 02130
(617)524-6686

Michigan
SEMCOSH (Southeast Michigan COSH)
2727 Second Street
Detroit, MI 48201
(303)961-3345

New York
ALCOSH (Allegheny Council on Occupational Safety and Health)
100 East Second Street
Jamestown, NY 14701
(716)488-0720

CYNCOSH (Central New York COSH)
615 West Genessee Street
Syracuse, NY 13204
(315)471-6187

NYCOSH (New York COSH)
275 Seventh Avenue
New York, NY 10001
(212)627-3900

ROCOSH (Rochester COSH)
797 Elmwood Avenue
Rochester, NY 14620
(716)244-0420

WYNCOSH (Western New York COSH)
2495 Main Street, Suite 438
Buffalo, NY 14214
(716)833-5416

North Carolina
NCOSH
P.O. Box 2514
Durham, NC 27715
(919)286-9249

Pennsylvania
PHILAPOSH (Philadelphia Project on Occupational Safety and Health)
3001 Walnut Street, 5th Floor
Philadelphia, PA 19104
(215)925-SAFE (7233)

Rhode Island
RICOSH
340 Lockwood Street
Providence, RI 02907
(401)751-2015

Tennessee
TNCOSH
1514 E. Magnolia, Suite 406
Knoxville, TN 37917
(615)5252-3147

Texas
5735 Regina
Beaumont, TX 77706
(409)898-1427

Wisconsin
WISCOSH (Wisconsin COSH)
1334 South 11th Street
Milwaukee, WI 53204
(414)643-0928

Canada
VanCOSH (Vancouver COSH)
616 East 10th Avenue
Vancouver, British Columbia V5T2A5

WOSH (Windsor Occupational Safety and Health Project)
1109 Tecumseh Road East
Windsor, Ontario N8W1B3
(519)254-4192

ORGANIZING TO PREVENT ASSAULT ON THE JOB

Unions have long led the fight for safe and healthy working conditions. That fight includes keeping our workplaces free of assaults. Union members in many different jobs have identified assault as a serious safety and health issue. Safe workplaces, free from assault, should be a goal of every union member. Like every union goal, our success depends on how well we organize.

1. How can you determine whether or not assault on the job is a problem at your workplace? *Talk, listen and encourage.*

Talk to your co-workers and find out if they share your concerns about safety. Develop a short survey (like the one in this chapter) to distribute in your facility. The SEIU Health and Safety Department can send you samples or help you design your own. Make sure you involve as many people in this activity as possible. Compile the results of your survey, consult with stewards and other union leaders, and then decide a group the most effective way to use these results.

One of the most important activities to get people involved in is documenting the problem. Urge members to document all assault incidents, and all the near-incidents, using the Incident Report Form in this Section. Review this data on a regular basis and update members every time there is an incident of assault.

Hold lunch-time meetings or form committees to research specific problems and develop solutions once the hazards have been identified. Keep members informed through the local union newsletter and on bulletin boards. Always encourage members to document incidents.

2. How do you solve the problem? *Develop a plan of action.*

A work environment in which people fear attack is NOT a healthy or safe place. Employers must recognize that it is their legal duty to protect workers from assault just as much as any other health and safety hazard. Assault is not something that "just happens." Taking the right steps, we can control and prevent violence in the workplace.

Use your Health and Safety Committee or form one to deal with on-the-job assault.

Keep members involved in developing the solution. Think through what steps you want management to take. When you have gathered your documentation and survey results, bring a group of workers to meet with management. Present your complaint, and ask them to correct it. If your demands are met, make sure you publicize your success.

3. What if management refuses to take action? *Grieve, Negotiate, Agitate, and Organize.*

If management refuses to respond to your demands, be prepared to take the following steps:

- File a grievance;
- Refuse to work alone or under certain conditions, but check with your local union leadership first;
- Contact OSHA and demand action;
- Publicize the problem using the media;
- Create a slogan campaign like "Understaffing Kills." Wear buttons, post notices and hand out flyers;
- Negotiate health and safety language in your next contract;
- Build coalitions and lobby for laws that provide proper protection for workers from being assaulted.

SEIU members and locals throughout the country that have grieved, negotiated, agitated and organized have succeeded in making their workplaces safer.

SEIU Success Stories

SEIU locals have fought for and won important protection for their members.

Staffing and a New Felony Law

When Georgia State Employees Union, Local 1985, organized a campaign to protest the 20 assaults in a three-month period at a youth development center, their actions resulted in a 10% staff increase, the allocation of $5 million to build a new facility for violent offenders, plus a new law which made assault on a staff member a felony.

Mandatory Training

After Local 1199NE in Connecticut discovered that 80% of all workers' compensation injuries were patient related—with most due to assaults—the union's actions led to the creation of a 21-hour mandatory training program for 9000 members across the state. At one mental health hospital, injuries due to assaults dropped 60% after the training and the implementation of an employee wellness program.

Better Patient to Staff Ratios

After several serious assault incidents at their mental health facility in West Virginia, Local 1199WO members filed a mass grievance. The outcome was an increase in staffing which changed the ratio of patient to provider from eleven to one to less than six to one.

ASSAULT ON THE JOB SURVEY—HOW SAFE IS YOUR WORK ENVIRONMENT?

Employee Commitment/Workplace Policy

Does your employer...

1. Place a high priority on eliminating hazards associated with assault on the job?
☐ Yes ☐ No

2. Have a policy that places employee safety on the same level of importance as patient/client safety?
☐ Yes ☐ No

3. Discuss assault openly at Health and Safety Committee meetings?
☐ Yes ☐ No

4. Investigate and document all instances of assault and/or harassment?
☐ Yes ☐ No

5. Have a written policy concerning assault on the job?
☐ Yes ☐ No

6. Involve employees in developing the policy?
☐ Yes ☐ No

7. Have a program to provide support for victims of assault?
☐ Yes ☐ No

8. Support employees who have been involved in an assault incident, rather than discipline them?
☐ Yes ☐ No

9. Provide legal counsel for assault victims?
☐ Yes ☐ No

10. Encourage reporting assault incidents to the police and support prosecution of offenders?
☐ Yes ☐ No

Staffing

1. Is staffing adequate?
☐ Yes ☐ No

2. Does your employer make sure you don't work alone?
☐ Yes ☐ No

3. Is the an adequate number of security staff?
☐ Yes ☐ No

4. Is back-up staff always scheduled?
☐ Yes ☐ No

Workplace Design

1. Are all work areas well-lit?
☐ Yes ☐ No

2. Are private washrooms provided for staff?
☐ Yes ☐ No

3. Is access to office areas/employees' work stations restricted to only authorized staff and clients?
☐ Yes ☐ No

4. Are there electronic alarm systems, closed-circuit TV or two-way radios?
☐ Yes ☐ No

5. Is furniture well placed so employees can't get trapped in a room with a client?
☐ Yes ☐ No

6. Are employees who do field work provided with personal alarm systems or beepers?
☐ Yes ☐ No

7. Are parking lots, garages, and other areas that employees need to walk through, secure and well-lighted?
☐ Yes ☐ No

Training

1. Do all employees receive adequate training on how to protect themselves from being assaulted on the job?
☐ Yes ☐ No

2. Were employees involved in developing the training programs?
☐ Yes ☐ No

3. Do all new staff receive training upon hire?
☐ Yes ☐ No

4. Do back-up staff receive training?
☐ Yes ☐ No

SAMPLE ASSAULT INCIDENT REPORT FORM

A sample of the form for you to complete and send it to your union representative so that we can (1) accurately record the incidents of assault and/or harassment that occur (2) notify and/or follow up with your employer regarding the problem and (3) plan strategies to prevent these problems from recurring:

1. Date of incident: mo ___ day ___ yr ___

2. Name, or pseudonym, of member(s): ______________________________

3. Work location: ______________________________

4. Local union: ______________________________

5. Please describe the incident: ______________________________

6. Where did the incident occur? ______________________________

7. Did the incident involve a weapon? ______________________________

8. Were you injured? ______________________________

9. To what extent? ______________________________

10. Did you lose any work days? How many? ______________________________

11. Have you applied for worker's compensation? ______________________________

12. Was the person who assaulted you a co-worker, supervisor, patient, client (or someone else)? ______________________________

13. Were you singled out, or was the assault directed at more than one individual?

____

14. Were you alone when the incident occurred? If yes, why? ______________

15. Did you have any reason to believe an incident might occur? If yes, why?

____

16. Did you report the incident to your supervisor? ______________________________

17. Have you filed a police report? ______________________________

18. If so, was the attacker charged? ______________________________

19. To your knowledge, has this type of incident ever happened before to your co-workers? ______________________________

20. Have you had any counseling or support since the incident? ______________

21. Do you have any thoughts on how this incident could have been avoided? What should your employer do to avoid similar incidents in the future? ______________

Remember, the more statistics we have on violent encounters in the workplace, the more compelling our arguments are for demanding improved workplace design, adequate staffing levels, better security and ongoing training.

Action Taken by Local

Talked with management (date): ______________________________

Grievance filed (date): ______________________________

OSHA complaint filed: ______________________________

Other actions: ______________________________

Comments: ______________________________

SAMPLE GRIEVANCE/PETITION FORM

Filing a group grievance or presenting management with a petition of protest is a good way to bring the issue of on-the-job assault to your employer's attention. If you have negotiated contract language that protects members from on-the-job assault, then use the grievance process to get some action. If you don't have contract language, or you want to include employees who may not be in your union (they may be managers or members of another union), then the group petition can be an effective action.

Here is a sample mass grievance/petition. You can adapt this to meet the specific needs of your campaign.

SEIU Local 9999
Group Grievance/Petition

To: Director(s) or person(s) in charge

From: The undersigned members of Local 9999

Date: January 1, 1994

Statement of Problem: The Employer is failing to provide a safe and healthy workplace, free from recognized hazards associated with assault on the job. (If you have contract language, cite the article.)

Relief Sought: Provide a safe and healthy work environment (as per Article __of the contract). Install alarms and other appropriate monitoring equipment in areas where the public has access, or where employees may be working alone. Develop a comprehensive policy to protect workers from being assaulted on the job, including the provision of adequate staffing, incident reporting procedures, the provision of appropriate and adequate training for employees who are at risk of assault, and support services for victims of assault.

Name	Title

TRAINING EXERCISES

Case Studies

In each of the case studies below, you play the role of union activists. Work together in small groups to discuss what actions you would take to address these problems concerning assault on the job.

1. Joe, a mental health aide, must often work alone due to cutbacks in staffing. Recently, he was attacked by a patient, and suffered cuts, bruises, and serious back injury. Joe is the fourth member of your union who has been assaulted by a patient in the last two months.

2. Donna works at the front desk in your social service agency. The clients, who are often frustrated and impatient with the system, often take their anger out on Donna. The abuse, hostility and threats of assault have taken their toll on her, as she continually talks about her fear of someone attacking her with a weapon.

3. You and your co-workers do field work, and most of the time you work alone. Years ago that was fine, but is seems the streets have gotten more dangerous, and you all feel very uncomfortable traveling alone. No one has gotten hurt yet, but there have been several near-misses, and everyone wants the union to do something about it.

4. Maria works in the nursing home where you work. One of her patients continually hassles her with insults and sexual advances. She has also been harassed by the son of another patient on that same floor. She recently asked to be transferred so she could avoid these situations, but her request was denied. She is now suffering frequent headaches and has trouble sleeping.

5. You and your co-workers drive to work, and you park your car in the building's parking lot. There have been several incidents in the last few months, and one co-worker was recently mugged as she approached her car.

6. The emergency room in your hospital has recently been the sight of two serious assaults on workers. Staff wants something done about it, but the employer keeps saying that they can't predict when violence is going to break out.

Response to Assault on the Job

The purpose of this exercise is:

- To discuss *what is currently being done* to protect workers from on-the-job assault;

- To discuss *what should be done to* protect workers from being assaulted on the job;
- For you and your co-workers to being to *develop an action* plan to deal with the issue of on-the-job assault.

How to Complete the Exercise

1. In small groups, participants should relate incidents of assault that have occurred on the job. All incidents should be listed on a chalkboard or chart paper. (Participants may want to look at "Case Studies" in this section for other examples of on-the-job assault.)

2. After all incidents have been listed, participants should complete the "Assault Response Chart" by answering the two questions for each group of people listed.

3. Based on your responses, have the group set some priorities for actions to take to address the issue of assault on the job.

Step 1: Discussing the incidents of assault at your workplace

Example:
Incident
Joan, a social service worker, was assaulted by one of her clients. She missed a week of work, and she's afraid to come back, because she fears the client will assault her again.

Discuss and list the incidents that have occurred among your members.

Step 2: What are these people doing about it?

Assault Responses

How do they respond, and what are they doing about it?	How should they respond, and what should they be doing about it?
Workers:	
______________________	______________________
______________________	______________________
Management:	
______________________	______________________
______________________	______________________
The Union:	
______________________	______________________
______________________	______________________

OSHA/Other Government Agencies:

Researchers/Academics:

Step 3: Setting priorities

Based on your discussion, list the actions that you and your co-workers will take to reduce the risk of assault on the job.

CHAPTER 4

Tuberculosis Engineering Controls

INTRODUCTION

Since the recent increases in health care worker conversions to TB and in some cases death, the need to understand and control the transmigration of droplet nuclei has become of paramount importance. Engineering controls, air changes per hour, negative pressure, HEPA filtration, ultraviolet germicidal irradiation, capture at the source, enclosure devices are now industry vocabularies that need to be understood by a variety of scientific disciplines in the health care setting. Selection and implementation of control options has become labor intensive and in some cases controversial. Preparing a health care facility to meet the standards for a TB Control Program is multi-dimensional as there needs to be scientific understanding of the aerosolization of TB, control procedures, budgetary considerations, air volume measurements of areas in the hospitals where suspected or identified cases or high risk medical procedures are to be done, types of engineering controls selected for the various areas of concern, maintenance of engineering controls selected for the various areas of concern, etc. Different scientific disciplines within each medical facility must work together harmoniously in order to achieve optimum results, for example, Infection Control, Engineers, Industrial Hygienists, Nursing, Administrators. In some states regulatory codes are in conflict with TB control guidelines and over time these conflicts must be resolved to mutual satisfaction.

The purpose of this section is to provide some additional information and design characteristics for engineering controls for TB. The reader is also referred to the CDC Guidelines, Part II, Department of Health and Human Services, Draft Guidelines for Preventing the Transmission of TB in Health Care Facilities for recommendations and additional references for engineering controls. This section contains: Part 1. An industrial hygiene approach to tuberculosis control; Part 2. Portable HEPA Filtration for TB Isolation in Hospitals and Clinics; Part 3. Preventing TB in the Workplace: What Did We Learn From HIV? Policies Regarding the HIV-Infected Health Care Worker: NOTE: At the writing of the CDC Guidelines there was no challenge data confirming the clearance efficiency of HEPA filters. The data presented here, though not yet in peer review but tested by independent laboratories funded by manufacturers, nonetheless give some positive indications as to the functionality and clearance rates of these systems; Part 4. Reducing the Spread of Tuberculosis in Your Workplace.

PART 1

An Industrial Hygiene Approach to Tuberculosis Control

Lorraine M. Conroy and John E. Franke

INTRODUCTION

Industrial hygienists address occupational health problems using a three step approach: recognition, evaluation and control. The approach involves answering three questions: 1) What are the potential health hazards? 2) Are the potential hazards truly hazardous? and 3) How can the hazard be eliminated or minimized? The toxic hazard of a material is defined as the likelihood of injury of a person by other than mechanical means.[1] The evaluation of the potential hazard is usually done by conducting air sampling or exposure monitoring and comparing the exposure with some established guideline or standard.

In the case of *Mycobacterium tuberculosis* (MTb), there are no established exposure limits. There is also no practical method for evaluating the MTb exposure with air sampling. Several studies indicate that a single bacteria particle can cause disease.[2–5] The evaluation of the hazard potential is influenced by the background incidence of disease in the population served by the institution, the number of TB cases treated by the institution, the tuberculin skin test (TST) conversion rate in the workforce of the institution, types of treatments and other aerosol-generating processes performed in the institution, and engineering and administrative controls in place at the institution.

Source identification and isolation are the key steps in controlling transmission of tuberculosis (TB) in health care facilities.[6] The central theme of this chapter is the application of basic industrial hygiene principles to the control of the TB transmission hazard. Industrial hygiene methods could enhance infection control efforts to identify TB sources in a facility. The application of ventilation to isolate known or suspected TB sources would also be improved by following industrial hygiene principles.

HAZARD CHARACTERIZATION

The number of reported cases of TB in the U.S. in 1991 was 26,283, 2% higher than in 1990.[7–8] There had been an annual decline in the number of cases of approximately 5% since the 1950s.[7] In 1958, Dublin[9] reported the death rate in the U.S. from TB as less than 8 per 100,000, a 96% reduction from 1900. This led him to conclude "We cannot say exactly when control will be complete, but there is every indication that it will be some time in the course of the next 20 years."[9]

1-56670-083-3/94/$0.00+$.50

There was a 6–7% annual decline in TB cases from 1981–1984. Using the trend from 1981–1984 to estimate the expected number of cases for 1985–1991, it is estimated that more than 39,000 excess cases of TB occurred between 1985 and 1991.[7]

Persons living in the same household, those who travel in the same vehicle[10] and those who share air with an infectious person through a common ventilation system for a prolonged time[11–12] are at risk of acquiring TB infection. Ventilation systems have contributed to the transmission of TB.[2] Twenty-seven of 67 susceptible office workers became infected following 160 hours of exposure to air shared by an infectious office worker in the same building.[13]

Several studies have documented higher than expected TST conversion rates in hospital personnel.[14–23] At least one case resulted in occupational transmission of active disease.[22] Procedures such as bronchoscopy, endotracheal intubation and suctioning with mechanical ventilation, open abscess irrigation and autopsy have been implicated in nosocomial transmission.[14,17,21]

The Centers for Disease Control and Prevention (CDC) and others have investigated several outbreaks of multiple drug-resistant TB (MDR-TB) in New York and Florida.[23–26] One of these investigations indicated that two categories of factors contributed to the outbreaks.[23] The first category included delays in diagnosing TB in HIV-infected persons and delays in recognizing drug resistance. The second category included delays in acid fast bacilli (AFB) isolation on admission or readmission, maintaining AFB isolation for an inadequate amount of time, lapses in AFB isolation, such as open doors and patients leaving isolation rooms, AFB isolation rooms with inadequate negative pressure and inadequate numbers of AFB isolation rooms. The outcome of these factors can be improved using industrial hygiene principles outlined in this chapter.

REGULATIONS AND GUIDELINES

In the U.S., the Occupational Safety and Health Administration promulgates and enforces standards for protecting the health of American workers. The authority for standard setting was established with the passage of the Occupational Safety and Health Act of 1970 (OSH Act).[27] The Act requires employers to "provide a safe and healthful workplace free from recognized hazards."[27] This is often referred to as the "general duty" clause of the OSH Act. The OSH Act also prescribes a formal standard setting procedure.[27] Health and safety standards established under the OSH Act generally have two forms. Many chemical agents are regulated under the Air Contaminants Standard.[28] This standard sets maximum exposure limits called Permissible Exposure Limits (PEL) for several hundred chemical agents. Several substance specific standards have also been promulgated. These standards set a PEL, but also establish other requirements of the employer such as training, medical surveillance and recordkeeping.

The PEL must be met through a combination of engineering and work practice controls. If engineering controls do not reduce the exposure below the PEL or while engineering controls are being implemented, respiratory protection may be used to achieve compliance with the standard. The second type of standard does not establish a maximum exposure limit but outlines requirements of the employer necessary to reduce the exposure to the lowest feasible level. An example of this type of standard is the Bloodborne Pathogens Standard.[29] The Bloodborne Pathogens Standard requires employers to implement a written exposure control plan; prepare an exposure determination; use a combination of universal precautions, engineering and work practice controls, and personal protective equipment to minimize or eliminate exposure; implement proper disposal and labeling procedures; provide hepatitis B vaccinations; and provide information and training on the risks of exposure to bloodborne pathogens, procedures and practices necessary to prevent exposure, and the requirements of the standard.[29]

At the present time no OSHA standard specifically regulates exposure to MTb. However, OSHA Region 2 has published enforcement guidelines for occupational exposure to tuberculosis.[30] The guidelines describe when the general duty clause, Section 5(a)(1) of the OSH Act, may be cited. Citations can only be issued to employers whose employees work on a regular basis in health care settings, correctional institutions, homeless shelters, long-term care facilities and drug treatment centers. The document lists examples of feasible and useful abatement methods for TB control and states that "the non-use of any of these methods is likely to result in the continued existence of a serious hazard and may, therefore, allow citation under 5(a)(1)." The methods are: 1) medical screening of employees; 2) work removal of employees who have current pulmonary or laryngeal TB until adequate treatment is instituted, their cough is resolved, or until a physician certifies that the person is no longer infectious; 3) training and education of employees about the hazards and control of tuberculosis; 4) respiratory isolation of infectious TB patients in negative pressure rooms exhausted to the outside; and 5) respiratory protection use by the patient, if possible, and the employee during patient transport.

The document also describes other OSHA standards that may apply. These include recordkeeping[31] and respiratory protection.[32] Additionally, OSHA Region 2 considers tuberculosis infections (positive skin test) and tuberculosis disease as recordable on the OSHA 200 log where tuberculosis has been identified as a hazard.[30]

In 1993 two hospitals in Madison, Wisconsin were inspected and citations for failing to adequately protect workers from TB were issued.[33] The first hospital was cited under Section 5 (a)(1) of the OSH Act for failing to provide medical surveillance and failing to ensure that an isolation room was properly ventilated to operate under negative pressure. The hospital was also cited for four violations of 1910.134 (respiratory protection) including using respirators that were not NIOSH approved, allowing workers with beards, sideburns, and skullcaps to wear respirators and not having or maintaining a written respiratory protection program.[33]

The second hospital was a Veterans Hospital. OSHA has the authority to inspect federal facilities and to issue citations but cannot assess penalties. The hospital was cited under Section 1-201(a) of Executive Order 12196, which is equivalent to the general duty clause of the OSH Act. The violations in this case were due to deficiencies in medical surveillance and failing to record an active case of TB on the injury and illness log.[33]

CDC has published several sets of guidelines for preventing tuberculosis in health care settings,[34–35] migrant farm workers,[10] long-term care facilities[12] and correctional institutions.[11] The most recent draft of the guidelines for health care settings was published in the Federal Register in October 1993.[36]

The current CDC guidelines have the following steps: 1) implement control measures that follow an established hierarchy of administrative and engineering controls and personal respiratory protection; 2) perform a risk assessment at each health care facility and develop a written TB control plan; 3) provide early identification and management of persons with TB; 4) implement a medical surveillance program for employees using purified protein derivative (PPD) skin testing; and 5) educate, train and communicate with health care workers about the risks of TB and the measures used to prevent MTb exposure.[36]

Regulations and guidelines have also been issued by state agencies and professional associations. The state of California, for example, has proposed regulations for occupational tuberculosis control that specify an exposure control plan, TB surveillance and employee notification, medical evaluation and preventive therapy, methods of exposure control, training and recordkeeping.[37] The American Society for Hospital Engineering has guidelines that discuss risk assessment and management for TB control in hospitals.[38] The American Society for Heating, Ventilating and Air Conditioning Engineers has general guidelines for health facilities that give design criteria and mention TB in the discussion.[39]

In addition to occupational health requirements, state and local building codes must be considered when designing or implementing control strategies for MTb. The recommendation of keeping TB patients in isolation rooms under negative pressure relative to the corridor violates state and local codes in many areas. A variance may be needed or the use of an anteroom may be required to meet both the CDC guidelines and building codes.

Source Characterization

Mycobacterium tuberculosis is a rod-shaped bacteria which varies in width from 0.2 to 0.6 µm and from 0.5 to 4.0 µm in length.[40–41] The bacteria are expelled from infected persons through coughing, sneezing, talking and singing[2–3] and become aerosolized as droplets.[34,7] The largest droplets (e.g., exceeding 100 µm) settle onto surfaces and are removed from the air.[42] Droplets less than 100 µm evaporate to form stable, nearly spherical, droplet nuclei in the 1–4 µm size range.[42] A study by Loudon and Roberts[43] indicated that 30% of droplet nuclei, by number, resulting from coughing were less than 3 µm in diameter.

The droplets are small enough that room air currents keep them airborne and spread them throughout a room or building.[44] Their small mass means that the droplet nuclei have negligible inertia and are unable to travel through air on their own. Instead they must follow the burst of air released by a cough, for example, and then follow room air currents when the air burst slows down. Air currents always exist in rooms and random room air velocities are typically in the range of 20–40 feet per minute (fpm).

The fact that infectious particles must follow the air currents that they encounter has important implications for control. Isolation design criteria aimed at controlling the contaminated air currents have a good chance to control the airborne infection hazard. Characterization of the room environment is, therefore, as important as characterizing the droplet nuclei source when specifying controls.

Because room air is not quiescent, droplet nuclei can remain airborne for prolonged periods of time (hours, at least).[45] Anyone who breathes air that contains these droplet nuclei can become infected with TB.[45] After inhalation, droplet nuclei can penetrate to the alveolar region of the lung, where infection is initiated.[40–41]

A study by Kent[46] indicated that the tuberculosis bacterium is highly resistant to environmental stresses, probably survives for extended time in the environment and could be resuspended in an infective state from settled dust. The viability of the resuspended particles depends, in part, on the moisture retained in the dust and the temperature and relative humidity of the room environment.[47]

It appears that a single *Mycobacterium tuberculosis* cell is adequate for infection.[2–5] It also appears that everyone who converts to a positive tuberculosis skin test acts as a low level source of infection, at least temporarily.[2,48]

Several studies indicate that the risk of infection is a function of several factors including concentration of droplet nuclei, cumulative time that air containing droplet nuclei is breathed and the worker's pulmonary ventilation rate.[49] Harris and McClement[50] describe the risk of airborne transmission to be a function of several factors including rate and concentration of expelled organisms, the physical state of the airborne discharge and the volume and rate of exchange of the air in the space where the bacilli are ejected. They state, however, that the most important risk factor is the length of time an individual shares a volume of air with an infectious case of tuberculosis. CDC lists the following environmental factors that enhance transmission: contact between susceptible persons and an infectious patient in relatively small, enclosed spaces; inadequate ventilation that results in insufficient dilution or removal of infectious droplet nuclei; and recirculation of air containing infectious droplet nuclei.[35]

Droplet nuclei concentration is higher near the source, especially after a burst release such as a cough. The droplets diffuse throughout the space with time, reducing this concentration gradient until another burst is released. Therefore, the risk of exposure to infectious droplets is highest and most variable near the source.[51] Controlling the length of time a worker shares this near-field air volume with a patient is crucial. Isolating this air space should also be the starting point for source control and ventilation design.

Since droplet nuclei follow the local air currents, it is possible to evaluate contaminated air flow patterns in existing buildings with smoke tube and tracer gas methods. The smoke tube can visually simulate the travel of a burst of droplets, for example, from the head position of a patient bed. The test can be used to specify work positions and local exhaust hoods. Smoke tubes release a momentarily irritating acid aerosol and caution must be used to avoid eye contact with and inhalation of the smoke.

Tracer gases are non-reactive, non-toxic gases that do not normally exist in the test space and are measurable over a wide range of low concentrations.[52] Tracer gases can be released in ways that simulate a source emission. Monitoring the air at critical exposure locations gives information about the transport of droplet nuclei in rooms and buildings (e.g., reception desks in clinic waiting rooms). Tracer gas testing is usually performed by specialists.

SOURCE CONTROL

The hierarchy of control strategies for any hazardous substance is 1) control at the source; 2) control between the source and the worker; and 3) control at the worker. Examples of control at the source are substitution with a less hazardous substance or process, source isolation and local exhaust ventilation. Examples of control between the source and worker are dilution ventilation, shielding and use of UV lights. An example of control at the worker is the use of personal respiratory protection.

Ventilation

Ventilation is one of the most important engineering techniques available for maintaining and improving workplace environments. Dilution or general ventilation refers to the dilution of contaminated air with uncontaminated air in a general area, room or building. Local exhaust ventilation refers to the capture of pollutants at the source. Local exhaust ventilation is preferred for controlling atmospheric concentrations of airborne hazards because the capture and control of contaminants can be complete and workers' exposure can be prevented. With general ventilation the contaminant concentration is diluted, but exposure is never completely prevented. Since one infectious droplet can transmit the disease, local exhaust controls should be considered before relying on dilution ventilation methods.

Other advantages of local exhaust ventilation are: 1) less exhaust air is required for equivalent control, making possible lower operating costs and 2) the contaminant is controlled by a smaller air volume, thus reducing costs of associated air cleaning.

A local exhaust ventilation system consists of hoods or enclosures, ductwork leading to an exhaust fan, often an air cleaning device and a discharge point. There are several general guidelines for design and operation of local exhaust ventilation systems.

1. The hood should physically enclose the source as completely as practicable.
2. The contaminated air should be captured with sufficient velocity so that it is always directed into the enclosure.
3. The system should be designed to direct the contaminated air away from workers' breathing zones.
4. The system should be operated so that workers are not placed in the air flow path between the source and the hood.
5. Sufficient make-up air must be supplied to replace the exhausted air at the design pressure differential between the room and the adjoining spaces.
6. The contaminated exhaust air should be discharged away from building air inlets and occupied areas.

There are three general categories of hood types: enclosures, receiving hoods and exterior hoods. Enclosures surround the point of emission or contaminant generation, either completely or partially. Complete enclosures require the lowest exhaust rate of the three hood types.

Receiving hoods are those hoods which use some characteristic of the process to help air contaminants flow into the hood. Two examples of receiving hoods are grinding wheel hoods and canopy hoods for heated processes. Grinding wheels release particles with a high velocity. Placing the hood directly in the path of the high velocity particles aids in capture. With hot sources, the contaminant is released, upward from the source, with a high velocity due to the buoyancy of heated air. Placing the canopy hood directly above the source also aids in contaminant capture.

Exterior hoods differ from enclosures in that they must capture contaminants being generated at a point outside the hood. They differ from receiving hoods in that they must capture contaminants without the aid of supplemental forces. Exterior hoods are sensitive to external conditions, especially crossdrafts, which may interfere with their "reach." Even slight drafts can cause some exterior hoods to become ineffective. Exterior hoods are used when processes require easy access to the source and enclosures obstruct performance of the job. For a given contaminant capture efficiency, exterior hoods require the most airflow.

The capture efficiency of local exhaust hoods depends on many factors including: physical state of the contaminant (gas/vapor or particle); temperature of source; direction and velocity of contaminant release; distance of the hood from the source (especially important with exterior hoods); drafts (significantly affect exterior hood performance but can also affect booths and partial enclosures); and worker activity, such as reaching into the zone of influence of the hood.

The current design method for local exhaust systems is given in the Industrial Ventilation Manual.[53] The first step is to eliminate or minimize air movements in the area of the hood and locate the hood around or as close as practicable to the source. The capture velocity is then determined. The Manual[53] defines capture velocity as "the air velocity at any point in front of the hood or at the hood opening necessary to overcome opposing air currents and to capture the contaminated air at that point by causing it to flow into the hood." The value of capture velocity depends on the velocity of the contaminant at release and the magnitude of room air currents. The CDC recommends a capture velocity of 200 fpm for exterior-type hoods. The recommendations also include having the patient face directly into the hood opening so that coughing or sneezing will be directed into the hood.[36]

Several researchers[54–63] have developed empirical expressions for centerline velocity as a function of hood shape, air flow into the hood, distance of the source from the hood and hood area. The airflow necessary to obtain the desired capture velocity is calculated using one of these empirical expressions. The most commonly used expressions are those of DallaValle[54] and Silverman[55–57] and are given by Equations (1)–(4).

Plain round or rectangular hood:

$$Q = v_c\ (10x^2 + A) \tag{1}$$

Flanged round or rectangular hood:

$$Q = 0.75v_c(10x^2 + A) \tag{2}$$

Plain slot hood:

$$Q = 3.7v_cLx \tag{3}$$

Flanged slot hood:

$$Q = 2.8v_cLx \tag{4}$$

where:

Q = hood air flow (volume/time)
v_c = desired capture velocity (length/time)
x = distance between source and hood (length)
A = hood area (length squared)
L = slot length (length)
and a slot hood is defined as having a length to width ratio ≥ 5.

Flanges are flat plates attached to the hood, usually parallel to the hood face, which limit the flow of air from behind the hood. Flanging improves the efficiency of exterior hoods by forcing air to flow from the zone directly in front of the hood, where the air is contaminated and drawing less air from behind the hood, where the air is not contaminated.

For booth type hoods, the required hood flow is calculated by multiplying the capture velocity by the area of the booth opening.

$$Q = vA \tag{5}$$

where:

Q = hood air flow (volume/time)
v = velocity at the booth opening (length/time), and
A = area of booth opening (length squared).

Figure 1[36] shows a booth used for sputum induction used for MTb control. Other examples of local exhaust ventilation systems used in hospitals are waste anesthetic gas scavenging systems and laser plume extraction systems.

In addition to the design procedure outlined above, the Industrial Ventilation Manual[53] gives design plates for hoods for many different industrial operations. The design plates are taken from designs used in actual installations of local exhaust ventilation systems. The manual cautions that modifications to the design plates may be necessary for special conditions, such as drafts. The manual also cautions against the design data being indiscriminately applied to highly toxic materials.

Many of the design plates given in the manual[53] could be used as a starting point for design of hoods for MTb control. The efficacy of the designs for MTb control would have to be validated before the designs could be widely used. Some examples of design plates that might be considered for MTb control are a welding hood, for cough-inducing procedures such as bronchoscopy or intubation, and a table slot hood for autopsy procedures. Laboratories handling infectious samples can refer to biological safety cabinet designs.[64]

Capture efficiency is a more quantitative index of hood performance than centerline capture velocity. Capture efficiency is defined as the fraction of contaminant generated that is captured directly by the hood. Roach[65] recognized that velocity is not the only determinant of exhaust effectiveness. Using dimensional analysis, capture efficiency, being dimensionless, must be equated with some other non-dimensional expression. Some physical quantity or quantities other than air velocity must be involved. Ellenbecker et al.[66] have shown capture efficiency to be a function of hood air flow, hood area, distance from the hood, crossdraft velocity and source temperature.

Several investigators have studied capture efficiency of exhaust hoods.[67–77] Many of these studies have been laboratory studies under controlled conditions. Conroy et al.[76] and Prodans et al.[77] have conducted field-based studies of local exhaust hood performance for hoods used for control of vapor degreasing solvents. Further development and validation of local exhaust hood designs for MTb control is needed using a capture efficiency approach.

All local exhaust systems need to be tested initially and periodically to ensure that they are operating as designed. One design criteria is the hood flow rate. Air velocity at some point in the system is measured and the hood air flow is calculated using Equation (5). The velocity can be measured in the duct

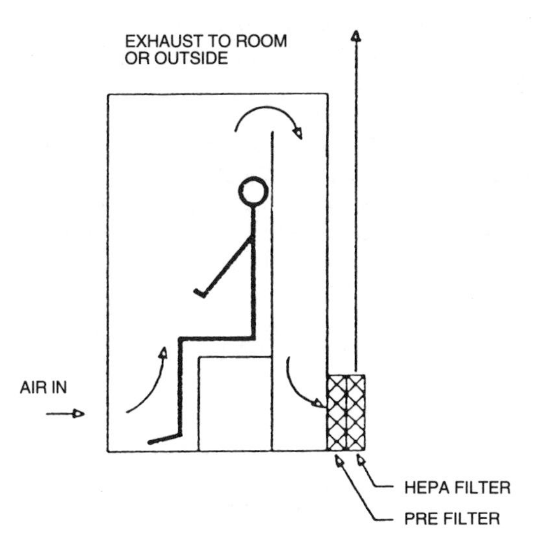

Figure 1. Sputum induction booth. *Source:* "Draft Guidelines for Preventing the Transmission of Tuberculosis in Health-Care Facilities, Second Edition; Notice of Comment Period" *Federal Register* 58:195 (October 12, 1993). pp. 52809–52854.[36]

downstream of the hood using a Pitot tube and manometer or at the face of the hood using a thermoanemometer, swinging vane anemometer or rotating vane anemometer. The cross-sectional area of the measurement location is also needed to calculate hood air flow.

Another design criteria is capture efficiency. Capture efficiency can be measured using a tracer gas. A tracer gas is released at a known rate at the source location and the concentration of the tracer gas is measured in the duct downstream of the hood. The hood air flow is also measured. The capture efficiency is calculated using Equation (6).

$$\eta = \frac{C_d Q}{S} \tag{6}$$

where:

η = capture efficiency (dimensionless)
C_d = concentration of tracer gas in duct (mass/volume)
Q = hood air flow (volume/time)
S = tracer gas release rate (mass/time)

A description of these instruments and procedures used for testing local exhaust systems is given in Chapter 9 of the Industrial Ventilation Manual[53] and Chapters 3, 5 and 13 of Burgess et al.[78]

EXAMPLE 1: The booth shown in Figure 1 is used for sputum induction from suspected or known TB patients. The booth opening is 84 × 32 in. with the door open. With the door closed there is a 3/4 × 32 in. gap below the door. a) With the door closed, what hood air flow is needed to maintain a capture velocity of 200 fpm through the booth opening? b) What hood air flow is would be necessary if the door is left open?

Use Equation (5):

$$Q = vA \tag{7}$$

Door Closed

$$A = (3/4 \times 32 \text{ in.})\left(\frac{1 \text{ f}^2}{12^2 \text{ in.}^2}\right) = 0.1667 \text{ f}^2$$

$$v = 200 \text{ fpm}$$

$$Q = vA = 200 \text{ fpm} \times 0.1667 \text{ f}^2 = 33 \text{ cfm}$$

Door Open

$$A = (84 \times 32 \text{ in.})\left(\frac{1 \text{ f}^2}{12^2 \text{ in.}^2}\right) = 18.67 \text{ f}^2$$

$$v = 200 \text{ fpm}$$

$$Q = vA = 200 \text{ fpm} \times 18.67 \text{ f}^2 = 3733 \text{ cfm}$$

Designing the booth for the closed door position would be the more economical choice. c) If the closed door design is installed, what capture velocity would result if a staff member left the door open during sputum induction?

Rearranging Equation (5) to solve for v gives:

$$v = \frac{Q}{A} = \frac{33 \text{ cfm}}{18.67 \text{ f}^2} \approx 2 \text{ fpm}$$

Isolation of the patient in the booth would be lost because the air velocity through the opening is less than disturbing room air currents typically found in rooms.

Directional Air Flow

Directional air flow uses directed air movement without benefit of enclosures to isolate workers from MTb sources. This concept has many potential applications in health care facilities. However, it is in the second tier of ventilation control designs after local exhaust.

Directional air flow systems use velocity and bulk air movement to direct contaminated air away from health care workers. The most studied application of directional air flow is the cleanroom. The technology has been applied successfully in critical health care settings such as operating rooms. Although federal guidelines regulate cleanroom performance, worker protection is not considered in cleanroom design standards.[53] The purpose of the room is protection of a product or patient.

Nevertheless, some of the cleanroom features can be applied to droplet nuclei control. The unidirectional flow system moves filtered, recirculated air in either a vertical or horizontal direction.[64] Because they distribute large air volumes across opposing room surfaces, these designs provide constant velocity, bulk air flows through the rooms in predictable paths that small particles must follow. The health care worker can be protected if their movements do not place them downstream of the patient or source.

Cleanrooms have several other features which need consideration when applying the designs to TB control:

1. Excess supply air keeps out dust from adjoining spaces.
2. A series of prefilters and high efficiency filters reduces the number concentration of dust particles to very low levels.
3. The life of the filters is extended by maximizing the recirculated air flow and minimizing the fresh outdoor air flow.
4. A unidirectional control air velocity of about 100 feet per minute is usually attained across the entire room.
5. Preventive maintenance time is high.
6. Cleanroom initial costs are high.

For application to MTb control, the first feature, excess supply air, should generally be avoided. The workers and other patients need protection from the TB patient. The system can be balanced to provide an excess of exhaust air to isolate the room from the adjoining ones. If the room has a positive pressure to protect an immune-compromised patient, for example, a negative pressure anteroom can be added to isolate the room from the adjoining spaces. However, workers entering the anteroom will need to be protected just as if they were in the patient room.

The second feature, filtration, is not needed for one-pass air flow rooms. However, the air flow needed for a 100 fpm control velocity is large and the energy cost is high. For example, an isolation room that is 10-feet wide and 10-feet high needs a 10,000 cfm capacity system [Q=(10 f × 10 f × 100 fpm) using Equation (5)]. If this were a one-pass system, the loss of conditioned air would be significant.

The third feature, recirculated air with filtration, would have a higher equipment expense but lower operating expense associated with energy savings. The example above would still need a 10,000 cfm fan. However, only a small percentage of this flow would need to be exhausted with proper fresh air flow for comfort.

Cleanroom systems are not practical for most TB control situations. However, there may be some critical applications in health care facilities where the high exposure hazard justifies the extra expense. For example, emergency rooms in hospitals that serve patient populations with high risks of TB infection may need rooms or curtained areas with reasonably good droplet nuclei control. Modified, unidirectional flow cleanroom designs could provide that control in ways that do not obstruct emergency room activities.

Multidirectional flow rooms are more commonly used for TB control. If the clean supply air is introduced at a position where the health care workers normally stand, the exposure time and risk in a patient isolation room, for example, would be reduced. Although the short-circuited air flow impairs the mixing needed for good dilution of air contaminants, good positioning of the worker (near the supply diffuser) and the patient (near the exhaust grill) could improve the exposure risk.

Figure 2 shows the concept proposed in the CDC guidelines.[36] Directional flow only exists very close to the supply and exhaust points in the room because the control air flows and velocities are much less than ones used in cleanrooms. This setup hopes that directional air flow patterns are established to transport droplet nuclei away from the worker. However, thermal air currents and air flows through window cracks can easily upset these patterns. The bulk air flow patterns over long averaging times may indeed be away from workers in these rooms. Whether or not the air flow patterns provide adequate protection from short term emission activities like coughing has not been validated.

EXAMPLE 2: A patient bed in a 2000 cubic foot room is located near the exhaust grill (Figure 2a). The 18-inch (1.5 f) square grill is 3-feet from the center of the bed. a) What air flow is needed to attain a 50 fpm control velocity at this point? b) If the room cross-sectional area is 100 f^2 and the supply air comes from the opposite wall, what is the average air velocity in the room?

a) Use Equation (2) for a flanged hood (the wall acts as a flange).

$$Q = 0.75v_c(10x^2 + A) = 0.75(50 \text{ fpm})(10(3^2 \text{ f}^2) + 1.5^2 \text{ f}^2) = 3459 \text{ cfm}$$

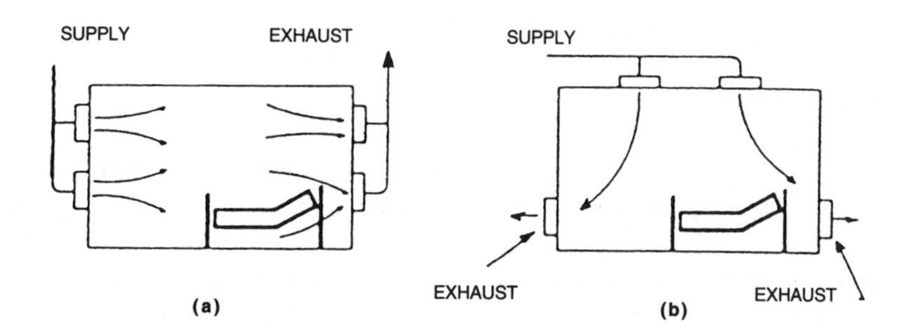

Figure 2. Directional air flow patterns. *Source:* "Draft Guidelines for Preventing the Transmission of Tuberculosis in Health-Care Facilities, Second Edition; Notice of Comment Period" *Federal Register* 58:195 (October 12, 1993). pp. 52809–52854.[36]

b) Use Equation (5)

$$v = \frac{Q}{A} = \frac{3459 \text{ cfm}}{100 \text{ f}^2} = 35 \text{ fpm}$$

Although this system could provide reasonably good directional air flow control, the expense associated with the fan and filters would be much more than normally invested in an isolation room.

Another directional air flow strategy is the air shower. This concept places an air supply hood over a fixed work location. The hood should have side curtains for better directional control and a thermostat for thermal comfort. Typical air velocities are in the range of 100 to 700 fpm. The hood provides a supplied air "island"[79] that workers can use to isolate themselves from contaminated air. Good communication about the use of the air shower is needed for workers to properly use this and any other directional air flow control.

A disadvantage of cleanrooms, and of directional air flow systems in general, is discomfort associated with drafts. Thermal comfort is mainly a function of room temperature, air velocity, air turbulence, the temperature difference between the room and the drafts and the part of the body in contact with the draft.[80]

Another factor to consider is the effect of drafts on the eyes. Occupant complaints about indoor air quality are sometimes associated with exposure of the eyes to draft velocities greater than 100 fpm.[81–82] The effects include eye irritation, contact lens problems, dry eyes and headache.

Control with Concentration Reduction

Dilution Ventilation

The use of dilution ventilation is most effective for low generation rates of non-toxic or low toxicity gaseous or vaporous contaminants where there is source-worker separation. Rather than isolate the worker from the patient, dilution ventilation reduces the residence time of droplet nuclei in the worker's air space. Prohibitively high air flow rates are needed for high generation rates and/or high toxicity contaminants. The worker will be exposed before the contaminant is diluted if the worker is very close to the source.

The current design method for general ventilation uses a completely mixed space mass balance approach. The model assumes perfect and instantaneous mixing of contaminants, no concentration of contaminant in the supply air, and a constant generation rate. A safety/mixing factor is introduced to account for deviations from these assumptions.[53] The design calculation uses Equation (7).

$$t_2 - t_1 = \frac{KV}{Q} \ln \frac{G - \frac{C_1 Q}{K}}{G - \frac{C_2 Q}{K}} \tag{7}$$

where:

t_1 = initial time
t_2 = time of interest
C_1 = contaminant concentration at time t_1
C_2 = contaminant concentration at time t_2
Q = air flow through space
V = volume of space
G = contaminant generation rate, and
K = safety/mixing factor.

At steady state, Equation (7) reduces to:

$$C_{ss} = \frac{GK}{Q} \tag{8}$$

where:

C_{ss} = steady state contaminant concentration.

In order to calculate the required air flow (Q), an estimate of the contaminant generation rate (G) and the acceptable contaminant concentration (C_{ss}) is needed. In the case of MTb, both of these estimates are unavailable. Generation rates are difficult to measure and will vary from person to person, with various activities (e.g., talking, coughing, sneezing will all result in different generation rates), and with various procedures (e.g., sputum induction, bronchoscopy, etc.). An acceptable exposure concentration has not been determined for MTb. Again there will be high variability in this parameter due to the infectivity of the droplet and the susceptibility of the exposed individual.

The use of Equation (8) for MTb control is limited because steady-state concentrations are rarely, if ever, achieved. The most common generation source is a person coughing. In that case, there will be a sudden large increase in the concentration of droplet nuclei with subsequent decay when the person stops coughing. Conceptually, an average concentration (the integral of the concentration during the sudden increase and subsequent decay) would be the appropriate concentration for use in designing the ventilation system.

In addition to estimates of the generation rate and acceptable concentration, the mixing factor must be determined or estimated. The mixing factor describes the effectiveness of the supplied ventilation in reducing contaminant concentrations. The Industrial Ventilation Manual[53] gives estimates of mixing factors for several room configurations. The mixing factor can also be estimated through experimental measurement of tracer gas concentration decay and mechanical ventilation rates. A tracer gas is released throughout the space of interest and the concentration is measured at a location representative of the concentration in the space, e.g., an exhaust grill. The decay of the tracer gas concentration with time and measured ventilation rate and space volume are used in Equation (7) to calculate the mixing factor, K.

EXAMPLE 3: A patient room has a volume of 1500 f^3. The mechanical exhaust ventilation is 45 cfm. The mechanical supply ventilation is 30 cfm and the air flow under the door to the room is 15 cfm. Tracer gas concentration as a function of time was as follows:

Time	Elapsed Time, min	Concentration, parts per billion (ppb)
1518	0	8.12
1532	14	7.49
1538	20	5.76
1552	34	4.27

What is the mixing factor for this room? Assume that there is no other air infiltration or exfiltration.

Following the release of the tracer gas, the generation rate is 0 and Equation (7) reduces to:

$$t_2 - t_1 = \frac{KV}{Q}\ln\frac{C_1}{C_2}$$

which can be rearranged to:

$$\ln C_2 = \ln C_1 - \frac{Q}{KV}\Delta t$$

Plotting ln (C_2) versus the change in time (Δt) results in a line with an intercept equal to ln (C_1) and a slope equal to (-Q/(KV)). This is shown in Figure 3. The slope of the "best fit" line through the data is -0.0085 min^{-1}. The mixing factor, K, is calculated as follows:

$$K = -\frac{1}{slope}\frac{Q}{V} = -\frac{1}{-0.0085\ min^{-1}} \times \frac{45\ cfm}{1500\ f^3} = 3.5$$

The current recommendations for ventilation rates are given in Table 1.[39] The recommendations in Table 2 are based on comfort criteria and experience and will not necessarily provide adequate protection from MTb. The recommendations should be used as minimum design criteria and higher ventilation rates will result in lower contaminant concentrations. Using Equation (8) shows that an infinite air flow is necessary to maintain the concentration of MTb at zero if there is any generation of particles. Without an estimate of generation rate and acceptable concentration, it is impossible to calculate the required air flow.

One calculation that is possible is the amount of time needed to reduce the concentration to some specified fraction of the initial concentration once generation has ceased, for example, when an infected person is no longer in the space. Equation (7) can be rearranged to give:

$$t_2 = -\frac{VK}{Q}\ln\frac{C_2}{C_1} \qquad (9)$$

Table 2[36] gives the number of minutes necessary to reduce the concentration by 90, 99, and 99.9% for various air exchange rates. The time given in Table 2 must be multiplied by the appropriate mixing factor for the space.

EXAMPLE 4: A sputum-induction booth is shown in Figure 4. a) What value of K (mixing factor) would you assign for this space? How can you determine K? b) If

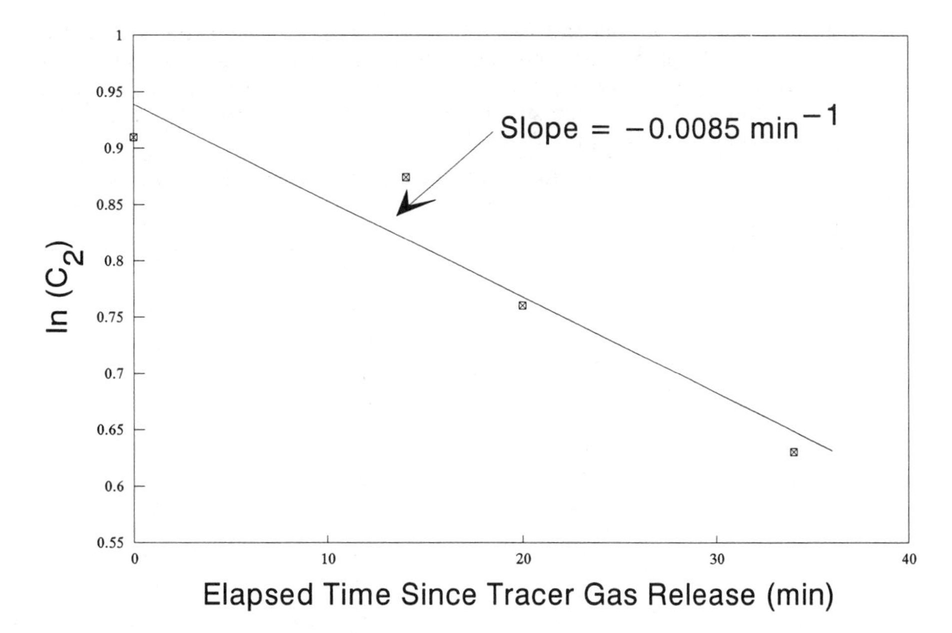

Figure 3. Tracer gas decay curve.

Table 1. Recommended ventilation rates. ***Source: Heating, Ventilating, and Air-Conditioning Applications,*** **American Society of Heating, Refrigerating and Air-Conditioning Engineers, Inc., Atlanta, GA (1991).**[39]

		Air Changes per Hour	
Function	**Pressure**	**Outside Air**	**Total**
Operating	Positive	15	15–25
Recovery	Equal	2	6
Trauma	Positive	5	12
Intensive Care Unit	Positive	2	6
Isolation	Negative, Positive	2	6
Isolation Anteroom	Negative, Positive	2	10
Bacteriology Laboratory	Negative	2	6
Treatment	Negative, Positive	2	6

Table 2. Air changes per hour and time in minutes for removal efficiencies of 90, 99, and 99.9% of airborne contaminants. *Source:* "Draft Guidelines for Preventing the Transmission of Tuberculosis in Health-Care Facilities, Second Edition; Notice of Comment Period" *Federal Register* 58:195 (October 12, 1993). pp. 52809–52854.[36]

Air Changes per Hour	Minutes Required for a Removal Efficiency of:		
	90%	99%	99.9%
1	138	276	414
2	69	138	207
3	46	92	138
4	35	69	104
5	28	55	83
6	23	46	69
7	20	39	59
8	17	35	52
9	15	31	46
10	14	28	41
11	13	25	38
12	12	23	35
13	11	21	32
14	10	20	30
15	9	18	28
16	9	17	26
17	8	16	24
18	8	15	23
19	7	15	22
20	7	14	21
25	6	11	17
30	5	9	14
35	4	8	12
40	3	7	10
45	3	6	9
50	3	6	8

the air flow through the space, Q, is 41 cfm, what is the air change rate? c) For the K value assigned in step a and the air change rate in step b, how long would be needed between users for a 90% reduction in concentration? d) How long for a 99.9% reduction?

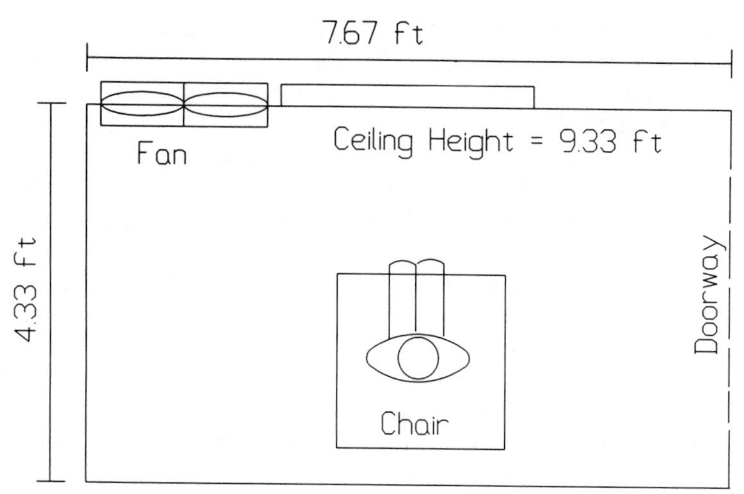

Figure 4. Sputum-induction room.

a) Using reference 53, p 2–4, the situation that most resembles this situation has K = 2.5 as a minimum. For MTb control you might want to add in a safety factor for toxicity and increase K to 3–5. For this example, pick K=4. You could determine K experimentally by releasing a tracer gas and measuring its decay.
b) The air change rate is the air flow through space divided by the space volume.

$$V = 4.33\ f \times 7.67\ f \times 9.33\ f = 310\ f^3$$

$$\text{air change rate} = \frac{Q}{V} = \frac{41\ \text{cfm}}{310\ f^3} \times \frac{60\ \text{min}}{\text{hr}} \approx 8 \text{ air changes per hour (ach)}$$

c) For 90% reduction and 8 ach, Table 2 gives t = 17 min. The total time necessary would be t × K = 17 min × 4 = 68 min. For 99.9% reduction and 8 ach, Table 2 gives t = 52 min. The total time would be 52 min × 4 = 208 min or 3.5 hr. For a busy emergency room or clinic, 3.5 hr or even 68 min might be too long. To reduce the amount of time, the air flow could be increased, the volume of the space reduced, or the mixing characteristics of the space improved.

Air Filtration

Another method of concentration reduction is the use of filters to remove droplet nuclei from the air. High efficiency particulate air (HEPA) filters are filters that are 99.97% efficient for particles 0.3 µm in diameter. Ultra low penetration air filters (ULPA) are 99.999% efficient for particles 0.3 µm in diameter. The filter efficiency increases with increase in particle size. Since droplet nuclei which carry MTb are approximately 1–4 μm, a HEPA filter can be expected to be at least 99.97% efficient. However, the efficiency of HEPA filters for MTb has not been validated.

HEPA filters can be applied to TB control in two ways: 1) duct filtration and 2) room air filtration. Duct filtration places the filter in the exhaust duct of the space. All the air being exhausted from the space passes through the filter before discharge to the outside or recirculation to occupied spaces. In-room filtration involves the use of a self-contained unit to reduce the concentration in the room. The unit will have a fan which pulls room air through the filter before discharging the air back into the room. The success of both filtration methods depends on all of the contaminated air in the space passing through the filter, which in turn depends on the mixing characteristics of the room.

The use of portable filtration units, for reducing particle concentrations, is limited by the number of air changes that can be generated by the fan. Larger flow rates through a filter with constant cross-sectional area will result in higher filter face velocities, which may reduce the filter efficiency. Larger flow rates also require larger fans which may make the units too large for the space and/or unacceptably noisy. Additionally, the inlet and outlet locations on portable units are relatively close to each other. This may mean that the units are repeatedly cleaning the same air, i.e., the relative location of the inlet and discharge may result in air "short-circuiting" the space. More research in actual settings is needed on the effects of these units on room particle concentrations and mixing characteristics. Limited research by the manufacturers suggests that these units may be beneficial in reducing particle concentrations (unpublished data).

The CDC guidelines specify recirculation of HEPA filtered air as a supplemental control, with local and one pass dilution ventilation used as the primary control. Air that cannot be discharged to the outside away from air intakes or occupied spaces should be exhausted through a HEPA filter. The guidelines further state that "in any application, HEPA filters need to be carefully installed and meticulously maintained to ensure adequate function."[36]

EXAMPLE 5: This example is meant to illustrate the effect of a HEPA filtration unit on the contaminant concentration reduction in an existing isolation room. In order to completely solve the example, an estimate of the generation rate is needed. For this example, assume G = 18000 particles/min while the patient is coughing and G = 90 particles/min while the patient is not coughing.[43] The true generation rates for a patient coughing and not coughing are unknown and highly variable.

CASE 1: Isolation Room without HEPA filtration unit.

Room volume = 1350 f^3, mechanical exhaust ventilation rate = 90 cfm, mechanical supply ventilation rate = 90 cfm, air leakage under door from pressurized anteroom = 7.7 cfm, approximate mixing factor = 4, supply air is 50% outside air.

a) What is the air change rate?

$$\text{exhaust air flow} = \frac{90 \text{ cfm}}{1350 \text{ f}^3} \times \frac{60 \text{ min}}{\text{hr}} = 4 \text{ ach}$$

b) What is the outdoor air change rate?

$$\text{outdoor air flow} = \frac{(0.5)(97.7 \text{ cfm})}{1350 \text{ f}^3} \times \frac{60 \text{ min}}{\text{hr}} = 2.2 \text{ ach}$$

CASE 2: Isolation room with HEPA filtration unit.

Room volume = 1350 f^3, mechanical exhaust ventilation rate = 90 cfm, mechanical supply ventilation rate = 90 cfm, air leakage under door from pressurized anteroom = 7.7 cfm, approximate mixing factor = 4, supply air is 50% outside air, HEPA filtration flow = 400 cfm, HEPA filter efficiency = 99.97%.

c) What is the new air change rate?

With the HEPA unit the total air flow rate is increased according to:

$$Q = Q_E + FQ_H$$

where Q_E = mechanical exhaust rate, Q_H = HEPA unit exhaust rate, and F = filter efficiency.

$$\text{exhaust air flow} = \frac{90 \text{ cfm} + (0.9997)(400 \text{ cfm})}{1350 \text{ f}^3} \times \frac{60 \text{ min}}{\text{hr}} = 22 \text{ ach}$$

d) What is the outdoor air change rate with the unit operating?

$$\text{outdoor air flow} = \frac{(0.5)(97.7 \text{ cfm})}{1350 \text{ f}^3} \times \frac{60 \text{ min}}{\text{hr}} = 2.2 \text{ ach}$$

e) How would the room concentration change with the introduction of the HEPA filtration unit?

Rearranging Equation (7) to include the recirculation of air and to solve for concentration results in:[83]

$$C_2 = \frac{KG}{Q_E + FQ_H}\left(1 - \exp\left(\frac{(Q_E + FQ_H)t}{KV}\right)\right) + C_1 \exp\left(\frac{(Q_E + FQ_H)t}{KV}\right) \quad (10)$$

For simplicity, assume the patient coughs for the first 5 min of each hour. Figure 5 shows the concentration curve as a function of time, using Equation (10), for each case. The example assumes that the mixing factor is not affected by the HEPA filtration unit. If the HEPA filtration unit is only affecting a pocket of air in its general vicinity and not cleaning the rest of the room then the mixing factor would be different and the curve would have to adjusted.

This example illustrates the theoretical effect of the addition of a HEPA filtration unit. These units have not been thoroughly tested in actual workplace settings and more research is needed to understand their performance.

Ultraviolet Irradiation

The third method of reducing the concentration of infectious MTb droplet nuclei is with the use of ultraviolet germicidal irradiation (UVGI). Ultraviolet (UV) radiation is in the 100–400 nanometer (nm) wavelength range of the electromagnetic radiation spectrum. The UV spectrum is divided into three regions. The UV-A range is from 320–400 nm. The UV-B range is from 290–320 nm. The UV-C range is from 100–290 nm. Commercially available germicidal lamps are low pressure mercury-vapor lamps which are operated in the UV-C range with a predominate wavelength of 254 nm.

UVGI used in exhaust ducts has been shown to be effective in disinfecting air of TB bacilli in experiments with guinea pigs.[84] Other studies have shown the effectiveness of UVGI in reducing transmission of other infections in hospitals,[85] classrooms[86–88] and military housing.[89] The use of UVGI has been suggested by CDC as a supplemental control measure. A recent study where UVGI was evaluated in an outpatient waiting room showed a 14–19% reduction in culturable airborne bacteria.[90] The researchers in that study concluded that environmental factors such as open doors and windows and the mechanical ventilation in the space may limit the effectiveness of UVGI as a control measure for airborne bacteria.

As with HEPA filtration, two methods of UVGI may be used: 1) duct irradiation and 2) upper air irradiation. Unlike HEPA filtration the UV lamps cannot be located in the occupied zone of the room because of the health hazards associated with UV exposure.

Duct irradiation involves placing the UV lamps in the exhaust duct to disinfect the air before it is recirculated. If properly designed and maintained, high UV

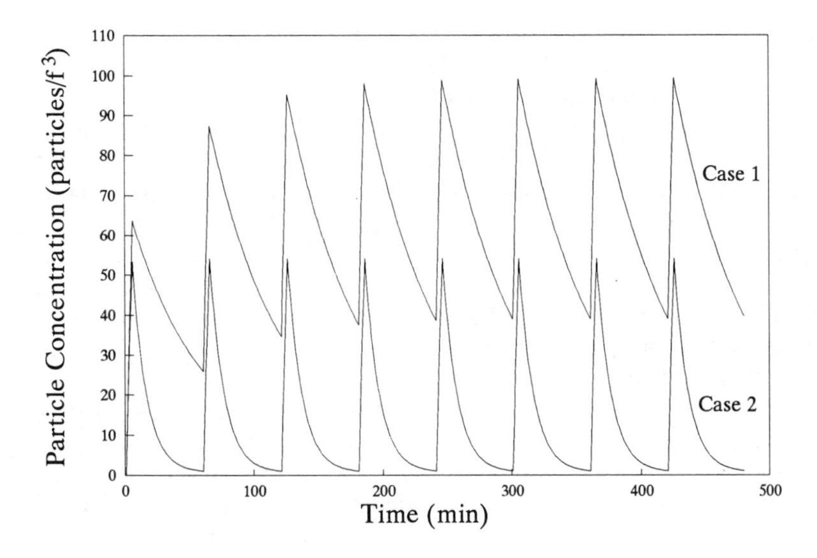

Case 1: No HEPA Filtration Unit   Case 2: HEPA Filtration Unit

Figure 5. Concentration versus time with and without the use of a HEPA filtration unit.

intensities can be used in the duct without exposure to humans, except during maintenance activities. To prevent exposure to maintenance personnel, the system should be designed to prevent access to the duct until the UV lamps have been turned off.

The CDC guidelines[36] outline the situations where duct irradiation may be used and situations where it should not be used. Duct irradiation may be used for isolation and treatment rooms to recirculate the air back into the isolation or treatment room and for other patient rooms and general use areas, such as waiting areas and emergency rooms, where there may be unrecognized TB. Duct irradiation is not recommended as a substitute for HEPA filtration for recirculation from isolation rooms to other areas of the facility or when discharge to the outside is not possible.

Upper room air irradiation involves suspending UV lamps from the ceiling or mounting them on the walls. The lamp must be high enough to prevent eye or skin exposure to persons in the room. The bottom of the lamp must be shielded and the ceiling and walls should be nonreflective surfaces. Upper room air irradiation depends on contaminated air in the lower part of the room moving to the upper part and remaining there long enough to kill the bacteria. This in turn, depends on the mixing characteristics of the room. Unlike dilution ventilation and HEPA filtration, there is a tradeoff in effectiveness with increasing air change rates and mixing. Good mixing improves the effectiveness of UVGI if there is adequate residence time of the air in the upper portion of the room. High air change rates may not allow for adequate residence time in the upper portion of the room.

UV radiation has been shown to cause keratoconjunctivitis and skin erythema.[91] Broad spectrum UV has been associated with increased risk of skin cancer[92–93] and the International Agency for Research on Cancer (IARC) has listed UV-C as a Group 2A (probable human) carcinogen.[93] NIOSH and the American Conference of Governmental Industrial Hygienists (ACGIH) have published exposure guidelines for UV radiation. The NIOSH Recommended Exposure Limit (REL)[91] and the ACGIH Threshold Limit Value (TLV)[94] are essentially the same. The guidelines are intended to protect nearly all workers from the acute effects of UV exposure (keratoconjunctivitis and skin erythema). The exposure guidelines are not intended to protect against long term effects such as skin cancer or to protect individuals who may be photosensitized due to disease or exposure to photosensitizing chemicals.

The REL and TLV limit the amount of time an individual may be exposed to a certain UV intensity or conversely the UV intensity for a given time period. Table 3 shows the allowable time for various UV intensities. The values given are for UV at a wavelength of 270 nm, considered to be the most hazardous. For other wavelengths, the intensity values in Table 3 should be divided by S_λ, given in Table 4. For UV irradiation at a wavelength of 254 nm, the allowable eight-hour exposure is 0.2 $\mu W/cm^2$.

In addition to limiting the exposure to 0.2 $\mu W/cm^2$ for an eight-hour average, CDC[36] recommends that when UVGI is used in a facility: 1) employees should be trained in the general principles of UVGI, the potential hazardous effects of UVGI, the potential for photosensitivity from certain medical conditions or pharmaceuticals, and general maintenance procedures for UVGI fixtures; 2) warning signs (Figure 6) should be posted on UV lamps and at accesses to ducts where UV lamps are used; 3) a preventive maintenance program should be implemented; and 4) a regularly scheduled evaluation of UV exposure to hospital personnel and patients should be conducted. There are direct reading instruments that can be used for measuring UV intensities in a room.

Patient Isolation Rooms

Negative pressure isolation rooms protect the occupants of adjoining rooms and spaces. People who enter the isolation room are not isolated from the TB patient unless the room has local exhaust or directional air flow controls. Instead, the exposure hazard is reduced by diluting the droplet nuclei concentration in the room.

Existing health care facilities serving high risk populations have difficulty providing the number of isolation rooms needed to care for suspected and confirmed cases of tuberculosis. Many facilities do not have isolation rooms that meet the current requirements for dilution air flow (Table 1), air flow into the room at all points (negative pressure) and air flow distribution within the room (Figure 2). Some issues that need to be considered when specifying improvements or new construction of the rooms are:

Table 3. **Maximum permissible exposure times for UV radiation.** ***Source:*** **"Draft Guidelines for Preventing the Transmission of Tuberculosis in Health-Care Facilities, Second Edition; Notice of Comment Period"** ***Federal Register*** **58:195 (October 12, 1993). pp. 52809–52854.[36]**

Duration of Exposure	Effective Irradiance, μW/cm^2
8 hrs	0.1
4 hrs	0.2
2 hrs	0.4
1 hr	0.8
30 min	1.7
15 min	3.3
10 min	5.0
5 min	10.0
1 min	50.0
30 s	100.0

Table 4. **Relative spectral effectiveness for several wavelengths of UV radiation.** ***Source:*** **Centers for Disease Control, National Institute for Occupational Safety and Health:** ***Criteria for a Recommended Standard-Occupational Exposure to Ultraviolet Radiation.*** **U.S. Department of Health, Education, and Welfare, NIOSH (publication no. (HSM) 73-110009), Washington, DC (1973).[91]**

Wavelength, nm	Relative Spectral Effectiveness, S_λ
240	0.300
250	0.430
254	0.500
255	0.520
260	0.650
265	0.810
270	1.000
275	0.960
280	0.880
285	0.770
290	0.640

1. How many rooms are needed for acute care patients with suspected TB?
2. What control performance criteria will be used to design the acute care isolation rooms?
3. Does the emergency room need isolation controls?

CAUTION ULTRAVIOLET ENERGY PROTECT EYES AND SKIN

CAUTION ULTRAVIOLET ENERGY TURN OFF LAMPS BEFORE ENTERING ROOM

Figure 6. Example warning signs for UV radiation. *Source:* "Draft Guidelines for Preventing the Transmission of Tuberculosis in Health-Care Facilities, Second Edition; Notice of Comment Period" *Federal Register* 58:195 (October 12, 1993). pp. 52809–52854.[36]

4. What are the performance criteria for the emergency room?
5. How many rooms are needed for long term care TB patients and what are the performance criteria?
6. Where should the isolation rooms be located within the building?
7. What are the intended room layouts, equipment locations, room penetrations and window openings in the rooms?
8. What interim measures will be used and how will these controls be evaluated?
9. Are anterooms necessary for good isolation in these locations?
10. Are there local building codes or fire safety regulations that conflict with the isolation room specifications?
11. Is there a detrimental effect of the room air balances on the central ventilation system in the building?
12. How will the isolation room ventilation controls be monitored?
13. Are the room isolation components accessible for maintenance?
14. Who will maintain the systems?

The proposed CDC guidelines provide the minimum design criteria for isolation rooms.[36] They are based on American Institute of Architects (AIA) guidelines.[95] The following example presents these criteria as applied to a review of architectural drawings for a proposed renovation project.

EXAMPLE 6: A dental clinic serving clients with high risk of TB infection decided to convert its two dental operatories into negative pressure, isolation rooms. The patient waiting room also needed to be isolated from the rest of the facility. A review of the architectural drawings gave the following information.

The existing floor area for each room is 90 f^2 and the ceiling height is 9 feet. There are 1/2-inch openings along the bottoms of the 7×3-feet, sealed doors. The rooms will use existing supply air ducts connected to the central ventilation system. Each room will have one 175 cfm supply air diffuser near the entry.

Existing return air grills and ducts will be removed. Two 115 cfm exhaust grills will be installed in each room at a high and low location on the wall opposite from the door. A new exhaust fan will be installed on the roof 50 feet from air intakes. A UV light will be used in the exhaust duct to disinfect the discharged air. The design uses rectangular duct with several branches and elbows up to the exterior wall. Circular duct is specified for the vertical run up to the fan. The specified fan delivers 675 cfm at 0.625 inches of static pressure and 1165 rpm.

Evaluate the plans with respect to the proposed CDC guidelines.[36]

1. Guideline: Maintain 10% more exhaust than supply air flow or at least a 50-cfm difference.

Each room design specifies 175 cfm supply and 230 cfm exhaust. The difference in the balance is -30% or 55 cfm more exhaust.

2. Guideline: Maintain an air velocity of at least 100 fpm through openings around the door in its normal position.

If the door is normally shut, the air velocity through the bottom is 440 fpm (55 cfm/(0.125 f^2). If the door is ajar with a 1/2-inch opening along all sides, the air velocity is 66 fpm. If the door is wide open, the air velocity through the opening is only 3 fpm.

3. Guideline: Provide dilution air flow volumes of at least 6 air changes per hour, 2 of which must be outdoor air.

The total air change rate, based on the exhaust air flow, is:

$$\frac{230 \text{ cfm}}{(90 \text{ f}^2)(9 \text{ f})} \times \frac{60 \text{ min}}{\text{hr}} = 17 \text{ ach}$$

This meets the 6-ach guideline if the room air mixing factor is at least 2.8 (K=17/6). The locations of the supply air grill at the ceiling and high/low exhaust air grills across the rooms may indeed provide the necessary mixing.

The outdoor air change rate, based on the supply air flow and an assumed outdoor air percentage of 20%, is:

$$\frac{(0.2)(175 \text{ cfm})}{(90 \text{ f}^2)(9 \text{ f})} \times \frac{60 \text{ min}}{\text{hr}} = 2.5 \text{ ach}$$

4. Guideline: Discharge the exhaust air flow outdoors away from walkways and air intakes or filter it before recirculating.

The exhaust fan location on the roof is stated to be more than 30 feet away from intakes. The UV light in the duct will help clean the discharged air. However, the light may accumulate dust and lint and need frequent cleaning.

Maintaining the design exhaust air flows in the rooms (as well as the proper balance with supply air) is critical to the success of this design. The accuracy of the static pressure estimate for this system should be checked to verify the selected fan size. The fittings and use of rectangular duct suggest that the air leakage into the duct could be as high as 15% of the total air flow. Since the specified air balance is set at the minimum recommended difference of 50 cfm, there is no room for error. The designers should offer some assurance that the selected fan will deliver more than the designed air flow rate. Sealing the duct seams and fittings could reduce the leakage. Choosing a belt-driven fan would offer some adjustability of air flow.

Measuring air flows in existing isolation rooms should be part of the preventive maintenance program at the facility. Air flows from supply and exhaust air grills can be measured with a swinging vane anemometer attached to an air flow hood.[53] Flow from an obstructed diffuser should be checked with a smoke tube if a reliable measure of air flow cannot be made.

Air flow under the isolation room door can be measured with a tape measure and an air velocity device such as a thermoanemometer. The average of six or more velocity readings along the bottom are multiplied by the area of the opening to get air flow (Equation (5)). Air flow direction should be checked with a smoke tube at the bottom and around the door perimeter with the door in its normal position. The air flow direction between the anteroom and corridor should also be checked with smoke.

The room pressurization can be checked with an electronic manometer if the resolution is very sensitive (0.001 inches water) and the meter is calibrated. The meter senses pressure in the room and the hall with two probes and reports the differential pressure between the two spaces. The suggested measurement points are at the center of the door in the corridor and at the kick plate near the door bottom inside the room.

EXAMPLE 7: The following measurements were made in a patient isolation room. Compare them with the CDC guidelines.

The room layout has two beds, a bathroom and an anteroom. The room pressurization is switchable from positive to negative with a control in the anteroom. The negative pressure mode is achieved by increasing the supply air flow in the anteroom. Air flows in the isolation room are not changed by the pressure control. The facility ventilation system uses a minimum of 50% outdoor air.

The floor area is 150 f^2 and the ceiling is 9 f high. A large, hinged window is on the outdoor wall. There are 3 access panels to the false ceiling space above the beds and bathroom.

The room has three supply air diffusers above the beds and at the window. A curtain rail blocks one diffuser. There are exhaust grills above the door and in the bathroom. The 7 × 3.5 f door is gasketed and has a 0.25-inch opening at the bottom.

Air Flow Measurement				
Location	**Method**	**Velocity**	**Direction**	**Air Flow**
Bed 1 diffuser	Airflow hood		Into room	30 cfm
Bed 2 diffuser	Airflow hood		Into room	30 cfm
Window diffuser (by curtain rail)	Smoke tube		Into room	? (Assume 30 cfm)
Bath grill	Airflow hood		Out of room	50 cfm
Grill by door	Airflow hood		Out of room	40 cfm
Room door at bottom	Thermoanemometer (6 point traverse)	105 fpm	Into room	
Anteroom diffuser	Airflow hood		Into anteroom	80 cfm
Anteroom grill	Airflow hood		Out of anteroom	30 cfm
Hall door at bottom	Smoke tube		Out of anteroom	
Both sides of room door	Differential pressure gauge		Into room	-0.002"H2O

1. Guideline: Patients with known or suspected TB should be placed in a private room.

The isolation room in this example has two beds, which would not meet the guideline.

2. Guideline: Maintain 10% more exhaust than supply air flow or at least a 50 cfm difference.

$$\text{Exhaust air flow} = 50 \text{ cfm} + 40 \text{ cfm} = 90 \text{ cfm}$$

$$\text{Area under door} = \frac{0.25 \text{ in}}{12 \frac{\text{in}}{\text{f}}} \times 3.5 \text{ f} = 0.073 \text{ f}^2$$

$$\text{Air flow under door} = 105 \text{ fpm} \times 0.073 \text{ f}^2 = 7.7 \text{ cfm}$$

$$\text{Mechanical supply air flow} = 30 \text{ cfm} + 30 \text{ cfm} + (30 \text{ cfm}) = 90 \text{ cfm}$$

$$\text{Total supply air flow} = 90 \text{ cfm} + 7.7 \text{ cfm} = 97.7 \text{ cfm}$$

The mechanical exhaust and supply air flows in the room are equal. Although the 50-cfm excess exhaust flow guideline is not met, the smoke tube test and differential pressure measurement at the door indicate that the isolation room is in a negative pressure mode compared to the anteroom. This is because the anteroom is pressurized by excess supply air flow. The exhaust grill near the door will promote the flow of contaminated air toward it. Opening either door disrupts the pressure difference with the room so that a slug of contaminated air can escape into the corridor.

Closing the air balance in existing rooms is difficult because of unknown infiltration air flows through the window cracks and room wall penetrations. Air flows around closed windows and penetrations can be significant. They cannot be measured accurately with conventional instruments. Smoke testing the ceiling access panels is needed in this room to detect contamination of the false ceiling space.

3. Guideline: Maintain an air velocity of at least 100 fpm through openings around the door in its normal position.

The closed door air velocity is 105 fpm into the room. Although this meets the guideline, a better guideline is 120 fpm because of the measurement method error (about 20%).

4. Guideline: Provide dilution air flow volumes of at least six air changes per hour, two of which must be outdoor air.

$$\text{Room volume} = 150 \text{ f}^2 \times 9 \text{ f} = 1350 \text{ f}^3$$

$$\text{Exhaust air flow} = \frac{90 \text{ cfm}}{1350 \text{ f}^3} \times \frac{60 \text{ min}}{\text{hr}} = 4 \text{ ach}$$

$$\text{Outdoor air flow} = \frac{(0.5)(97.7 \text{ cfm})}{1350 \text{ f}^3} \times \frac{60 \text{ min}}{\text{hr}} = 2.2 \text{ ach}$$

The total air flow is less than 6 ach. The outdoor air flow is more than the 2-ach guideline. Increasing the exhaust air flow in the room to 140 cfm so that it exceeds the supply air flow by 50 cfm would increase the total air change rate to:

$$\text{Exhaust air flow} = \frac{140 \text{ cfm}}{1350 \text{ f}^3} \times \frac{60 \text{ min}}{\text{hr}} = 6.2 \text{ ach}$$

The balance between supply and exhaust air in the anteroom should also be adjusted to provide an excess of 50 cfm supply air. This will promote air flow into the isolation room without promoting air flow into the corridor.

Upgrading the system in this way (increasing exhaust flow to 140 cfm) would help meet guidelines 1 and 2. Guideline 3 would be met only if the room had near perfect mixing of the air. The location of supply air diffusers over the beds and exhaust grills near the door is the opposite of the recommended mixing pattern (Figure 2).

RESEARCH NEEDS

This chapter described the application of basic industrial hygiene principles to the control of the TB transmission hazard. Several areas needing research were also identified. The first was the need to develop state and local codes which are consistent with infection control methods while still meeting fire safety and other considerations.

In order to design effective control systems, the droplet nuclei generation rate for various activities must be known. Droplet nuclei generation rates and emission factors need to be developed and validated for representative particles or surrogate microorganisms. Additionally, it is important to determine if there is an acceptable exposure concentration that can be used for design.

The application of directional air flow to MTb control should be investigated and validated in real situations. There is also a need for experimental determination and validation of the "near field" of the source. At what distance from the source does the droplet nuclei concentration equal the background concentration?

The design and validation of local exhaust hoods for MTb control needs to undertaken. The research should evaluate design factors such as capture velocity and efficiency. Acceptance of the hoods by the workers using them should also be optimized.

With the design of isolation rooms, several questions need to be addressed. How many isolation rooms are necessary in a facility? Where should clinic and isolation rooms be located in the facility to minimize the hazard to other occupants of the hospital? What is the minimum air flow difference necessary to achieve effective isolation? What is an adequate negative pressure? What is an adequate control velocity at the openings to the room? What factors (room size, furniture placement, etc.) affect the necessary air flow difference, negative pressure, and control velocity?

The performance of portable HEPA filtration units and UVGI in actual facilities needs to be validated. The hazard reduction and associated costs need to be quantified for these controls. Using them to reduce reentry hazard of air discharged outdoors is another application that needs validation.

This is only a partial list of research needs. Scientists are once again investigating some of these items. Development and maintenance of engineering controls for infectious diseases is a continuing challenge. Hopefully, motivation and resources needed by the healthcare community to control the renewed tuberculosis hazard will be realized.

REFERENCES

1. Peterson JE: *Industrial Health.* American Conference of Governmental Industrial Hygienists, Inc., Cincinnati, OH (1991).
2. Houk VN: Spread of tuberculosis via recirculated air in a naval vessel: the Byrd study. In *Airborne Contagion,* RB Kundsin, ed. Annals of the New York Academy of Sciences, Vol. 353. New York (1980).
3. Riley RL: Indoor airborne infection. *Environ Int* **8**:317–320 (1982).
4. Nardell EA: Dodging droplet nuclei, reducing the probability of nosocomial tuberculosis transmission in the AIDS era. (editorial) *Am Rev Respir Dis* **142**:501–503 (1990).
5. Bloom BR, Murray CJL: Tuberculosis: commentary on a reemergent killer. *Science* **257**:1055–1064 (1992).
6. Melius J: Source Characterization and Control, presented at Workshop on Engineering Controls for Preventing Airborne Infections in Workers in Healthcare and Related Facilities, CDC, NIOSH, July 14, 1993.
7. American Thoracic Society: Control of tuberculosis in the United States. *Am Rev Respir Dis* **146**:1623–1633 (1992).
8. Centers for Disease Control: Summary of notifiable diseases, United States. *MMWR* **40**(53):57 (1992).
9. Dublin LI: The course of tuberculosis mortality and morbidity in the United States. *Am J Pub Health* 48(11):1439–1448 (1958).
10. Centers for Disease Control: Prevention and control of tuberculosis in migrant farm workers. *MMWR* **41** (No.RR-10) (1992).
11. Centers for Disease Control: Prevention and control of tuberculosis in correctional institutions: recommendations of the Advisory Committee for the Elimination of Tuberculosis. *MMWR* **38**:313–320, 325 (1989).
12. Centers for Disease Control: Prevention and control of tuberculosis in facilities providing long-term care to the elderly. *MMWR* **39** (No. RR-10) (1990).
13. Nardell EA, Keegan J, Cheney SA, Etkind SC: Airborne infection-theoretical limits of protection achievable by building ventilation. *Am Rev Respir Dis* **144**:302–306 (1991).
14. Catanzaro A: Nosocomial tuberculosis. *Am Rev Respir Dis* **125**:559–562 (1982).
15. Brennan C, Muder RR, Muraca PW: Occult endemic tuberculosis in a chronic care facility. *Infect Control Hosp Epidemiol* **9**(12):548–552 (1988).
16. Malaskey C, Jordan T, Potulski F, Reichman LB: Occupational tuberculous infections among pulmonary physicians in training. *Am Rev Respir Dis* **142**: 505–507 (1990).
17. Centers for Disease Control: Mycobacterium tuberculosis transmission in a health clinic-Florida, 1988. *MMWR* **38**(15):256–258, 263–264 (1989).
18. Craven, RB, Wenzel RP, Atuk NO: Minimizing tuberculosis risk to hospital personnel and students exposed to unsuspected disease. *Annals Int Med* **82**: 628–632 (1975).

19. Dooley SW, Villarino ME, Lawrence M, Salinas L, Amil S, Rullan JV, Jarvis WR, Bloch AB, Cauthen GM: Nosocomial transmission of tuberculosis in a hospital unit for HIV-infected patients. *JAMA* **267**(19):2632–2634 (1992).
20. Ehrenkranz NJ, Kicklighter JL: Tuberculosis outbreak in a general hospital: evidence for airborne spread of infection. *Annals Int Med* **77**(3):377–382 (1972).
21. Frampton MW: An outbreak of tuberculosis among hospital personnel caring for a patient with a skin ulcer. *Annals Int Med* **117**(4):312–313 (1992).
22. Haley CE, McDonald RC, Rossi L, Jones WD, Haley RW, Luby JP: Tuberculosis epidemic among hospital personnel. *Infect Control Hosp Epidemiol* **10**(5):204–210 (1989).
23. Centers for Disease Control: Nosocomial transmission of multidrug-resistant tuberculosis among HIV-infected persons-Florida and New York, 1988–1991. *MMWR* **40**(34):585–591 (1991).
24. Edlin BR, Tokars JI, Griego MH, Crawford JT, Williams J, Sordillo EM, Ong KR, Kilburn JO, Dooley SW, Castro KG, Jarvis WR, Holmberg SD: An outbreak of multidrug-resistant tuberculosis among hospitalized patients with the acquired immunodeficiency syndrome. *N Engl J Med* **326**(23): 1514–1521 (1992).
25. Fishl MA, Uttamchandani RB, Daikos GL, Poblete RB, Moreno JN, Reyes RR, Boota AM, Thompson LM, Cleary TJ, Lai S: An outbreak of tuberculosis caused by multiple-drug-resistant tubercle bacilli among patients with HIV infection. *Annals Int Med* **117**(3):177–183 (1992).
26. Pearson ML, Jereb JA, Frieden TR, Crawford JT, Davis BJ, Dooley SW, Jarvis WR: Nosocomial transmission of multidrug-resistant Mycobacterium tuberculosis. *Annals Int Med* **117**(3):191–196 (1992).
27. Public Law 91-596. 91st Congress, S.2193, December 29, 1970.
28. "Air Contaminants," *Code of Federal Regulations* Title 29, Pt. 1910.1000.
29. "Bloodborne Pathogens Final Rule" *Federal Register* 56:235 (December 6, 1991). pp. 64003–64182.
30. Occupational Safety and Health Administration Region 2: Enforcement Guidelines for Occupational Exposure to Tuberculosis, May 1992.
31. "Access to Employee Exposure and Medical Records," *Code of Federal Regulations* Title 29, Pt. 1910.20.
32. "Respiratory Protection," *Code of Federal Regulations* Title 29, Pt. 1910.134.
33. TB control plans focus of settlements between OSHA, two Wisconsin hospitals. *Occupational Safety and Health Reporter,* Vol. 23, No. 18, The Bureau of National Affairs, Inc., Washington, DC, September 29, 1993.
34. Centers for Disease Control: Prevention of TB Transmission in Hospitals. U.S. Department of Health and Human Services, Public Health Service, CDC, DHHS Publ. No. (CDC) 82-8371, Atlanta GA, October 1982.
35. Centers for Disease Control: Updated Guidelines for Preventing Transmission of Tuberculosis in Health Care Settings with Special Emphasis on HIV Related Issues. Draft 5, January 15, 1993.
36. "Draft Guidelines for Preventing the Transmission of Tuberculosis in Health-Care Facilities, Second Edition; Notice of Comment Period" *Federal Register* 58:195 (October 12, 1993). pp. 52809–52854.
37. *Title 8, California Code of Regulations.* Div. 1, Ch. 4, Subchapter 7, Group 16 Control of Hazardous Substances. Article 109, Section 5197, Occupational Tuberculosis Control, Draft III (March, 1993).

38. Gershom RM, McArthur BR, Thompson E, Grimes MJ: TB Control in the Hospital Environment, In *Healthcare Facilities Management Series,* American Society for Hospital Engineering of the American Hospital Association, Chicago (1993).
39. *Heating, Ventilating, and Air-Conditioning Applications,* American Society of Heating, Refrigerating and Air-Conditioning Engineers, Inc., Atlanta, GA (1991).
40. Smith D: Mycobacterium tuberculosis and tuberculosis. In *Zinsser Microbiology,* 15th ed. Joklik WK and Smith DT, eds. Ashton-Century-Crofts, New York (1972).
41. Breed RS, Murray EGD, Smith NR: *Bergey's Manual of Determinative Bacteriology,* 7th ed. The Williams & Wilkins Company, Baltimore, MD (1957).
42. Riley RL, O'Grady F: *Airborne Infection—Transmission and Control.* Macmillan, New York (1961).
43. Loudon RG, Roberts RM: Droplet expulsion from the respiratory tract. *Am Rev Respir Dis* **95**:435–442 (1967).
44. Wells WF: Aerodynamics of droplet nuclei. In *Airborne Contagion and Air Hygiene.* Harvard University Press, Cambridge, MA (1955).
45. Centers for Disease Control: National action plan to combat multidrug-resistant tuberculosis: recommendations of the CDC TB Task Force. *MMWR* **41**(No. RR-11) (1992).
46. Kent DC: Tuberculosis as a military epidemic disease and its control by the Navy Tuberculosis Control Program. *Dis Chest* **52**:588–594 (1967).
47. Cole EC: Aerosol Characterization, presented at Workshop on Engineering Controls for Preventing Airborne Infections in Workers in Healthcare and Related Facilities, CDC, NIOSH, July 14, 1993.
48. Kent DC, Reid D, Sokolowski JW, Houk VN: Tuberculin conversion: the iceberg of tuberculous pathogenesis. *Arch Environ Health* **14**:580–584 (1967).
49. Centers for Disease Control, National Institute for Occupational Safety and Health: NIOSH Recommended Guidelines for Personal Respiratory Protection of Workers in Health-Care Facilities Potentially Exposed to Tuberculosis. U.S. Department of Health and Human Services, Public Health Service, CDC, NIOSH, Atlanta, GA, September 14, 1992.
50. Harris HW, McClement JH: Pulmonary tuberculosis. In *Infectious Diseases-A Modern Treatise of Infectious Processes,* 3rd ed., Hoeprich PD, ed., Harper & Row, Philadelphia (1983).
51. Woods JE, Braymen DT, Rasmussen RW, Reynolds GL, Montag GM: Ventilation requirement in hospital operating rooms, Parts I and II. *ASHRAE Trans. 92, part* 2 (1986).
52. Grimsrud DT, Sherman MH, Janssen JE, Pearman AN, Harrje DT: An intercomparison of tracer gases used for air infiltration measurements. *ASHRAE Trans.* 86, part 1, pp. 258–267 (1980).
53. *Industrial Ventilation—A Manual of Recommended Practice.* 21st ed. American Conference of Governmental Industrial Hygienists, Committee on Industrial Ventilation, Lansing, MI (1992).
54. DallaValle JM: Studies in the Design of Local Exhaust Hoods. Doctoral Thesis, Harvard University (1930).
55. Silverman L: Fundamental factors in the design of lateral exhaust hoods for industrial tanks. *The J Ind Hyg and Tox* **23**(5):187–266 (1941).
56. Silverman L: Centerline velocity characteristics of round openings under suction. *The J of Ind Hyg and Tox* **24**(9):259–266 (1942).
57. Silverman L: Velocity characteristics of narrow exhaust slots. *The J Ind Hyg and Tox* **24**(9):267–276 (1942).

58. Garrison RP: Nozzle Performance and Design for High Velocity/Low Volume Exhaust Ventilation. Doctoral Thesis, University of Michigan (1977).
59. Garrison RP: Centerline velocity gradients for plain and flanged local exhaust inlets. *Am Ind Hyg Assoc J* **42**(10):739–746 (1981).
60. Garrison RP: Velocity calculation for local exhaust inlets-empirical design equations. *Am Ind Hyg Assoc J* **44**(12):937–940 (1983).
61. Fletcher B: Centerline velocity characteristics of rectangular unflanged hoods and slots under suction. *Ann Occup Hyg* **20**:141–146 (1977).
62. Fletcher B: Effect of flanges on the velocity in front of exhaust hoods. *Ann Occup Hyg* **21**:265–269 (1978).
63. Fletcher B and Johnson AE: Velocity profiles around hoods and slots and the effects of the adjacent plane. *Ann Occup Hyg* **25**:365–372 (1982).
64. *Procedural Standards for Certified Testing of Clean Rooms.* National Environmental Balancing Bureau, Vienna, VA, pp. 4.14 (1988).
65. Roach, SA: On the role of turbulent diffusion in ventilation. *Ann Occup Hyg* **24**:105–132 (1981).
66. Ellenbecker MJ, Gempel RF, Burgess WA: Capture efficiency of local exhaust ventilation systems. *Am Ind Hyg Assoc J* **44**(10):752–755 (1983).
67. NIOSH Research Report: *Ventilation Requirements for Grinding, Buffing, and Polishing Operations.* National Institute for Occupational Safety and Health, Pub. No.: (NIOSH) 75–107 (1974).
68. Jansson A: Capture efficiencies of local exhausts for hand grinding, drilling, and welding. *Staub-Reinhalt.* **40**:111–113 (1980).
69. Rake BW: Influence of crossdrafts on the performance of a biological safety cabinet. *Appl and Env Microbiology.* **36**(2):278–283 (1978).
70. Fuller FH, Etchells AW: The rating of laboratory hood performance. *ASHRAE Journal.* 49–53 (1979).
71. Fletcher B, Johnson AE: The capture efficiency of local exhaust ventilation hoods and the role of capture velocity. *Ventilation '85*. HD Goodfellow, ed. Elsevier Science, Amsterdam (1986).
72. Flynn MR, Ellenbecker MJ: Capture efficiency of flanged circular exhaust hoods. *Ann Occup Hyg* **30**(4):497–513 (1986).
73. Flynn MR, Ellenbecker MJ: The potential flow solution for air flow into a flanged circular hood. *Am Ind Hyg Assoc J* **46**(6):318–322 (1985).
74. Conroy LM, Ellenbecker MJ: Capture efficiency of flanged slot hoods under the influence of a uniform crossdraft: model development and validation. *Appl Ind Hyg* **4**: 135–142 (1989).
75. Conroy LM, Ellenbecker MJ: Capture efficiency of flanged slot hoods and area sources under the influence of a uniform crossdraft. In *Ventilation '88,* Vincent JH, ed. Pergamon Press, Oxford, England, pp. 41–46 (1989).
76. Conroy LM, Prodans RS, Fergon SM, Lachman M, Yu X, Franke JE, Barbiaux M: Field study of vapor degreaser local exhaust hood performance. Presented at Ventilation '91, Cincinnati, OH, September 19, 1991.
77. Prodans RS, Conroy LM, Fergon SM, Lachman M, Franke JE, Barbiaux M: Field study of local exhaust performance for hoods exhausting vapor degreasers. Presented at the American Industrial Hygiene Conference, Salt Lake City, UT, May 21, 1991.
78. Burgess WA, Ellenbecker MJ, Treitman RD: *Ventilation for Control of the Work Environment.* John Wiley & Sons, New York, NY (1989).
79. Burton DJ: *Industrial Ventilation Work Book.* 2nd ed. IVE, Inc., Bountiful, UT (1992).

80. *Fundamentals*. Chapter 8. American Society of Heating, Refrigerating and Air-Conditioning Engineers, Inc., Atlanta, GA (1993).
81. Franck C: Eye symptoms and signs in buildings with indoor climate problems (office eye syndrome). *ACTA Ophthalmologica* **64**:306–311 (1986).
82. Wyon NM, Wyon DP: Measurement of acute response to draught in the eye. *ACTA Ophthalmologica* **65**:385–392 (1987).
83. Wadden RA, Scheff PA: *Indoor Air Pollution*. John Wiley & Sons, New York, NY (1983).
84. Riley RL, Wells WK, Mills CC, Nyka W, McLean RL: Air hygiene in tuberculosis: quantitative studies of infectivity and control in a pilot ward. *Am Rev Tuberc* **75**:420–431 (1957).
85. McLean RL: General discussion: the mechanism of spread of Asian influenza. *Am Rev Respir Dis* **83**:36–38 (1961).
86. Wells WF, Wells MW, Wilder TS: The environmental control of epidemic contagion. I. An epidemiologic study of radiant disinfection of air in day schools. *Am J Hyg* **35**: 97–121 (1942).
87. Wells WF, Holla WA: Ventilation in the flow of measles and chicken pox through a community. Progress report, January 1, 1946 to June 15, 1949. Airborne Infection Study, Westchester County Department of Health. *JAMA* **142**:1337–1344 (1950).
88. Perkins JE, Bahlke AM, Silverman HF: Effect of ultra-violet irradiation of classrooms on spread of measles in large rural central schools. *Am J Pub Health* **37**:529–537 (1947).
89. Willmon TL, Hollaender A, Langmuir AD: Studies of the control of acute respiratory diseases among naval recruits. I. A review of a four-year experience with ultraviolet irradiation and dust suppressive measures, 1943–1947. *Am J Hyg* **48**:227–232 (1948).
90. Macher JM, Alevantis LE, Chang L-L, Liu K-S: Effect of ultraviolet germicidal lamps on airborne microorganisms in an outpatient waiting room. *App Occup Environ Hyg* **7**(8):505–513 (1992).
91. Centers for Disease Control, National Institute for Occupational Safety and Health: *Criteria for a Recommended Standard-Occupational Exposure to Ultraviolet Radiation*. U.S. Department of Health, Education, and Welfare, NIOSH (publication no. (HSM) 73-110009), Washington, DC (1973).
92. Urbach F: The biological effects of ultraviolet radiation (with emphasis on the skin). In: *Proceedings of the 1st International Conference Sponsored Jointly by the Skin and Cancer Hospital, Temple University Health Sciences Center and the International Society of Biometeorology*, F. Urbach, ed. Pergamon Press, Oxford, England, (1969).
93. *IARC Monographs on the Evaluation of Carcinogenic Risks to Humans: Solar and Ultraviolet Radiation*, Vol. 55, World Health Organization, International Agency for Research on Cancer, Lyon, France (1992).
94. *Threshold Limit Values for Chemical Substances and Physical Agents*. American Conference of Governmental Industrial Hygienists, Inc., Cincinnati, OH (1993).
95. AIA Committee on Architecture for Health: *Guidelines for Construction and Equipment of Hospital and Medical Facilities*. American Institute of Architects Press, Washington, DC (1993).

PART 2

Portable HEPA Filtration for TB Isolation in Hospitals and Clinics

Byron S. Tepper

INTRODUCTION

High efficiency particulate air (HEPA) filters, also referred to as ultra-high efficiency air filters or absolute filters, have been generally accepted as capable of removing viable and nonviable particulates from air streams, producing ultra-clean, microbiologically sterile air. HEPA filters were developed during the 1940s and 1950s by the U.S. Army Chemical Corps and Naval Research Laboratories and the Atomic Energy Commission to provide respiratory protection from biological warfare agents for military personnel and to contain radioactive dust and other airborne radioactive particulates generated in the developing nuclear materials industry. Since then HEPA filters have been used in a large number of medical, research, electronic, pharmaceutical and industrial applications where clean air is essential to the work or where emissions of toxic particulates or hazardous biologic agents must be controlled.

A knowledge of the construction, performance, and mechanics of HEPA filtration is essential to the successful application of this filter in contamination control. Familiarity with the mechanisms of filtration and the methods used to test their efficiency are also necessary to dispel the erroneous assumption that HEPA filters only remove airborne particles down to a diameter of 0.3 micron.

MECHANICS OF FILTRATION

HEPA filters are composed of continuous sheets of borosilicate glass filter paper. Although glass may provide the best mesh of fibers of circular cross-sections and a small diameter, cellulose, asbestos, and other mineral fibers have been used, and today, many HEPA filters use fibers of plastic materials. The mesh of fibers, often referred to as a "depth filter," offers little resistance to air flow since it consists principally of empty spaces. The fibers, however, present an enormous surface area of the fibers within the matrix for adsorption of particles by mechanisms which will be described below. HEPA filters are not sieve or screen type filters as many erroneously assume. Sieve filters contain pores or holes which are smaller than the smallest particle to be eliminated; the larger particles are entrapped in the pore, the smaller pass through the pore. Particles are almost entirely trapped on the surface of sieve filters. Resistance to air flow is very high, increasing as the pores become occluded until the pores are totally occluded. The sieve filter obviously has little application in the bulk cleaning of air.

1-56670-083-3/94/$0.00+$.50

By definition, a HEPA filter has an efficiency of 99.97% for particles 0.3 micron in diameter. The 0.3-micron particle was selected for filter challenge because theoretical studies[1] have shown that filtration efficiency should be at a minimum for that size particle and that efficiency increases for particles smaller or larger than 0.3 micron (Figure 1). Concern has often been expressed that viruses, which are much smaller than 0.3 micron, will readily pass through a HEPA filter. Figure 1 indicates that the filtration efficiency should be greater than 0.03% penetration (99.97% efficiency) for the test particles; actual tests with T1 coliphage, which has a diameter of 0.1 micron, have recorded penetrations in the order of 0.0002 to 0.002 % (99.9998% to 99.998% efficiency). It is unlikely that aerosols are generated that are pure monodispersates of viruses. In nature viruses are usually associated with body fluids (saliva, sputum, blood, urine) which, when dried, contribute to the size of the particle. Until there is evidence to the contrary, it is assumed that virus aerosols will produce droplet nuclei similar in size and nature to bacterial aerosols.

Airborne particles can be divided into two classes, those that are large enough to settle rapidly near the source and those in the form of droplet nuclei which remain suspended in air and behave as a gas. In HEPA filtration, the large particles may actually be sieved and remain loosely attached to the surface of the filter rather than firmly attached to the fibers. There are at least 5 mechanisms by which the small particles are collected on a single fiber[3]: a) inertial impaction, b) diffusion, c) direct interception, d) sedimentation and e) electrostatic attraction.

The air flowing to and around a single fiber is shown in Figure 2a. Note that the airstream bends as it goes past the fiber. Inertial impaction occurs when a relatively large particle (0.5 micron or above) carried in the air fails to follow the air stream around the fiber. Such particles, having a specific gravity greater than air, follow a trajectory in the direction of the obstruction and are impacted and adhere to the fiber by van der Waals forces. The role of inertial impaction in filtration increases markedly as the air velocity increases.

Particles too small to be influenced by the inertial effect but small enough to exhibit Brownian motion may diffuse across air stream lines with "a high probability" of impaction on a filter fiber (Figure 2b). This diffusion mechanism should be most efficient for the removal of particles the size of nondispersed viruses. Increasing air velocity can smooth out Brownian motion and shorten the course of the particle through the filter bed, reducing the chance of collision.

Direct interception operates in the case of particles too large to exhibit Brownian motion and too small to be captured by the inertial effect (Figure 2c). Such particles tend to remain in the stream lines and are the particle size that most readily penetrates the filter bed. Test methods involving 0.2- to 0.3-micron particles rely on this principle and, hence, are a useful indication of the maximum penetration (lowest efficiency) to be expected for the working air velocity of a given filter.

Sedimentation in accordance with Stokes Law is of relatively little importance under normal conditions as it will only apply to heavy particles. However, as flow decreases sedimentation may become increasingly important for particles as small

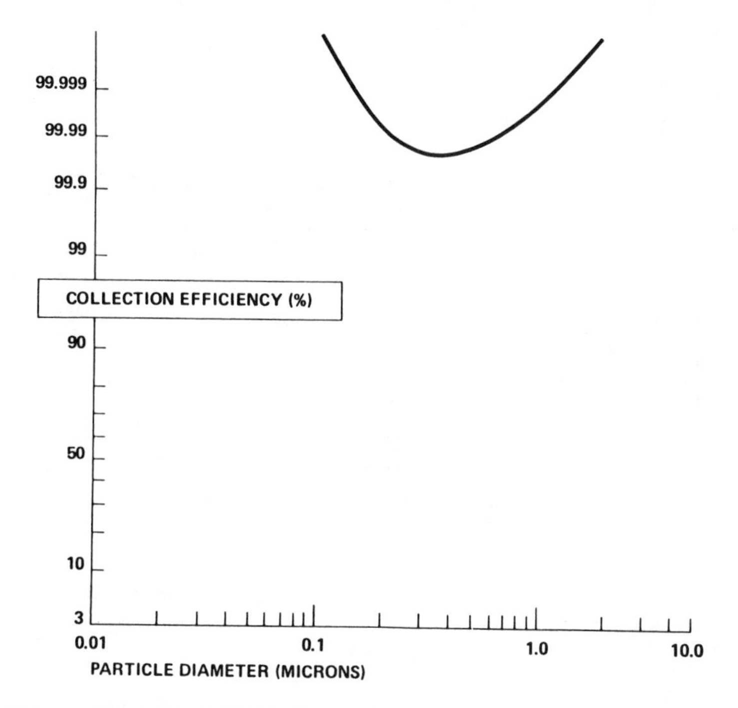

Figure 1. Theoretical HEPA filter collection efficiency. *Source:* National Cancer Institute. A Workshop for Certification of Biological Safety Cabinets. Office of Biohazards and Environmental Control, Publication no. BH 74-01-11, Rockville Bio-Engineering Services, Dow Chemical U.S.A.; 8–15, 1974.[2]

as 0.5 micron in diameter. Electrostatic attraction is of relatively minor importance since current filter materials fail to develop a significant charge. The role of electrostatic charge may change as newer plastic fibers are used in HEPA filters.

The contribution of each mechanism of filtration to the overall efficiency of a HEPA filter is shown in Figure 3.

HEPA FILTRATION IN HEALTHCARE FACILITIES

It is not surprising that techniques using HEPA-filtered, sterile air and laminar airflow were tested in the early 1960s for the control of airborne infections and for protective isolation of patients in the health care environment.

No area of a hospital requires more careful control of environmental contaminants than does the surgical suite. The laminar airflow concept developed for industrial cleanroom use has been successful in reducing airborne contaminants generated by the activities of the surgical team in the vicinity of the operating

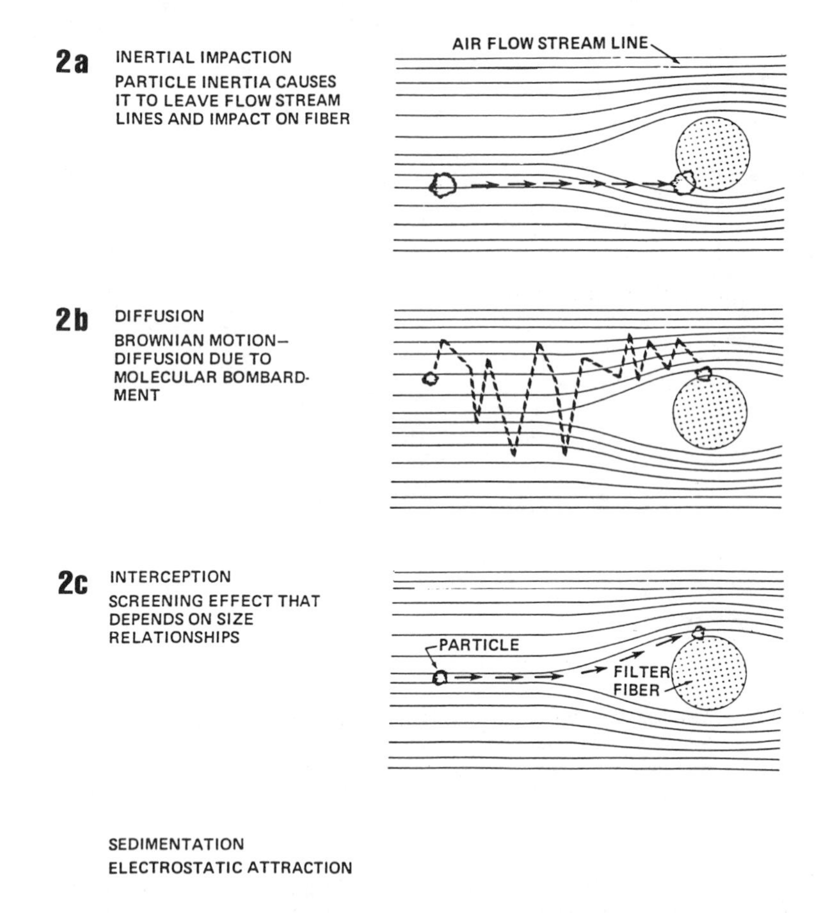

Figure 2. Air filtration theory particle collection mechanisms. Adapted from National Cancer Institute. *Source:* National Cancer Institute. A Workshop for Certification of Biological Safety Cabinets. Office of Biohazards and Environmental Control, Publication no. BH 74-01-11, Rockville Bio-Engineering Services, Dow Chemical U.S.A.; 8–15, 1974.[2]

table. Laminar airflow in surgical operating rooms is defined as ultra-clean (HEPA-filtered) airflow that is predominantly unidirectional, either vertical or horizonal, when not obstructed. The unidirectional laminar airflow pattern is commonly attained at a velocity of 90±20 fpm. In a commonly used alternative to laminar airflow in the operating room, HEPA-filtered supply air is delivered from the ceiling, with downward movement to several exhaust inlets located low on opposite walls. This is probably the most effective air movement pattern for maintaining the concentration of contamination in the operating room at acceptable levels.

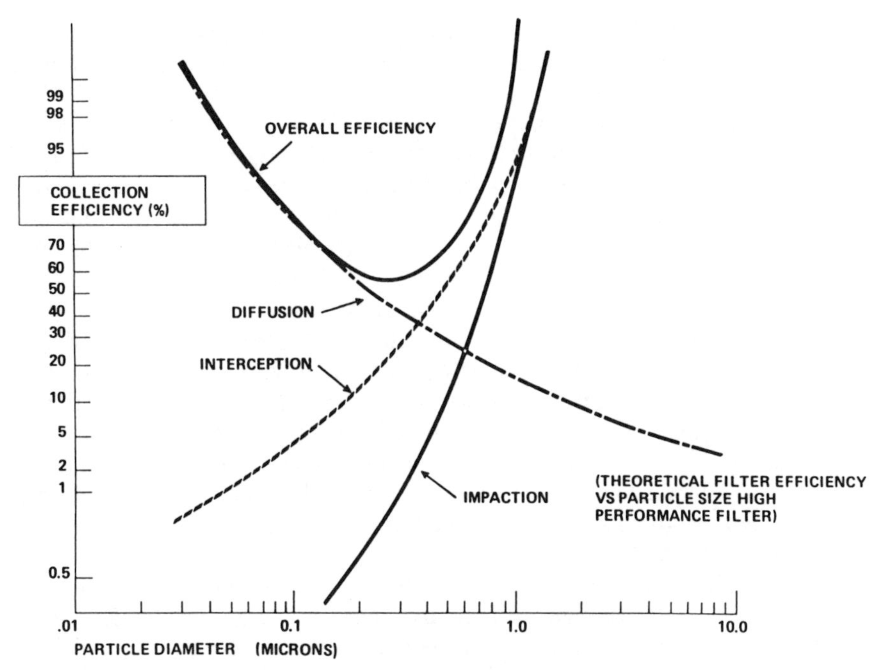

Figure 3. Relative effect of particle collection mechanisms.

HEPA-filtered supply air systems have also been used and are recommended by ASHRAE[4] for rooms used for clinical treatment of patients with a high susceptibility to infection, such as leukemia, burns, bone marrow transplant, organ transplant or acquired immunodeficiency syndrome (AIDS). Laminar airflow systems have been used to protect patients during and after treatments which suppress immunity. Some physicians prefer to house patients in rooms with air changes of 15 ACH (air changes per hour) or greater in which sterile air (HEPA-filtered), supplied by nonaspirating diffusers, is drawn over the patient and returned near the floor or at the door to the room.

Stand-alone, bench-type laminar airflow units routinely provide protective sterile air environments in hospital pharmacies, tissue banks, and blood banks. Biological safety cabinets, which combine HEPA filtration and controlled airflow, have found a place in hospital laboratories not only to protect microbial cultures from contamination but, most importantly, to protect the laboratory workers from exposure to the potential pathogens in the diagnostic specimens they handle daily.

RESPIRATORY ISOLATION FOR TUBERCULOSIS

The resurgence of tuberculosis (TB), compounded by nosocomial outbreaks of multiple drug-resistant tuberculosis, (MDR-TB) has posed an immediate challenge to hospital engineers. Whereas hospital wound infection and, possibly, infection of patients undergoing chemotherapy may be the result of exposures to relatively large and heavy particulates averaging 13 micron, TB is transmitted primarily by airborne droplet nucleic 1 to 5 microns in diameter produced when infected individuals cough, sneeze or speak. Such particles are respirable and are retained in deep pulmonary spaces. For TB, a single organism deposited in the lungs may be all that is needed to cause infection. There is no doubt that the risk of acquiring tuberculosis in a medical environment is almost exclusively a function of the concentration of the infectious particles in the air. Air control is of obvious importance for eliminating or reducing the airborne contaminants; this is accomplished by dilution ventilation and local exhaust ventilation.

In general ventilation systems rely on dilution ventilation to control airborne contaminants, the contaminated air is continually exhausted while room air is replaced and mixed with uncontaminated air. The resultant gradient reduction of airborne contaminants is related to the number of air changes per hour; the greater the number of air changes the more rapid the reduction in airborne contaminants. In dealing with particulate contaminants, i.e. TB droplet nuclei, the uncontaminated air can be fresh (outdoor) air or air that has been "cleaned" by passage through an appropriate filter.

Current CDC guidelines for the prevention of transmission of tuberculosis in health care facilities[5] recommend that any patient suspected or known to have infectious TB should be placed in TB isolation in a private room with a minimum of 6 ACH. The recent draft revision of the guidelines[6] suggests that airflows greater than 6 ACH, up to 37 ACH, would be expected to result in a greater dilution of droplet nuclei. In addition to the high ventilation rates, other specifications for isolation include: a) maintenance of the room under negative pressure (airflow into the room) to prevent airborne contaminants from escaping the isolation room, b) airflow patterns in the room designed to assure proper mixing to prevent stagnation or "short circuiting" of air, c) air from the isolation room should be exhausted to the outside (single pass air) and dispersed so that it is not entrained in the building or neighboring building fresh air supplies; if re-entrainment cannot be avoided the exhaust must be HEPA filtered before discharge and d) if the hospital building recirculates air, HEPA filters should be used to remove contaminants before the air is returned to the general ventilation system. Similar isolation ventilation is recommended for areas where high risk patients are examined and treated. Such areas include triage rooms, waiting rooms, examining rooms in the emergency department, and ambulatory care areas including radiology suites. Isolation ventilation is most important where high risk procedures, i.e., sputum induction, bronchoscopy, and pentamidine or other aerosol therapy are performed.

A recent evaluation of isolation facilities in seven hospitals in a midwestern metropolitan[7] area showed that: a) there were very few rooms designed to have suitable air change and negative pressure ventilation suitable for respiratory isolation, b) only three hospitals had intensive care respiratory isolation rooms, c) none had isolation rooms in the emergency department, and most important, d) only 55% of the isolation rooms demonstrated negative pressure airflow with the doors closed. The survey documented that not only was there an inadequate number of respiratory isolation rooms, but that those that were designed for the purpose were poorly maintained. The term "adequate" is defined by the CDC[5] as "enough TB isolation rooms to appropriately isolate all patients with suspected or confirmed active TB. This number should be derived from the risk assessment of the health care facility." For some metropolitan hospitals with AIDS treatment programs this could mean 10 or more TB isolation rooms; all acute care hospitals should have at least one. The cost of providing isolation rooms with single pass air exhausted directly to the outside may be prohibitive in this era of healthcare cost containment, especially since the risk/benefit analyses have not been done. Retrofitting of existing buildings could even be more difficult considering that most HVAC systems in older buildings are not adequately sized to provide and condition (heat, cool, dehumidify) the increased volumes of air required. Exhausting air from a room is only a minor engineering problem; providing the conditioned supply or make-up air, without disturbing the air balance in the rest of the facility, is a major problem.

Recirculation of HEPA-filtered air to other areas of the facility or recirculation of HEPA-filtered air back into the isolation room are safe alternatives which could conserve energy and increase the air changes of uncontaminated air without disturbing air balance. Fixed systems are not without problems. Filter housings must be located as close as possible to the isolation room to minimize the amount of potentially contaminated exhaust or return ductwork. They must be accessible for decontamination, maintenance, and filter changes. Fans may have to provide the static pressure required to overcome the resistance to airflow caused by the filter media; fixed fans are often noisy and may require sound suppression to avoid disturbing patients. Location of supply air diffusers and returns must be located to avoid dead air spaces and short circuiting to exhaust and return systems. Systems should be tested to assure that proper air mixing and dilution ventilation occur after installation and after furnishings which may obstruct airflow are provided.

PORTABLE IN-ROOM HEPA FILTRATION

Portable in-room HEPA filtration systems are available which can remove particulates and recirculate uncontaminated air into the isolation rooms. Portable filtration units are available that can recirculate 30,000 to 48,000 cfh, which can supplement the general ventilation air changes in a 1600 cubic feet isolation room by as much as 30 ACH. Although not all portable filter units are quiet, at least one portable HEPA filtration system can deliver 800 cfm of filtered air at the extremely low noise level of 55 dBA. The portable HEPA filtration systems have an advantage over the fixed systems in that they can be removed from the patient room and moved to a remote location for maintenance, decontamination, and filter changes. Portable HEPA filtration systems are available which can quickly convert a standard patient room into a TB isolation room by increasing the number of air changes per hour and, by using inexpensive duct work connections, can create a negative pressure (inward airflow) condition by delivering a fraction of the HEPA-filtered air to the building exhaust system, to air recirculation return ducts, and/or out the room window.

The efficiency of portable HEPA filters to remove particulates under actual conditions of use can be evaluated and, if not acceptable, the filter unit can be readily located to optimize the dilution ventilation. Lastly, portable HEPA filtration units are inexpensive; multiple units can be acquired for the cost of designing and installing fixed filtration systems.

The efficiency of a portable HEPA filtration unit, designed for use at the Johns Hopkins Hospital, has been tested in a typical patient room. The room measured 11ft W × 18ft D × 8ft H (1584 cubic feet) and had a ventilation rate of 8 ACH. The room was deliberately chosen because it was 20 cfm positive to the corridor. The filter unit, 3ft W × 2ft D × 6ft H, had an inlet at floor level and the supply at the top; the unit was designed to deliver 550 cfm or 21 ACH at a sound level of 55 dBA. The room was challenged with bis-(2-ethylhexyl) sebacate (dioctyl sebacate, DOS), average particle size 0.3 micron, at 250,000 particles per cubic foot. Particle counts were taken with and without the HEPA filtration. Figure 4 shows the decay curve, average of three runs, for the normal ventilation and the additional dilution ventilation provided by the HEPA filtration unit; the time to reduce particulate levels to ambient background was reduced from 27 to 7 minutes. Within the next 7 minutes, the ambient particle level was reduced by 60% and then a steady state was achieved between the dilution ventilation and the particles in the supply air. The two 7-minute segments represent two aspects of contamination control. The first segment represents the burst contamination, the cough or sneeze that adds a relatively large number of droplet nuclei to the environment. The decay curve shows rapid clearance. The second segment models the continuous generation of infectious droplet nuclei from a TB patient. The cough frequency at the time of admission to the hospital may average as many as 15 coughs per hour before treatment; the decline in cough frequency during treatment is rapid with most patients, reducing their count to half the initial value within two weeks of treatment with antituberculous drugs.[8] The decay curves show that the counts are reduced but are not cleared; a steady state is achieved.

Figures 5 and 6 demonstrate the air curtain effect of the filtration unit placed between the patient bed and the room door. When particles of DOS are generated at the bed headboard with the filter running, only a small percentage, approximately 1%, escape to the door side of the filter; the unit appears to establish an anteroom or airback effect in the patient room even in a room positive to the corridor. The air curtain theoretically will capture any droplet nuclei that would, without the air curtain, be entrained in the wake following attending staff who are leaving the room. In standard isolation rooms, only relatively small airflows are needed to control the direction of flow through the cracks around closed doors. The opening of a door instantaneously reduces any existing pressure between the separated areas to such a degree as to nullify the effectiveness of the pressure.[9] The air supply rates needed to control airflow direction through an open door are considerable and impractical. The air curtain produced by the portable HEPA filtration unit located near the door provides the added protection with the door open as well as closed, reducing transfer on entering or leaving the room.

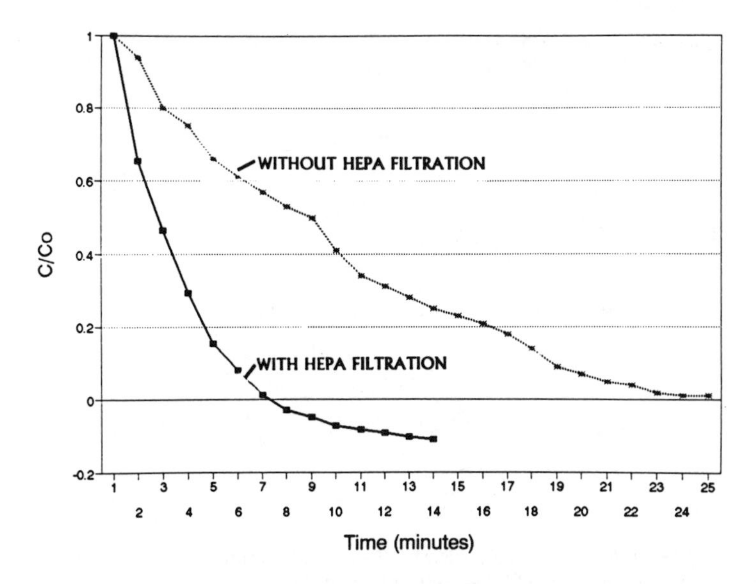

Figure 4. Clearance of particles generated in a patient room with the portable HEPA filtration unit off and with the filtration unit on.

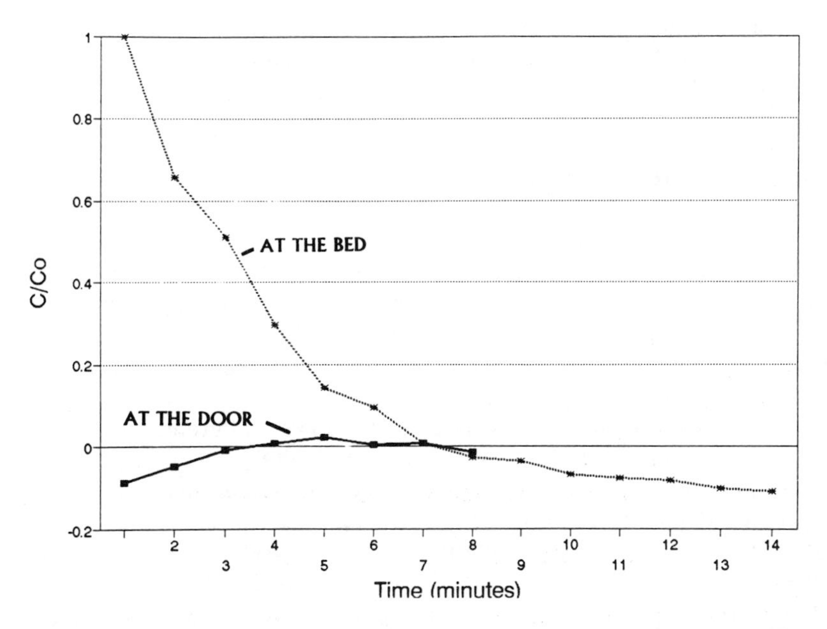

Figure 5. Clearance of particles generated in a patient room with the portable HEPA filtration unit on.

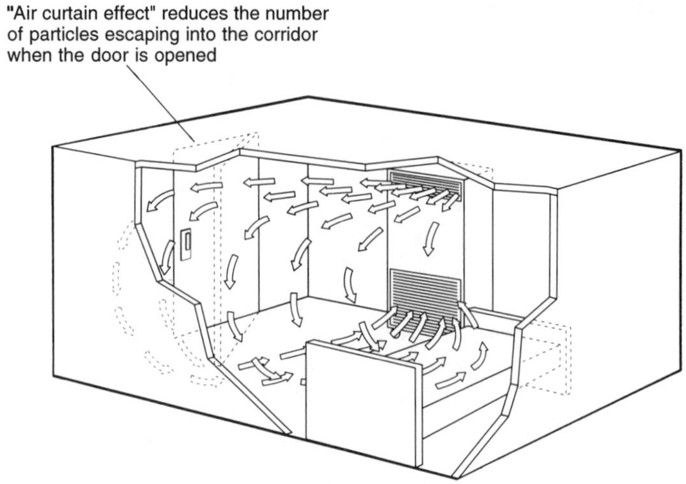

Figure 6. Placement of portable HEPA filtration unit in patient room which produces "air curtain effect." See Figure 5 for data.

The data in this model system show that HEPA filtration can rapidly reduce airborne contaminants, thus reducing the exposure of health care workers to infectious droplet nuclei.

RESPIRATORY PROTECTION

It should be apparent that dilution ventilation only reduces the risk of infection but never entirely eliminates it. The isolation room cannot protect the health care worker from infectious droplet nuclei generated while they are attending the patient. Respiratory protection is obviously required for all health care workers and visitors entering rooms where patients with known or suspected infectious TB are isolated. The type of respirator to be worn is controversial. Without entering and documenting the arguments, it appears that the minimal protection required will be a NIOSH-approved HEPA particulate air respirator. NIOSH-approved respirators providing greater protection will also be acceptable. This is the standard outlined in the October 8, 1993, OSHA directive on "Enforcement Policy and Procedures for Occupational Exposure to Tuberculosis."[10] Along with the requirement to use the HEPA respirators is the responsibility of health care facilities to be in full compliance with the OSHA standard 29 CFR 1910.134: Respiratory Protection, which includes medical examinations, training, and qualitative or quantitative fit testing of all employees required to wear respiratory protection.

SOURCE CONTROL

The most efficient control of airborne contaminants is source control, the capture of the contaminant at its source and its removal without exposing persons in the area.

Historically, studies on the transmission of tuberculosis have shown that providing patients with tissues to cover their mouths and noses when coughing or sneezing significantly reduces the number of droplet nuclei released to the environment. Similarly, providing patients surgical masks to be worn in common areas and during transport has been shown to reduce infectious particles. Since both of these controls provide only partial protection and since they require the cooperation of the patient, they must be supplemented with other control methods.

Local exhaust is a source-control technique which removes airborne contaminants at or near the patient who is the source of infectious droplet nuclei. This technique is especially important during the performance of medical procedures likely to generate aerosols containing infectious particles or during procedures likely to induce coughing or sneezing. These high risk procedures include endotracheal suctioning, bronchoscopy, and sputum induction; aerosol treatments such as pentamidine therapy; cough-inducing; and other aerosol-generating procedures. Hoods, booths or tents provided with exhaust systems can be used to capture and exhaust the droplet nuclei directly to the outside of the building. If re-entrainment is possible or when the exhaust cannot be directed to the outside, the exhaust containing the droplet nuclei and the medication should be discharged only after HEPA filtration.

At Johns Hopkins, the specially designed free-standing hood (reverse flow clean air bench) shown in Figure 7 has been used since 1974 in sputum induction to protect health care workers from exposure to potentially infectious aerosols generated during the procedure. More recently, the same hoods have been used for the control of health care worker exposure to fugitive pentamidine during aerosol administration.[11] Sampling of the area and in the breathing zone of health care personnel as positive for pentamidine during administration of the drug without the engineering control; samples were all below the analytical limits of detection when the engineering control was used with the hood velocity at 100 LFM at the exterior edge (Table 1). The exhaust from the hood which had its own HEPA filter was captured by a thimble connection and exhausted to the outside. The use of the thimble exhaust connection provided directional airflow from the treatment room door to the hood and then to the outside. The data show the effectiveness of the hood in capturing fugitive pentamidine and provide indirect evidence that the hood had been providing protection from infectious agents, including tuberculosis, since its use was initiated.

Booths and tents which totally enclose patients have been used during sputum induction and pentamidine administration procedures to capture droplet nuclei from infectious patients. These devices, usually connected to local exhaust, undoubtedly are effective; however, the effectiveness of such units has not been adequately evaluated.

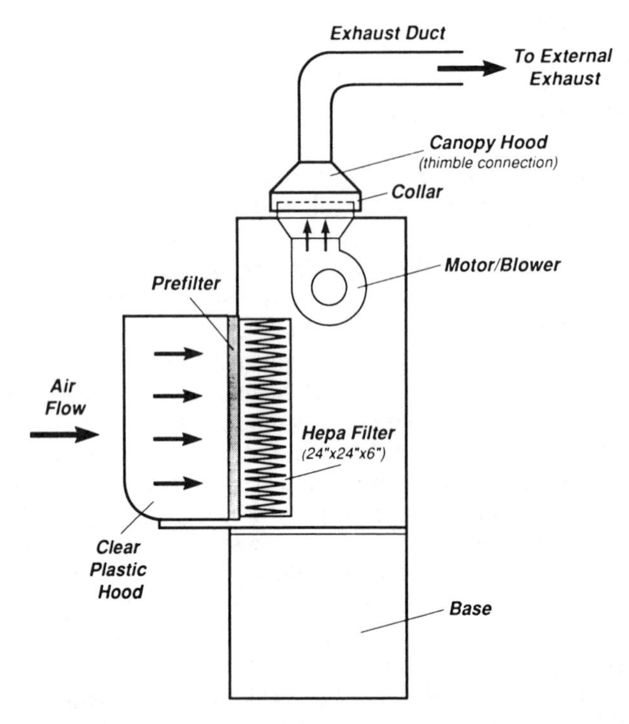

Figure 7. Design of a reversed flow, horizontal flow containment hood. The diagram shows the location of filters and the direction of airflow. *Source:* McDiarmid MA, Schaefer J, Richard CL, Chaisson RE, Tepper BS: Efficacy of engineering controls in reducing occupational exposure to aerosolized pentamidine. *Chest* 102-1764-1766, 1992.[11]

Table 1. Air sampling results for pentamidine.

	Equipment Controls	
	Without Hood	With Hood
Sample Type	No. of Samples >LD*	No. of Samples >LD
Area	11/13 (85%)	0/7 (0%)
Personal	5/8 (63%)	0/7 (0%)

* LD= limit of detection ~.00033 mg/m^3. Adapted from McDiarmid et al.[11]

DISCUSSION

Since 1988, outbreaks of TB, including MDR-TB involving over 200 patients, have been reported to the CDC. In outbreaks in seven hospitals and one prison at least 16 healthcare workers have developed MDR-TB infection; at least 5 healthcare workers have died. Nationwide, several hundred health care employees have become infected after workplace exposure and have required medical treatment. These events have sensitized the health care industry to the resurgence of TB. In 1990, in response to concerns about TB transmission, the CDC issued guidelines for preventing the transmission of TB in health care settings. The engineering controls, specifically the dilution ventilation and "negative pressure" isolation rooms specified in the guidelines have presented an engineering and financial challenge to many hospital administrations.

As recent as 1992, surveys have shown that 27% of acute care hospitals do not have isolation rooms that were designed to meet the CDC specifications, and the evidence suggests that if all the existing isolation rooms were tested, most would not meet their design specifications. It can be assumed further that the number and location of existing isolation rooms would not be appropriate or adequate for the preventive measures recommended in the CDC guidelines.

A survey of our isolation rooms showed results similar to the Fraser et al.[7] study. The isolation rooms in our newer buildings were designed for protective isolation (positive pressure). Over the years clinical services have been relocated so that many of the isolation facilities with anterooms and directional air flow were not in hospital spaces which patients with tuberculosis could occupy. Most of the isolation rooms in areas admitting patients at high risk for tuberculosis did not achieve the optimal combination of directional airflow, air changes, and external exhaust. In addition, there were no isolation rooms in the ER, ICU, radiology, or other areas at high risk for nosocomial transmission of TB. The renovation of existing isolation rooms and the addition of new rooms in appropriate locations was given high priority; however, design and construction will be extensive because of the age of existing buildings and inadequate HVAC systems. Adding isolation rooms to the already limited space in intensive care and emergency rooms has challenged our engineers and facility planners. It was obvious that we needed an interim environmental control strategy while optimizing our isolation room capacity.

The interim solution was to use portable, recirculating, high flow rate HEPA filter units in the existing isolation rooms and in patient rooms used for isolation. Operationally, each patient with possible pulmonary tuberculosis was placed in AFB isolation, which carries the requirement that a HEPA filtration unit be placed in the isolation room. The filters were placed in the rooms as shown in Figure 6. The requirement was also established that all staff wear a high filtration (dust/mist) mask when entering the rooms of patients on AFB isolation.

In one year, 284 patients were placed on AFB isolation. Of these, 10 patients were culture positive for *Mycobacterium tuberculosis*; 4 of the 10 were AFB smear positive and presumed to have transmissible TB. To date, we have not observed PPD conversions in any staff attending these patients.

In our hospital, the portable HEPA filtration units have also been used in bronchoscopy rooms, bronchoscopy recovery rooms, intensive care cubicles, renal dialysis cubicles, in the ER triage room and in a special examination/treatment room in the ER. As described earlier, we have used a containment hood for sputum indication and pentamidine administration for several years without associated occupational TB infection. Probably the longest interim use of the portable HEPA filtration units will be in outpatient waiting rooms in high risk areas such as the ER, AIDS clinics and substance abuse clinics. These areas which usually are large, open, and poorly ventilated have been supplied with multiple portable recirculating HEPA filtration units until such time as the areas are enclosed and the ventilation is upgraded.

Review of our data on AFB isolation reveals that over the course of the past year, there have been at least 5 days in which 10 or more patients have been on AFB isolation with the portable HEPA filtration units in place. The total number of AFB isolation rooms in areas which could be occupied by patients with suspected or confirmed TB has not yet been established, but undoubtedly there will be occasions where the number of patients with suspected TB exceeds the number of permanent isolation rooms. It appears that even after the upgrading and additions there still may be a niche for the portable HEPA filtration units.

REFERENCES

1. Langmuir I: *Filtration of Aerosols and Development of Filter Materials.* O.S.R.D. Report 865. Washington, DC, Office of Technical Services; 1943.
2. National Cancer Institute. A Workshop for Certification of Biological Safety Cabinets. Office of Biohazards and Environmental Control, Publication no. BH 74-01-11, Rockville Bio-Engineering Services, Dow Chemical U.S.A.; 8–15, 1974.
3. Darlow HM: Air filters for recirculating systems: minimum efficiency for bacteria and viruses. In: Hers JF Ph, Winkler KC (Eds.) *Airborne Transmission and Airborne Infection.* Utrecht, The Netherlands: Oosthoek Publishing Co.; 516–519, 1973.
4. ASHRAE. Health Facilities. In: *HVAC Applications,* 1991 ASHRAE Handbook Atlanta, GA; Chapter 7, 1991.
5. Centers for Disease Control and Prevention. Guidelines for preventing the transmission of tuberculosis in health-care settings with special focus on HIV related issues. *MMWR* 39(RR-17):1–29, 1990.
6. Centers for Disease Control and Prevention. Draft guidelines for preventing the transmission of tuberculosis in health-care facilities, Second edition; Notice of Comment Period. *Federal Register* 1993; 58 (no. 195):52810-52854.
7. Fraser VJ, Johnson, Johnson K, Primack J, Jones M, Medoff G, Dunagan WC: Evaluation of rooms with negative pressure ventilation used for respiratory isolation in seven midwestern hospitals. *Infect Control Hosp Epidemiol* 14:623–628, 1993.
8. Loudon RG, Spohn SK: Cough frequency and infectivity in patients with pulmonary tuberculosis. *Am Rev Resp Dis* 99:109–111, 1969.
9. Caplan KL: Ventilation and air conditioning. In: Bond RG, Michaelsen GS, DeRoos RL (Eds.) Environmental Health and Safety in Health-Care Facilities. New York, NY: Macmillan Publishing Co.; 66–121, 1973.
10. Occupational Safety and Health Administration. Memorandum for Regional Administrators: Enforcement Policy and Procedures for Occupational Exposure to Tuberculosis. October 8, 1993.
11. McDiarmid MA, Schaefer J, Richard CL, Chaisson RE, Tepper BS: Efficacy of engineering controls in reducing occupational exposure to aerosolized pentamidine. *Chest* 102-1764-1766, 1992.

PART 3

Preventing TB in the Workplace: What Did We Learn From HIV? Policies Regarding the HIV-Infected Health Care Worker

John Mehring

The HIV/AIDS epidemic brought about a revolutionary change in our practice of infection control in health care and other settings. With the adoption of universal precautions as a control methodology, the risk of exposure to infected blood and other body fluids or materials is acknowledged to exist potentially anywhere, rather than linked exclusively to a confirmed diagnosis.

However, we know that some tasks and procedures place workers at greater risk than others, and now efforts are being made to evaluate these tasks and procedures to modify them, to eliminate them, if feasible, or to institute safer devices or barriers with the goal of further decreasing the risk of exposure to potentially infectious materials.

The concept of applying universal precautions to patients and clients and other members of the public is on firm ground despite efforts to implement mandatory screening programs to determine HIV infection in individual patients and clients.

However, the concept of applying universal precautions to workers is on less firm ground and efforts have been made by various sectors of our society, including the Centers for Disease Control and Prevention (CDC), to discuss and implement mandatory, or less coercive, screening programs to determine HIV infection in individual health care workers and restrict infected health care workers' job duties and responsibilities.

Organizations in the field, such as labor unions, which have argued for increased infection control vis-a-vis patients and clients, agree that there is a concomitant need for increased infection control vis-a-vis workers, although there is significantly less risk of bloodborne infection from workers as opposed to patients and clients. We believe, however, that the same principles that apply in decreasing the risk to workers should apply to decreasing the risk to patients and clients: that is, new, more aggressive infection control methods must be researched and implemented in order to make the workplace safer for all parties. Again, these methods include evaluating the tasks and procedures which place these parties at greater risk of infection, and modifying them, or eliminating them, if feasible, or instituting safer devices or barriers with the goal of decreasing the risk of exposure to all potentially infectious materials.

While there is general support for a more aggressive approach to infection control to provide protection against bloodborne pathogens, this support has not been implemented in concrete ways. All three relevant federal government agencies, the CDC, the Food and Drug Administration (FDA) and the Occupation-

1-56670-083-3/94/$0.00+$.50

al Safety and Health Administration (OSHA) have been extremely conservative in recommending, supporting or requiring infection control efforts beyond elementary universal precautions. The CDC, which has been preoccupied over the past few years regarding the risk of infection posed by invasive procedures, has never recommended, ironically, the use of new procedures or devices to reduce this risk.

A dichotomy has thus developed: there appears to be a pronounced lack of will in these federal agencies and other quarters to go beyond what has already been achieved, while parties on the front lines shift attention away from an individual's infection status, and try to re-focus attention on a more advanced program of infection control that protects society across-the-board.

Regardless of our society's inability to move more quickly to a more effective infection control program, the fact remains that our progress in this area has been substantial. Our mileposts include adoption of universal precautions as a recommendation by the CDC in 1987, and the promulgation of the Bloodborne Pathogens Standard by federal OSHA in 1992. These actions by government have not only prevented exposure to potentially infected body fluids more effectively than previous infection control programs, but they also allowed health care workers and health care patients and clients to resist successfully repeated attempts to discriminate against them. In addition, the passage and implementation of the Americans with Disabilities Act of 1990 provided solid legal ground to protect HIV-infected health care workers and health care patients and clients from discrimination. This is a model that must be carried over and applied to the new struggle to control and prevent tuberculosis infection and disease, and protect HIV-infected healthcare workers and healthcare patients and clients from discrimination.

For a decade, HIV-infected healthcare workers were the subject of concerted efforts to remove them from the workplace to protect patients and clients from exposure to HIV infection. Now HIV-infected healthcare workers are being pressured to leave the healthcare workplace because of the risk they face from tuberculosis infection.

HIV-infected healthcare workers, and other immunocompromised workers, are at greater risk for acquiring tuberculosis infection when coming into contact with the airborne bacillus and developing active tuberculosis once they are infected. This risk is exacerbated by the emergence of multi-drug resistant tuberculosis, which has an extremely poor prognosis. HIV-infected healthcare workers are being counseled by some individuals and organizations to leave the healthcare workplace in order to decrease or eliminate this risk of tuberculosis infection and disease.

Labor unions which have grappled with this issue have decided against the medical removal of a class of individuals who can be medically protected in the workplace if an effective infection control program to prevent the transmission of airborne tuberculosis is instituted. Of course, we know these programs do not exist in most workplaces. Institutions have been slow to adopt the CDC's recommendations. Many employers are reluctant to commit financial resources to tuberculosis prevention or are opposed philosophically to an industrial hygiene approach to tuberculosis prevention. Federal OSHA has dragged its feet in moving toward an enforceable Prevention of Occupational Tuberculosis Standard. In the meantime, workers are fending for themselves. In this vacuum, confusion and inaction predominate, allowing advocates of the medical removal of HIV-infected healthcare workers to appear sensible and humanitarian.

What can HIV-infected healthcare workers do to protect their health and their employment rights? HIV-infected healthcare workers should assess their risk for tuberculosis infection, and they must be allowed, in principle and in practice, to reduce the risk of tuberculosis infection by asking for, and receiving, "reasonable accommodation," a right given them by the Americans with Disabilities Act of 1990. Nevertheless, most HIV-infected healthcare workers will remain on the job where they are currently, because that is what they want to do, or because they are ignorant of their infection status. That is, and should remain, their right. And just as we know that we could test everyone for HIV infection and still not know accurately what everyone's infection status is, we also know that the occupational transmission of tuberculosis will continue even when pressure is exerted to make the healthcare workforce "HIV-free." In terms of infection control, a medical removal program for HIV-infected workers is as productive an outcome as a mandatory HIV screening program.

If employers are committed to a safe work environment, they will give their employees the education and training to make their individual risk assessment, and they will provide an occupational environment that protects all workers and the public, across-the-board. The general application of industrial hygiene principles will protect these populations at risk for tuberculosis infection, just as the specific application of another industrial hygiene principle—universal precautions—lays the foundation for protecting these populations at risk for bloodborne infections.

For HIV-infected workers this means getting critical information and services in a way that guarantees their confidentiality. It is very important that immuno-compromised workers screen themselves carefully for tuberculosis infection and disease. Workers who may be immunocompromised include those who are HIV-infected, as well as those with leukemia, lymphoma or those who are using various drug or radiation treatments. Accurate screening for tuberculosis in immunocompromised persons is difficult to accomplish because the standard tuberculosis test, the PPD skin test, requires a healthy immune system to work correctly. A compromised immune system may not be able to mount an immune response, which means that the person is anergic, and therefore the PPD test result may not be accurate. A negative PPD skin test may actually be a false negative. The person may be infected with tuberculosis, but the test may not reveal that.

HIV-infected healthcare workers should test themselves for anergy. Testing for anergy involves using at least two other skin test antigens such as mumps antigen, tetanus toxoid and candida antigen. If a person does not react to these antigens, the person can be considered anergic. This person should be medically monitored for tuberculosis symptoms.

Practically speaking, an effective occupational tuberculosis prevention program can only be built when all workers are informed about the connection between immunocompromised status and the risk of tuberculosis infection, so that the target population of immunocompromised workers is reached successfully, and their privacy rights are respected. For example, employers should provide anergy testing in tandem with PPD testing or after a negative result has been obtained, to all employees who elect it, with no explanation required, as well as offering an outside referral.

All workers need to make informed decisions about their risk and be given the appropriate equipment to protect themselves. In the debate over appropriate respirator use, we must agree upon a minimum level of protection, while an additional level of protection must be provided to those workers who choose it, a choice which will be made based on workers' education and training and individual risk assessment.

HIV-infected healthcare workers will only be as protected to the degree all other healthcare workers are protected. We must learn the lessons from the HIV/AIDS epidemic and apply those lessons to the tuberculosis epidemic: (1) an exposure control program which casts a broad net of infection protection, incorporating an industrial hygiene hierarchy that goes beyond diagnosis, or infection status alone and (2) policies and procedures that protect every healthcare worker and patient from discrimination.

Organizations with TB Recommendations Specific to Immunocompromised Healthcare Workers
1. American Association of Physicians for Human Rights, "TB: Recommendations for the HIV-Infected Health Care Provider," 1992.
2. American Medical Association, "Multiple-Drug Resistant TB: A Multi-faceted Problem," 1992; "Update on TB," 1993.
3. American Nurses Association, "Position Statement on Tuberculosis and HIV," 1993.
4. Centers for Disease Control and Prevention, "Guidelines for Preventing the Transmission of TB in Health-Care Facilities, 2nd Edition" (Draft), 1993.
5. Labor Coalition to Fight TB in the Workplace: AFGE, AFSCME, AFT, District 1199, SEIU, "Fed-OSHA Petition for TB Prevention Standard," 1993.
6. State of New York Department of Health, "Recommendations for HIV-Infected Healthcare Workers Regarding Tuberculosis," (Draft), 1994.
Organizations with no TB Recommendations Specific to Immunocompromised Healthcare Workers
American Hospital Association
Association for Practitioners in Infection Control
Federal OSHA

Organization	Document	Year	Recommendations Specific to Immunocompromised Health Care Workers
American Association of Physicians for Human Rights	"TB: Recommendations for the HIV-Infected Health Care Provider"	1992	(1) Know HIV status, and receive regular medical care. (2) Perform PPD and anergy testing every 6 months. (3) Be well acquainted with signs and symptoms of TB. (4) HEPA respirators should be worn with known or suspected TB. (5) Must weigh personal risk of TB infection. A change in job setting or career may be appropriate. (6) HIV-infected providers who risk exposure to MDR-TB should strongly consider a change in job setting or career.
American Hospital Association	"Tuberculosis Control in Hospitals"	1992	None
American Medical Association	"Multiple-Drug Resistant TB: A Multifaceted Problem"	1992	(1) HIV-infected health care workers should be carefully apprised of their risk of clinical TB. Risks and benefits of caring for persons with active or suspected TB should be carefully considered.
	"Update on TB"	1993	(1) Powered air-purification respirators (PAPRS) may be useful for the protection of immunocompromised health care workers who care for patients with infectious TB.
American Nurses Association	"Position Statement on Tuberculosis and HIV"	1993	(1) HIV-positive nurse to know their TB status. (2) Self-limit their nursing practice based on a case-by-case assessment of their TB status. (3) Self-restrict their contact with patients, co-workers and visitors if symptoms associated with TB are present. (4) Adhere to prescribed medication regime for TB to decrease the opportunity for transmission of the disease.

Organization	Document	Year	Recommendations Specific to Immunocompromised Health Care Workers
Association for Practitioners in Infection Control	"Position Statement on TB Prevention and Control"	1992	None
Centers for Disease Control and Prevention	"Guidelines for Prevention the Transmission of Tuberculosis in Health-Care Facilities, 2nd Edition" (Draft)	1993	(1) All health care workers should know if they have a medical condition or are receiving a medical treatment that may lead to severely impaired cell-mediated immunity. (2) All healthcare workers should be counseled about the potential risks, in severely immunocompromised persons, associated with taking care of patients with some infectious diseases, including TB. (3) Severely immunosuppressed healthcare workers should avoid exposure to *M. tuberculosis.* Healthcare workers with severely impaired cell-mediated immunity (due to HIV infection or other causes) who may be exposed to *M. tuberculosis* should consider a change in job setting. (4) Employers should make reasonable attempts to offer alternative job assignments to an employee with a documented condition compromising cell-mediated immunity who works in a high-risk setting for TB. The facility should offer, but not compel, a work setting in which the healthcare worker would have the lowest possible risk of occupational exposure to *M. tuberculosis.*

Organization	Document	Year	Recommendations Specific to Immunocompromised Health Care Workers
Centers for Disease Control and Prevention (continued)	"Guidelines for Prevention the Transmission of Tuberculosis in Health-Care Facilities, 2nd Edition" (Draft) (continued)	1993	(5) All healthcare workers should be informed that immunosuppressed healthcare workers need to have appropriate followup and screening for infectious diseases, including TB. Healthcare workers who are known to be HIV-infected or otherwise severely immunosuppressed should be tested for cutaneous anergy at the time of PPD testing. Consideration should be given to retesting immunocompromised healthcare workers with PPD and anergy tests at least every six months because of the high risk of rapid progression to active TB should infection occur. (6) Information provided by healthcare workers regarding their immune status should be treated confidentially.
Federal OSHA	"Enforcement Policy and Procedures for Occupational Exposure to Tuberculosis"	1993	None
Labor Coalition to Fight TB in the Workplace: AFGE, AFSCME, AFT, District 1199, SEIU	"Fed-OSHA Petition for TB Prevention Standard"	1993	(1) Prior to the PPD test, all employees will be counseled regarding increased risk if immunocompromised. Upon the employee's request, the employees will be tested with at least two delayed-type skin-test antigens, in addition to the PPD. Those employees found to be anergic will receive a follow-up evaluation by an appropriate healthcare professional and monitored for development of TB-related symptoms. The PPD and anergy tests will be offered at least every six months for immunocompromised workers. (2) The employer must provide a confidential referral at the request of immunocompromised healthcare workers to an employee health professional who can confidentially counsel the employee on an individual basis regarding his/her risk of TB.

Organization	Document	Year	Recommendations Specific to Immunocompromised Health Care Workers
Labor Coalition to Fight TB in the Workplace: AFGE, AFSCME, AFT, District 1199, SEIU (continued)	"Fed-OSHA Petition for TB Prevention Standard"	1993	(3) Employers must inform all healthcare workers with severely cell-mediated immunity who may be exposed to *M. tuberculosis* about their rights to reasonable accommodations under the Americans with Disabilities Act (ADA), including voluntary transfers to work areas and activities in which there is the lowest possible risk of *M. tuberculosis*. (4) Employers must treat confidentially all information provided by a healthcare worker regarding their immune status. The employer must have written procedures on confidential handling of such information.
State of New York Department of Health	"Recommendations for HIV-Infected Health Care Workers Regarding TB" (Draft)	1994	(1) Employers should educate all workers at risk for occupational exposure to TB about the risks associated with TB and the precautions they should take to avoid exposure. That TB presents an especially serious risk to people who are immuno-compromised for any reason, including HIV infection, must be a part of the education message. Discussions about the relative risk of continued employment in a specific setting should be included. (2) All employees at risk for occupational exposure should be routinely assessed for TB infection. Immuno-suppressed individuals with TB infection will not develop a positive PPD, when cutaneous anergy is present. If there is a high likelihood that an anergic worker has been exposed to TB, then the worker and his or her health care provider should consider further diagnosis evaluations (such as chest X-ray) and possible chemoprophylaxis. (3) Employers should not mandate reassignment of immuno-suppressed workers for worksites with a high prevalence of TB solely because of the risk to the worker. If an immuno-suppressed worker requests reassignment or modification of duties to reduce risk, the employer should consider the situation carefully, and accommodate the worker's request whenever possible.

Organization	Document	Year	Recommendations Specific to Immuno-Compromised Healthcare Workers
State of New York Department of Health (continued)	"Recommendations for HIV-Infected Health Care Workers Regarding TB" (Draft)	1994	(4) Employers must ensure that workers with infectious TB be removed from the workplace. (5) All individuals who may be exposed to HIV through personal behavior, blood products or occupational exposure should be counseled to seek HIV testing so that they may benefit from medical management. Recent outbreaks of TB, including drug-resistant strains, make the reasons for knowing one's HIV status even more compelling. (6) Individuals who are infected with TB should be encouraged by their health care providers to undergo HIV testing...since their HIV status may influence the value and interpretation of the tuberculin skin (PPD) test as well as recommendations for preventive therapy and treatment of TB disease. (7) Individuals with HIV who are at risk for TB infection should be under the care of a physician. Decisions regarding diagnosis, prophylaxis and treatment should be made by the individual and his or her physician.

PART 4

Reducing the Spread of Tuberculosis in Your Workplace

Derrick Hodge and Daniel Kass

INTRODUCTION

In the last few years, rates of tuberculosis infection have climbed. In New York City, poverty, homelessness, conditions in homeless shelters, high rates of HIV and AIDS and overcrowded jails and prisons have all contributed to the epidemic. Some get TB because they have no home and are forced to spend night after night in crowded shelters. Others are sick with AIDS or weakened by HIV, making them more susceptible to active TB. Even housed and healthy people have become infected with TB from continued contact with others who are sick. Each of these people have something in common—they got TB after breathing in the air coughed out by someone who was sick with TB. If your work puts you in contact with people who may have active tuberculosis, then you might also be at risk for infection.

When people with active tuberculosis cough or sneeze, they may release small particles into the air called "droplet nuclei." People with active TB are infectious to others. People with TB infection, or those who do not have symptoms, are not infectious to others. Indoors, incoming sunlight and fresh air reduce the amount of particles that remain in the air and infect others. Since you usually cannot control the amount of sunlight at work, you need to focus on making sure there is a lot of fresh air to get rid of the particles in the air.

This fact sheet will help you to:

- look at your workplace for problems with the ventilation and quality of air which increase the risk of TB transmission, and
- try some low cost, easy-to-do solutions to improve the problems you identify in your survey.

This fact sheet recommends low- and no-cost changes to your workplace. Alone, these cannot eliminate the risk of TB transmission. To significantly reduce this risk, a combination of approaches is needed, which may include training your staff about TB prevention and control, improving or adding ventilation systems, installing ultraviolet lights and selecting the proper personal protective equipment, such as respirators. Most important, your agency needs to identify those with active TB promptly and help them get treatment.

The first part of this fact sheet tells you how to evaluate the ventilation and flow of air in your workplace.

1-56670-083-3/94/$0.00+$.50

CHECKING YOUR BUILDING'S VENTILATION

What Is Ventilation?

The purpose of a ventilation system is to move stale or contaminated air out of a space, and bring in fresh air. When the air is moved by a motor and fan through vents and ducts, it is called forced air ventilation. Even without forced air, air will naturally move in and out of doorways, windows and other openings. This is called natural ventilation (Figure 1). Natural ventilation can be improved by using portable fans and air conditioners to help move the air and add fresh air. This fact sheet will help you evaluate natural ventilation in your building and improve it, if necessary.

Good air quality depends on good ventilation. When there is close and continuous contact with a person who has active TB, poor ventilation can lead to TB transmission if:

1. there is little or no fresh air moving into the building, or
2. contaminated air is removed very slowly or not at all.

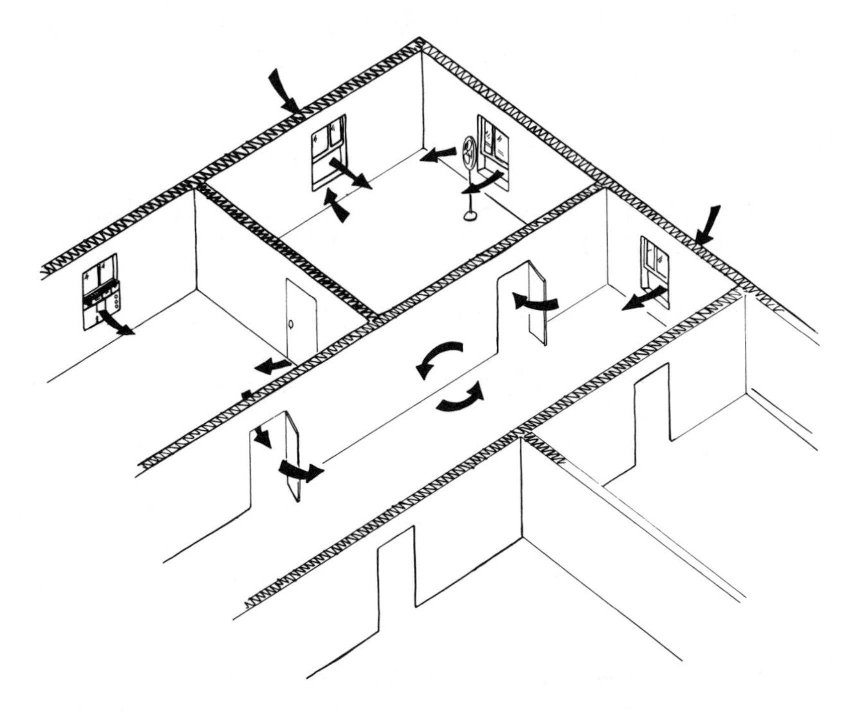

Figure 1. Air naturally moves through windows, doors and other openings. Some rooms also have window air conditioners and/or desktop, window or floor-standing fans.

To prevent TB transmission, and for general comfort, your natural ventilation must bring in an adequate amount of fresh air. The best way to do this, of course, is to leave an outdoor window open wide. If your office does not have a window, try to increase the amount of fresh, outside air moving through your work space.

AIR QUALITY SURVEY

It is difficult to test physically the amount of fresh air moving into a space from natural ventilation. One way to check is to see if you are experiencing any symptoms of working in a stuffy environment. Frequent headaches, drowsiness and fatigue are common among occupants of stuffy environments. In these environments, colds and flu may spread usually quickly, widely and linger longer. Allergies may develop or be aggravated, leading to frequent throat, eye and sinus irritation. Though each of us experience these symptoms at one time or another, stuffy, unhealthy environments may cause them to occur more frequently.

While stuffy air alone does not mean that there is a risk of TB transmission or infection, it may indicate that air movement needs to be improved.

Another way to assess and improve the air quality in your building is to look at how portable fans and air conditioners are used in your area. The following sections will help to determine the amount of fresh air and the proper movement of air in your environment.

Answer the following questions by checking Yes, No, or Not Applicable (N/A).

1. If there are windows in your office area, are they easy to open and close?
☐ Yes ☐ No ☐ N/A

2. Do odors or smells go away fairly quickly? ☐ Yes ☐ No ☐ N/A

3. If you have floor fans, window fans or table-top fans installed are they clean and do all of their features work (such as speeds and position adjustments)?
☐ Yes ☐ No ☐ N/A

4. If there is a window air conditioner, is its filter clean? ☐ Yes ☐ No ☐ N/A

5. If there is a window air conditioner, is the fresh air knob or vent open and working? ☐ Yes ☐ No ☐ N/A

6. How does the air move naturally in the area? To find out, you need to buy several incense sticks, and draw a simple floor-plan of the work area in question. The floor plan should show the location of doors, windows and where people are located or seated in the area. The following test involves making smoke and looking at the direction it moves in your area. Since incense gives off an odor, the

time it takes for the odor to travel and go away will help you figure out how the air is traveling throughout the building (see Figure 2).

Why is it useful to know the direction of air movement?

Knowing how the air flows in your work area can help you decide where to place fans, how to position desks and how to arrange seating. If someone is coughing, you want the air to carry the droplets from the cough away from other people, not toward them.

Knowing how the air is moving helps to:

1. Find areas where there is little or no air movement. Knowing this can help you make decisions about where and how to place fans and air conditioners to create an air flow.
2. Position fans so they are not blowing against the natural flow or air, but helping it. It is easier to add to the natural flow than to try to reverse it. When you do want to change the direction of the air, you'll need more powerful fans.
3. Decide where to place furniture and people.

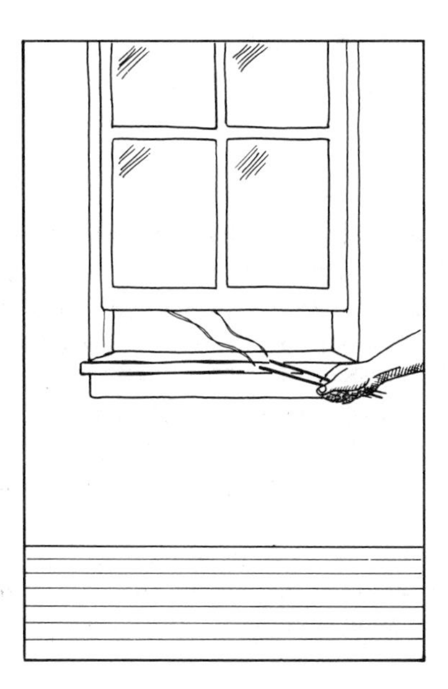

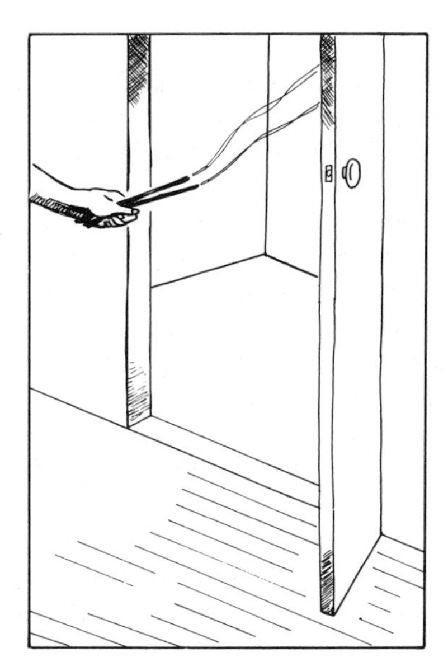

Figure 2. Light two incense sticks. After they start to burn, blow them out so that smoke rises. Hold them together horizontally near the base of each window and doorway in the room (if your window or doorway is usually wide-open or closed, keep it that way). Look to see which direction the smoke is blowing. Ask people in adjacent rooms to tell you when they begin to smell the incense.

Using the floor plan, draw arrows indicating the direction smoke is blowing from windows, doorways, hallways and in the middle of the room. You can also make a note of how strongly the air is flowing. (Use a single line for strong air flow, dotted line for light air flow. See Figure 3.) Adjacent areas where the incense smell travels first or is strongest probably receive more air from the testing area than others. Since windows usually stay open in summer and closed in winter, make sure you repeat this test in both seasons.

Figure 3 is one example of a completed air flow diagram.

HOW TO INTERPRET THE RESULTS OF THE AIR QUALITY SURVEY

If you and your co-workers answer no to one or more of the questions in the air quality survey, then the low-cost improvements on this page can help you find ways to improve air quality. If your air-flow floor plan from question 6 shows that you have poor air movement or that the air flow could carry droplet nuclei from one person directly to another, then use the next section to help you to position people and fans to make improvements.

LOW COST WAYS TO IMPROVE NATURAL VENTILATION TO REDUCE THE SPREAD OF TUBERCULOSIS

There are many ways to improve the ventilation of a building: an office can be remodeled, new openings can be cut into doors and walls, a forced air system can even be installed, but each of these strategies requires expert help and can be very expensive. Because of the cost, many building owners refuse to make these changes. Fortunately, there are some small, inexpensive improvements which will increase the amount of fresh air in the area where you work. This section offers some ideas about how to make these changes. Alone, these suggestions will not be enough to eliminate the risk of TB transmission, but they may help to reduce the risk.

Most of the suggestions below can be done by staff without the help of maintenance workers. You should consider all of the suggestions in the order they are listed. Together, they can help to improve air quality of your building.

Open Your Windows

Windows should be opened as much as possible. Opening a window is the simplest way to bring fresh air into a building. Some windows can be opened at the top and the bottom. If this is the case in your area, open both as much as

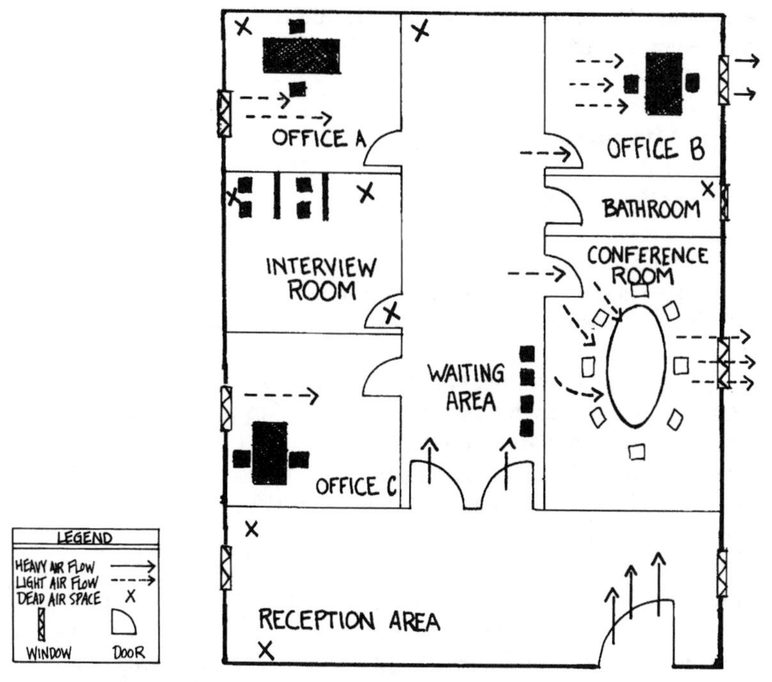

Figure 3. This completed floor plan shows air flow through doorways and windows, and across rooms and hallways. This facility has no forced air ventilation system and relies upon open windows and doorways for fresh air.

possible. If the windows are hard to open past a certain point, use a lubricant so they can be opened more easily. For wooden windows, try rubbing candle wax where the window slides against the frame. Metal parts can be lubricated with household oil, WD-40 or similar products. If these suggestions do not work, ask your maintenance staff to make the repairs.

In winter months, keep your windows as open as possible. If you cannot get enough heat to keep the windows open, there are two things you can do.

Make your heating system work better

If radiators are your main source of heat, ask your maintenance workers to service them. They may need to have water or air removed. This will help them produce more heat.

Add space heaters

Check with your building maintenance staff first to make sure the electrical system can handle the added load, otherwise, they could be fire hazards. Since space heaters dry the air, you might consider putting in a humidifier, or keeping a pan of water on top of a hot radiator. If you use a humidifier or pan of water, make sure you change the water frequently.

Install and Properly Position Fans

Fans can do three things:

1) move fresh air in
2) blow contaminated air out of a work area
3) circulate existing air to thin out the amount of TB droplets in an area.

All three, in combination, may reduce the risk of TB transmission. Without adding fresh air or removing contaminated air, a fan will only keep circuiting bad air and could do more harm than good.

There are three kinds of fans that can be added to your work area that can help improve the quality of air; window fans, floor-standing fans and portable table top fans (see Figure 4).

Fans blow air in one direction. Do not use fans just to create wind. Without a supply of fresh or clean air, fans will just recirculate any TB droplets which are present, possibly doing more harm than good. This section of the fact sheet describes ways to control the proper movement of air to reduce exposure to TB. Which direction do you want the fans to blow in your own area? Here is what you can do to help decide.

Window Fans and Floor-Standing Fans

Window fans and floor fans can either bring fresh outside air in, or blow "dirty" indoor air out. If there are people coughing in an area next to your office, then have fresh air come in from your window. If people are coughing directly in your work area, then have the window fan blow the air out. You always want to move dirty air away from other areas and directly to the outdoors.

Sometimes a window fan's air direction can be changed just by turning it around in the window. If you are going to buy a window fan, buy one that can easily reverse the direction of the air. A floor-standing fan can also be put near a window to blow air in or suck air out. Use the same rules as those for window fans, above.

Figures 5 through 8 show ways of using fans to control air flow when a waiting room is located in an office and when a waiting area is located in a hallway.

Portable Table Top Fans

Table top fans can also help control the direction of air. They can be useful if you are sitting with someone who may be coughing. If the person is facing you,

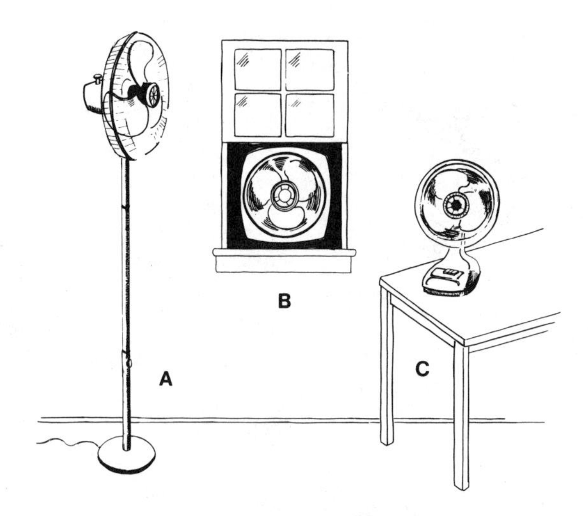

Figure 4. A. Floor-standing fan. B. Window fan. C. Portable table top fan.

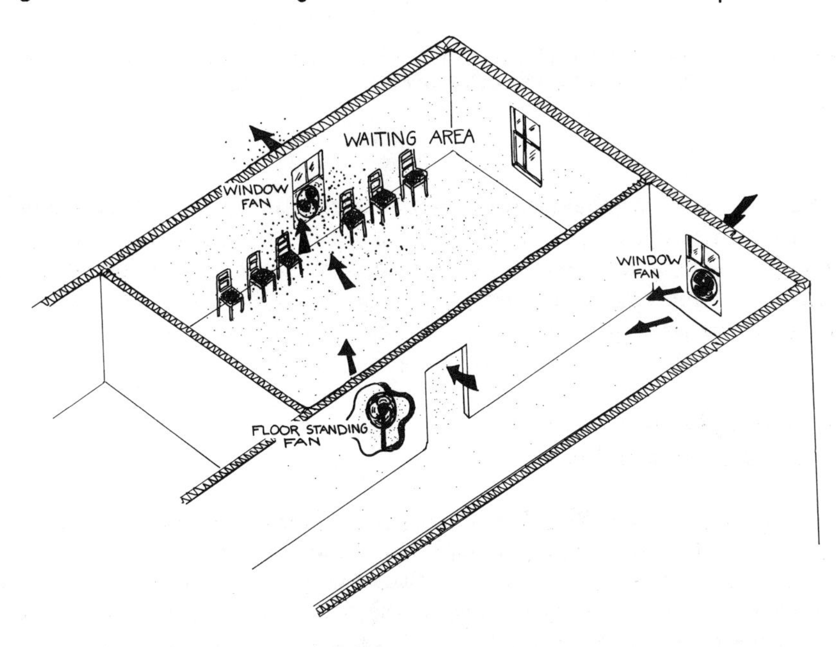

Figure 5. In this diagram, the dots represent air contaminated with TB droplets. To minimize contamination and exposure, a window fan draws air out of the waiting area, helped by a floor-standing fan. Another window fan brings fresh air into the hallway. Whenever a fan is added to force air out, make sure that you bring in more fresh air.

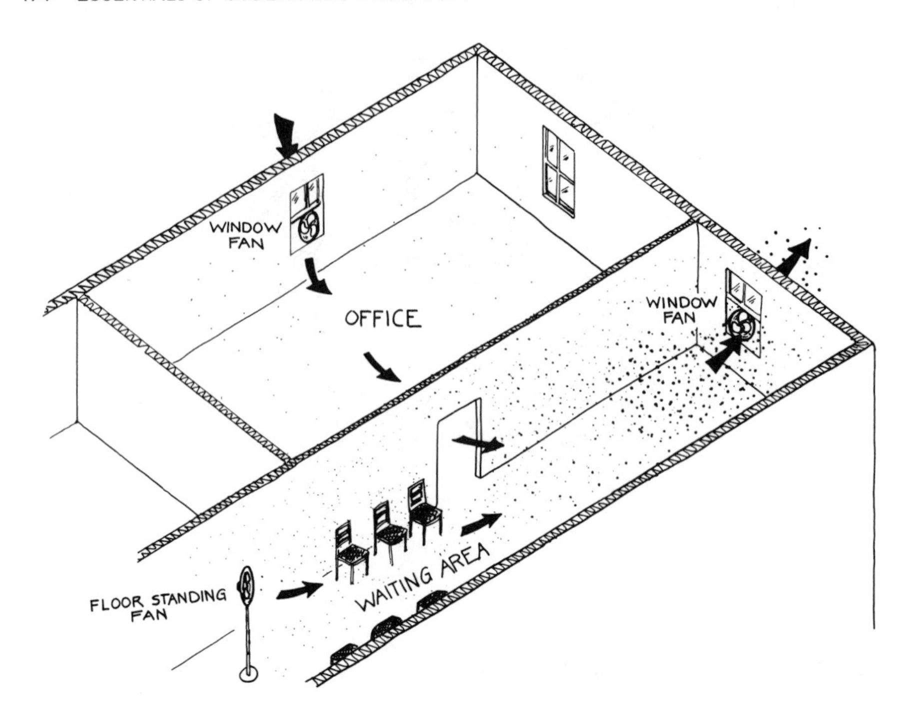

Figure 6. Here, a floor-standing fan and window fan move air out of the hallway. Fresh air is added though a window fan in the office.

have a fan positioned behind you, blowing air over your shoulder. The fan should blow the air diagonally and away from you. If this is uncomfortable for the person coughing, point the fan so it blows air between you. Make sure that the fan is not pointed at anyone else in the work area. Also, have the air flowing toward a window where air exits the building or toward an air purifier.

Clean and Maintain Your Air Conditioner and Fans

Window (see Figure 9) and wall-mounted air conditioners need to be checked to make sure they are blowing as much fresh air as they are able to.

You should regularly clean or replace the filter on the air conditioner. Remove the cover or face plate to get to the filter. Most can be gently rinsed. Also, make sure that the dial or sliding knob which adjusts the fresh air vent works. If the knob doesn't work, have someone fix it. Always keep the air conditioner set to provide the maximum amount of fresh air. If the air conditioner does not work after trying these suggestions, it may need professional servicing.

Figure 7. When two people sit across from one another, a fan can be positioned so that it blows air between the two people, directly toward an open window. If your fan oscillates back and forth, turn off that feature so that the fan always blows in the same direction. If you are buying a fan, make sure it can stay in the same position.

Use a Portable Room Air Cleaner

The most reliable ventilation strategy to lower the risk of getting TB at your workplace is to have as much fresh air circulating as possible at all times. In rooms without windows, or in winter months, an air cleaner may help.

Air cleaners filter out tiny particles in the air like smoke, pollen and bacteria. Though no one is yet sure, it is thought that some portable air cleaners can help filter out the "droplet nuclei" which spread TB. Air cleaners which may help eliminate TB from the air contain filters which are called "high-efficiency" or "HEPA" filters.

Room air cleaners (see Figure 10) can be placed in conference rooms, office rooms or any area where a large number of people gather at the same time. Larger models generally filter more air, at a faster rate, than smaller ones. Expect to pay $500 or more for a good one.

Here are some tips for shopping for an air purifier:

1. Buy the largest you can afford.
2. Buy the newest model—they tend to filter more air and smaller particles.
3. You may need more than one for large rooms.

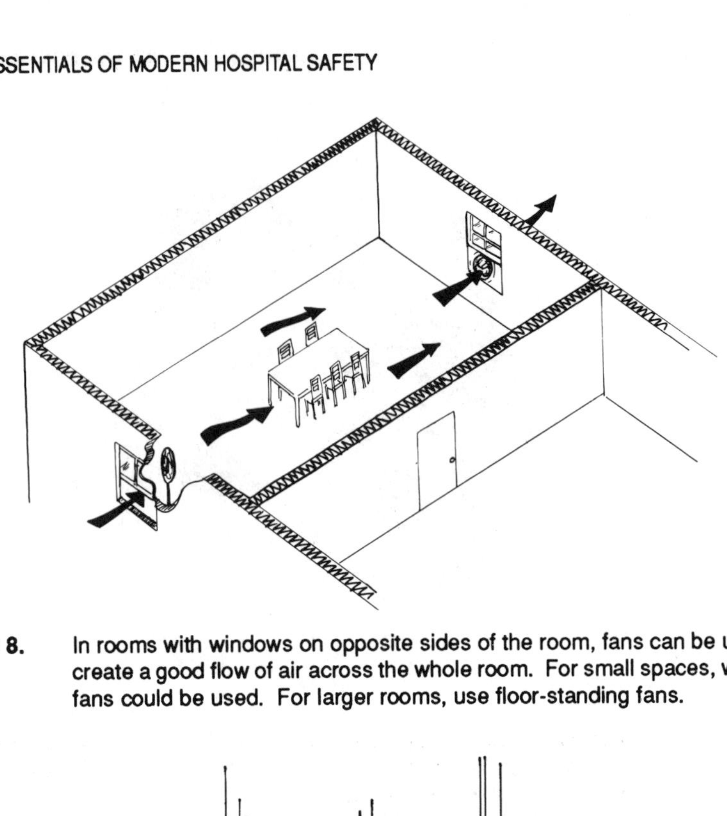

Figure 8. In rooms with windows on opposite sides of the room, fans can be used to create a good flow of air across the whole room. For small spaces, window fans could be used. For larger rooms, use floor-standing fans.

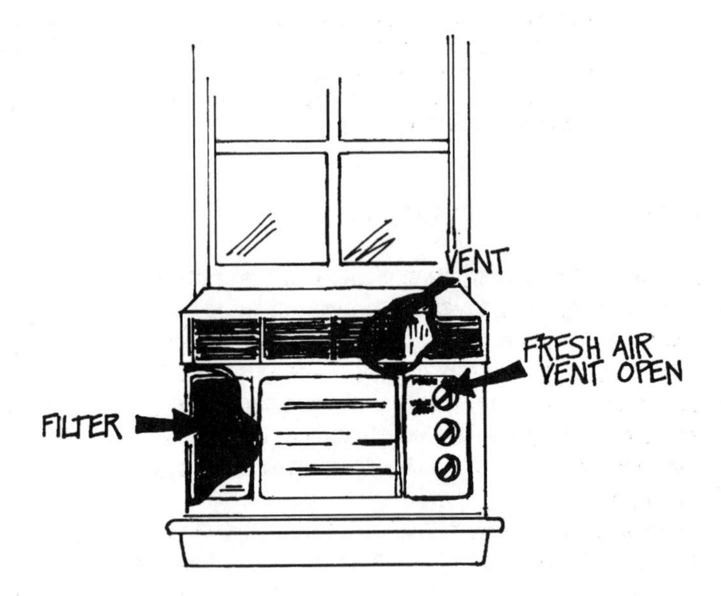

Figure 9. Typical window air conditioner.

4. Make sure the cleaner has filters which are labeled "high-efficiency" or "HEPA."
5. Have a supply of extra filters and change them often (follow manufacturer's directions).

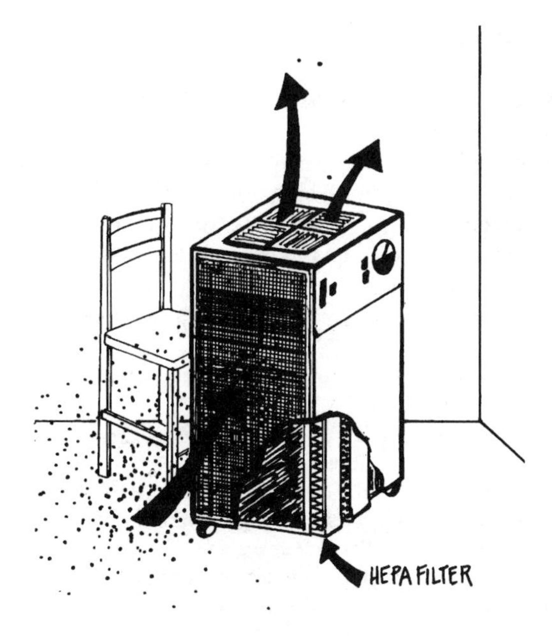

Figure 10. Typical portable room air conditioner.

CHANGE THE WAY YOU INTERACT WITH PEOPLE

Make Sure People Cover Their Mouths When Coughing or Sneezing

Have a good supply of tissue in your work areas at all times. Don't be shy about asking people to use them. Ask people to throw these tissues in a covered garbage can and make sure to empty it frequently.

Leave the Room Momentarily

If a person starts to sneeze repeatedly or cough heavily, you can hold your breath and calmly leave the area for a few minutes until the attack subsides and the air has a chance to clear.

Decrease Crowding as Much as Possible

Spread staff and clients out evenly throughout the building so fewer people are around one another at one time. If there are many people coming for some service, try to limit the number coming at any given time. Serve fewer people over a longer period of time rather than all at once.

Move Desks, Chairs and People

Put people where the air flow is best. Move people away from areas where there is little or no ventilation.

If you are meeting one on one with someone who may be coughing, have them positioned so that they are not facing you directly. Limit close contact with anyone who shows signs of TB. As noted before, getting prompt treatment for people with active TB is the best strategy for protecting clients and staff.

The floor plan in Figure 11a shows the results of an air flow survey using incense sticks in one facility. Notice the following problems:

1. The interview room has no window and no detectable air flow, potentially resulting in a concentration of airborne particles.
2. In Office A, there is a dead air space (an area where no air flows) where people sit at a table across from one another.
3. When people are sitting in the chairs in Office B, air flows from one to the other. If one coughs, the particles may go toward the other.
4. In the bathroom, there is a dead air space.
5. The waiting area is in a hallway. The air flow pattern may carry any airborne TB droplets into the Conference Room and other areas.

The floor plan in Figure 11b shows improvements. Using the suggestions offered in the previous section, several changes have been made.

1. The interview area has been moved to a room with better air flow.
2. Tables and chairs in Office A and Office B have been re-positioned so that air flows between people.
3. The waiting area has been moved from the hallway to the front of the facility.
4. Window exhaust fans have been added in the bathroom and in the new waiting area to draw air out of the room.

SUMMARY AND FOLLOWUP

Get Additional Help

Alone, the suggestions offered above will not solve complex problems in facilities where people with active tuberculosis reside, use or visit. Once you have completed the steps suggested here, or if you feel they do not apply to your facility, you should seek additional technical advice and assistance. Organize staff training on TB and help your agency establish TB protocols. Regulations may be forthcoming which require your facility to do more. Stay in touch with health professionals and TB advocates to keep up to date with recommendations and guidelines for air flow in facilities like yours.

For help in finding these resources, call the Hunter College Center for Occupational and Environmental Health or the Center on AIDS, Drugs and Community Health.

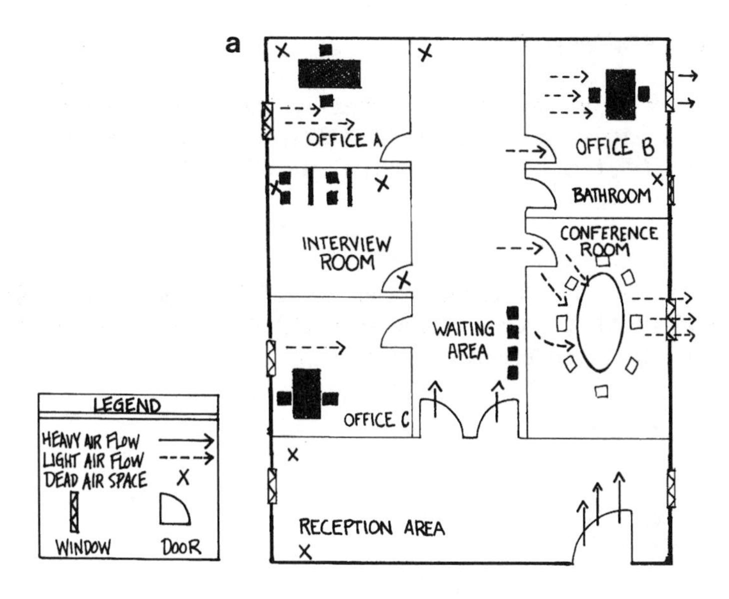

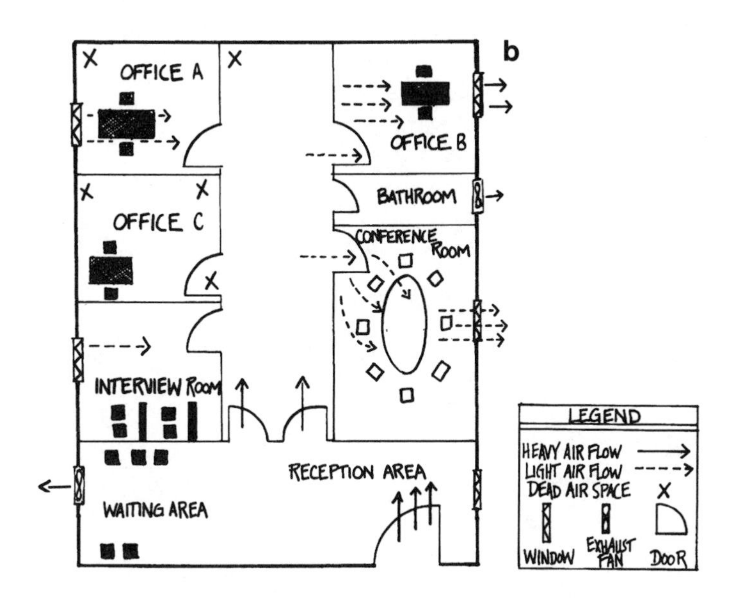

Figure 11. An example of workplace improvements which help to reduce the spread of tuberculosis: a) before improvements and, b) after improvements.

ACKNOWLEDGMENTS

This fact sheet was jointly prepared by the Hunter College Center for Occupational and Environmental Health ((212)481-8790) and the Hunter College Center on AIDS, Drugs and Community Health ((212)481-7672).

Source: The Center for Occupational Health and The Center on AIDS, Drugs and Community Health at Hunter College, 425 East 25th Street, New York, NY 10010 (212)481-8790. Reproduced with permission.

APPENDIX 4-1

Test Method

a. Perform rudimentary supply air volume measurements in 3D15 to determine nascent room air exchange rate.

b. Measure room/hallway positive pressure with doors and windows closed.

c. Conduct FS209E room qualification to establish background.

d. Elevate particulate load in room with Bolus and record concentration vs. time data to create clearance curve with machine off.

e. Turn air cleaning device on at T.B.D. air volume rate and record concentration vs time decay after Bolus AB. The room is quiescent to the hall today, positive pressure < 0.005 w.c. smoke reveals room slightly positive. HVAC air estimated to be approximately 150 CFM.

1. Test: 6/9/94, room 3D15 HVAC recovery test, no air scrubbing. *Count = 6 sec. @ 1.0 CFM = 0.1 Ft^3, Channel = > 1.0μm.

2.

Time (m)	Count*	Comment
0.0	1983	Background
0.5	1777	"
1.0	1686	"
1.5	1643	"
2.0	2029	Inj. Bolus
2.5	46128	Rising
3.0	110128	Rising
3.5	192962	Peak
4.0	181394	Decay
4.5	188361	"
5.0	191927	"
5.5	175957	"
6.0	174710	"
6.5	176260	"
7.0	167960	"
7.5	165877	"
8.0	150716	"
8.5	146112	"
9.0	139033	"
9.5	119913	"
10.0	120664	"
10.5	103928	"
11.0	97353	"
11.5	91656	"
12.0	88046	"
12.5	74029	"
13.0	61180	"
13.5	51444	"
14.0	49709	"
14.5	51665	"
15.0	47204	"
15.5	38419	"
16.0	36334	"
16.5	31325	"
17.0	32356	"
17.5	30267	"
18.0	26881	"
18.5	21615	"
19.0	20875	"
19.5	19097	"
20.0	18142	"

(continuation of SA/HVAC recovery without machine in place)

Time	Count	Comment
20.5	16515	Decay
21.0	14022	"
21.5	13451	"
22.0	13000	"
22.5	11055	"
23.0	10010	"
23.5	9925	"
24.0	8579	"
24.5	8549	"
25.0	8170	"
25.5	7980	"
26.0	7493	"
26.5	7804	"
27.0	7155	"
36.0	3498	<-- Note time change, this is
36.5	3563	start of next test run
37.0	3348	
37.5	3498	

1. Test: 6/9/94, room 3D15, recovery test with machine initially off, then switched on to 400 CFM. *Count = 6 sec. @ 1.0 CFM = 0.1 Ft^3, channel = > 1.0µm.

2.

Time(m)	Count*	Comment
0.0	3498	Background
0.5	3563	"
1.0	3348	"
1.5	3498	"
2.0	17651	Turn machine on add DOP;Peak decay
2.5	9416	Background
3.0	7700	"
3.5	7672	"
4.0	5964	"
4.5	5634	"
5.0	4723	"
5.5	4029	"
6.0	2933	"
6.5	2779	"
7.0	2061	"
7.5	1919	"
8.0	1751	"
8.5	1235	"
9.0	1022	"
9.5	830	"
10.0	762	"
10.5	628	"
11.0	540	"
11.5	419	"
12.0	502	"
12.5	342	"
13.0	275	"
13.5	260	"
14.0	222	"
14.5	212	"
15.0	No data	
15.5	164	"
16.0	155	"
16.5	139	"
17.0	85	"
17.5	146	"
18.0	100	"
		Floor of counts
18.5	85	"

1. Test 6/9/94, room 3D15, machine on at approximately 400 CFM throughout. *Count 6 sec, @ 1.0 CFM = 0.1 Ft^3, channel = > 1.0μm.

2.

Time	Count	Comment
0.0	164	Background
0.5	155	"
1.0	139	"
1.5	85	"
2.0	146	"
2.5	100	"
3.0	85	"
3.5	3032	Add DOP;Peak decay
4.0	49296	"
4.5	46944	"
5.0	33173	"
5.5	31736	"
6.0	19985	"
6.5	14639	"
7.0	10868	"
7.5	9165	"
8.0	8719	"
8.5	7368	"
9.0	5991	"
9.5	4983	"
10.0	4198	"
10.5	3469	"
11.0	3036	"
11.5	2356	"
12.0	2113	"
12.5	1714	"
13.0	1229	"
13.5	1078	"
14.0	865	"
14.5	742	"
15.0	577	"
15.5	471	"
16.0	424	"

1. Test: 6/9/94, room 3D15, repeat test with machine on at 400 CFM continuously. *Count = 6 sec @ 1.0 CFM = 1.0 CFM = 0.1 Ft^3, channel = > 1.0μm.

2.

Time	Count	Comment
0.0	577	Background
0.5	471	"
1.0	424	"
1.5	5370	Add Bolus
2.0	35539	Rising peak decay
2.5	124092	"
3.0	100655	"
3.5	83286	"
4.0	56860	"
4.5	45435	"
5.0	32057	"
5.5	21427	"
6.0	16271	"
6.5	12485	"
7.0	11821	"
7.5	8976	"
8.0	7674	"
8.5	6502	"
9.0	5464	"
10.0	3963	"
10.5	2963	"
11.0	2589	"
11.5	1985	"
12.0	1639	"
12.5	1296	"
13.0	1052	"
13.5	878	"
14.0	676	"
14.5	644	"

HEPA Test Mode 880 CFM

Time	Concentration
0.0	612300
0.5	596440
1.0	577300
1.5	545800
2.0	519320
2.5	458350
3.0	417870
3.5	378800
4.0	294170
4.5	235600
5.0	190900
5.5	146890
6.0	106150
6.5	75710
7.0	61530
7.5	43310
8.0	23640
8.5	29140
9.0	23570
9.5	19860
10.0	19670
10.5	17650
11.0	16260
11.5	16160
12.0	14380
12.5	16710

RS1000 HEPA CHALLENGE TEST

SAN FRANCISCO GENERAL HOSPITAL
MEDICAL CENTER

Technical Safety Services of San Ramon, CA was retained by San Francisco General Hospital to test the capacity of the portable RS 1000 HEPA recirculating negative air machine utilized in 10 TB isolation rooms. Clearance rates/decay rates were determined by challenging the machine with DOP, fogging the room with specific particle counts per cubic foot of air and timing the decay curve to background counts. The curve slope is considered a activity of time versus a reduction in quantified particle count in air.

Test equipment: Discreet particle counter: manufactured by particle measurement systems Mode 1# Lasaire 310: factory calibration date 4/94

Sample flow rate: 1 CFM

Room size: 1300 cubic feet

HVAC: Supply 150 CFM

Probe site: exhaust grille, ceiling supply diffuser

Machine speeds:

1. Trial test without machine operating.
2. Machine at 400 CFM setting actively used in isolation rooms as a balance between noise and clearance values.
3. Machine set at HEPA test mode 880 CFM.
4. A setting providing Micron range: 1 micron setting.

DISCUSSION OF RESULTS

The RS 1000 functioned as an air scrubber in the two basic modes tested, 400 CFM and 880 CFM (HEPA Test Mode). The unit was tested at these volumes with three challenge trials each and all decay slopes corresponded.

The first challenge test without the machine running indicated a 37.5 clearance rate or 16 minute clearance time to 90% of background. This corresponds to approximately 9 air changes per hour (see table, Federal Register, Air Changes Per Hour and Time, Vol. 58, No. 195), if one assumes a mixing factor of 1 (perfect mixing). However, the formula: Average CFM × 60 room volume of 400 CFM × 60, 1300 which equals 18 ach).

Three trials at 400 CFM showed a decay time of 5.5 minutes, 6.5 minutes and 4 minutes to decay to 90% of background. This range corresponds to approximately 20–25 air changes per hour when consulting Table S3 of the Federal Register, Vol. 58, No. 195. This in turn is slightly greater than the results with the air change formula due to better mixing rate.

Three trials at 880 CFM (HEPA Test Mode) showed the average recovery time to be 4.5 minutes. Adding the 880 CFM to the 150 CFM supplied by the house HVAC yields 46 air changes per hour. Applying this value to Table S3-1 suggests that the 90% recovery time should be 3 minutes, therefore the overall room performance does not compare well with the expected value. This can be explained by a less than 1 mixing value.

One other explanation is that the turbulence created by a high volume current creates a delay in decay, or that at the higher rate of air exchange, a homeostasis is reduced with performance. (See Table S3-1 Federal Register, Vol. 58, No. 195. Between 30 to 50 air changes. Little increase in performance is noted.

The RS 1000 which brings air in at the bottom through a 17" × 17" grille and exhausts air through the top, showed in all test modes to clear the air and mix the air extremely well in a 1300 cubic foot room. It proved that it can remove particles in an order of magnitude much greater than the central ventilation system due to a higher volume efficiency and better mixing factor.

Time	0	0.5	1	1.5	2	2.5	3	3.5	4	4.5	5	5.5	6
880 CFM	612300	596440	577300	545800	519320	458350	417870	378800	294170	235600	190900	148890	106150
Count	596300	580440	561300	529800	503320	442350	401870	362800	208170	219600	174900	130890	90150
Background	100	97.34	94.13	88.85	84.41	74.18	67.39	60.84	46.65	36.83	29.33	21.95	15.12
% Remain													
400 CFM (112K)	17651	9416	7700	7672	5964	5634	4728	4029	2933	2779	2061	1919	1751
Count	17566	9331	7615	7587	5879	5549	4638	3944	2848	2694	1976	1834	1666
Background	100	55.12	43.38	43.19	33.47	31.59	26.40	22.45	16.21	15.34	11.25	10.44	9.48
% Remain													
400 CFM (49K)	49296	46944	33173	31736	19985	14639	10868	9165	8719	7368	5991	4983	4198
Count	48872	46520	32749	31312	19561	14215	10444	9741	8295	6944	5567	4559	3774
Background	100	95.19	67.01	64.07	40.02	29.09	21.37	17.89	16.97	14.21	11.39	9.33	7.72
% Remain													
400 CFM (124K)	124092	100655	83286	56860	45435	32057	21427	16271	12485	11821	8976	7674	6502
Count	123448	100011	82642	56216	44791	31413	20783	15627	11841	11177	8332	7030	5058
Background	100	81.02	66.96	45.56	36.30	25.47	16.86	12.69	9.62	9.08	6.78	5.73	4.78
% Remain													
Control (193K)	192962	181394	188361	191927	175957	174710	176260	167960	165877	150716	146112	139033	119913
Count	189462	177894	184861	188427	172457	171210	172760	164460	162377	147216	142612	135533	116413
Background	100	93.69	97.57	99.45	91.02	90.37	91.18	86.80	85.70	77.70	75.27	71.54	61.44
% Remain													

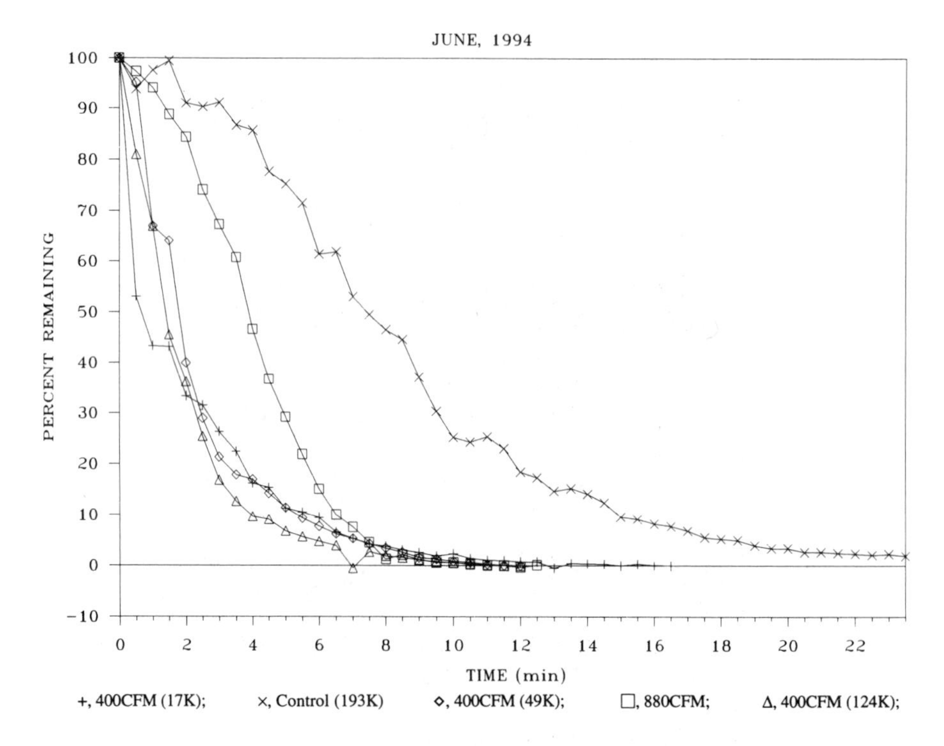

Figure 1. RS 1000 Challenge Test.

Ben Gonzales, PE, CSP, Technical Safety Services:
For: William Charney, DOH, Director Environmental Health, San Francisco General Hospital

APPENDIX 4-2

Challenge Data for Portable HEPA Filter Systems for TB

Olmsted Environmental Services

1.0 INTRODUCTION

Hazardous Material Technologies Corporation (HMTC) was retained by Biological Controls to test the effectiveness of the MICROCON® as an auxiliary ventilation system in filtering a simulated airborne pathogenic organisms such as TB bacillus and the AIDS virus. HMTC was asked to test the MICROCON®'s ability to control an airborne concentration of respirable particulate of 1–3 micron in diameter (see Appendix II for detail information of particulate characteristics).

The MICROCON® is a portable, self-contained free-standing air purification device. It is designed to function as an internal ventilation system capable of both filtration (99.97% of particulate ≥ 0.3 microns in diameter) and disinfection of pathogenic species. The unit can be operated at any one of three (3) flow rates.

The test method used consisted of generating an airborne particulate concentration within the test rooms, in the range of 10–20 mg/M^3. The goal of the test was to determine the ability of the MICROCON® to remove airborne particulate from a typical hospital environment under simulated operating conditions. All testing was performed in a soon to be renovated section of Horton Hospital, Middletown, New York, on December 22, 1992. The two test rooms used in the test were a special isolation room and a patient residence room.

2.0 SCOPE OF EVALUATION

The intent of this study was to evaluate by impartial objective means the ability of Biological Control's MICROCON® to filter a simulated human generated bio-aerosol intended to mimic the type of aerosol produced by a human cough or sneeze.

The particulate medium chosen to simulate the bio-aerosol was Arizona Road Dust. The test dust had a mean diameter of 0.76 microns. This medium was chosen because its aerodynamic properties are within the human respirable range. This medium also permits sufficient residency for consistent and reproducible measurement of airborne concentrations.

In general, the test consisted of generating an airborne dust concentration in the test room and measuring the change in the particulate concentration. The internal volume of the test isolation room and regular patient room were 1090 and 1612 ft^3, respectively.

Prior to the start of each test the particulate monitoring equipment was adjusted and a consistent airborne dust concentration generated. After an appropriate period of time to allow for stabilization of the dust concentration the MICROCON® was placed in central location (as possible) within the test room prior to the start of the test.

Once a stable airborne dust concentration was achieved and a baseline established the MICROCON® was then activated. The concentration of airborne dust within the test rooms was monitored using four (4) direct-reading portable aerosol monitors which measured airborne particulate levels at two levels in each of the two locations. The monitoring equipment chosen for monitoring particulate levels were battery powered, hand-held Haz-Dust™ Particulate Monitors manufactured by Environmental Devices Corp. The Haz-Dust™ monitor has a digital readout as well as a DC voltage output for data recording.

Four (4) of these monitors were used in each test. Two (2) test stands were used in each test cell and each contained two (2) monitors mounted at different heights. Monitors 1 and 3 were mounted on test stand number 1 with monitors 2 and 4 mounted on test stand number 2. Monitors 1 and 2 were mounted at a height of 66 inches with monitors 3 and 4 mounted at a height of 32 inches. The height of 66 inches was chosen as representative of the breathing zone height of the average standing adult. The height of 32 inches was chosen as an approximation of the height of a supine patient's breathing zone. The test stands were located as follows:

1) Test stand #1 - (holding monitors 1 and 3) was placed near the front side of the bed approximately 36 inches from the wall near the head of the bed.

2) Test stand #2 - (holding monitors 2 and 4) was placed in the corner, at the foot of the bed, near the window approximately 24 inches from the corner of the rooms.

The test protocol was designed to address possible increased settling attributable to the greater density of the test dust as compared to that of a bio-aerosol. Initially, in an effort to compensate for the lack of air currents necessary for uniform particle dispersion, one (1) 24-inch diameter fan was employed to distribute the test dust and to produce a more uniform concentration of the test dust within the test room. The primary reasons for the diminution of the test dust concentrations appeared to be impact losses on fan blades, associated surfaces, and gravity-induced settling or drop-out. Another source of loss was the scavenging effect produced by static charges on the room's interior surfaces.

As a result of scavenging, drop-out and impact losses the quantity of test dust required to produce a stable concentration of 10–20 mg/M^3 in each test room was somewhat greater than calculations indicated. A hand-held air-powered nebulizer was used to disperse the test dust in each test cell. Once the dust loading requirements were met, it took five (5) to seven (7) minutes to achieve a stable and acceptably uniform test dust concentration within the desired range.

3.0 DISCUSSION OF TEST

Unless otherwise noted all tests were performed with the MICROCON®'s flow setting selector in the medium position. The tests began after a four (4) to five (5) minute settling period, to allow the dust concentration to stabilize. After this settling period the MICROCON® unit was activated. During this stabilization period no additional mixing was provided other than that produced by the room's own dynamics.

Chart 1 - Background Dust Levels - Room 336

This test represents the background or residual airborne dust present in the test room prior to the introduction of the test dust. This dust is undoubtedly the result of lack of use because this section of the hospital was not being used.

Chart 2 - Settling Profile - Room 336

This test represents normal settling characteristics of the dust used in the tests, as well as any inherent flow dynamics of the test room. An airborne dust concentration was generated, mixed and then allowed to settle naturally.

Chart 3 - Test Number 1 - Room 336

This test shows the performance of the MICROCON® (air machine) in the test room with an elevated level of dust present. After the initial dust application it was noted that additional openings through which air could flow into the room needed to be sealed if a true test of the MICROCON® was to be obtained. After these openings were sealed a second dust application was performed (see Chart 3).

Chart 4 - Test Number 2 - Room 336

In this test all room HVAC vents and openings were opened (this condition is more typical of normal usage conditions).

Chart 5 - Test Number 3 - Room 336

The test conditions were the same as those indicated for Test Number 2, but the MICROCON® was operated with the flow setting on low.

Chart 6 - Test Number 4 - Room 335

The test conditions were the same as those indicated for Test Number 2.

Chart 7 - Test Number 5 - Room 335

The test conditions were the same as those indicated for Test Number 2, but the MICROCON® was operated with the flow setting on low.

4.0 DISCUSSION OF FINDINGS/RESULTS

An analysis of the purge test data indicates that the MICROCON® is effective in capturing and removing significant quantities of respirable airborne particulate from the immediate surrounding area. The effectiveness of the disinfection option of the unit was not evaluated as part of the purge testing reported herein.

The test data collected consisted of airborne particulate concentrations measured as a function of location and elevation (within the test room) and time. These data are presented in graph form in Appendix I. A total of five (5) purge tests were conducted. These included operating the MICROCON® alone and in tandem with the test room's HVAC system. The MICROCON® appeared to function most efficiently when operated with a room's own HVAC system but is also capable of operating well even without such aid. This was most likely due to the added turbulence and air movement provided by the HVAC system. It would also appear that the MICROCON® benefits when it is placed in a location where it can take best advantage of the induced air currents produced by the HVAC system when these currents do not comprise the design effectiveness of the unit.

It should be noted when reviewing the data continued in the charted test data in accordance with the Haz-Dust™ manufacturer's directions all monitors and subsequent readings are subject to a background correction of approximately 1–2% of the highest dust concentrations encountered. This correction is a function of the optics used in the monitors and *is non-correctable by zero correction.* Hence, each final reading should be viewed accordingly.

Because we were asked to perform the evaluation tests under the most stringent conditions possible, the reported data were not adjusted in any way for background buildup. When the background buildup is taken into account and the final readings so corrected *the actual levels or percentage of total dust removed by the MICROCON® (Air Machine) during the test is even greater than indicated.*

The lack of zero correction and the high levels of airborne dust used during these tests were of particular concern with regard to the data depicted by sensors number 4, 3, 3 and 3 on Charts Number 4 (Test No. 2), Chart Number 5 (Test No. 3), Chart Number 6 (Test No. 4) and Chart Number 7 (Test No. 5), respectively. These sensor read-outs show clear evidence of this phenomena. If the background corrections recommended by the manufacturer are performed on all data the end-point or final readings at these locations would be more consistent with those observed for the other sensor readings obtained during the respective test.

5.0 CONCLUSIONS

The test data indicate that the MICROCON® performs its intended task "removal of respirable size airborne particulate from the air within its effective area of influence." The size of the effective area of influence is to a large extent dependent upon a number of factors including drafts, obstructions and machine operating speed. All these factors can influence the aerodynamic behavior of airborne particulate and should be considered in any proposed application(s). If these factors are adequately addressed and the manufacturer's instructions followed then the MICROCON® should perform as indicated.

Appendix 4-2 Part I
Purge Test Charts 1–7

Background Dust Levels - Room 336

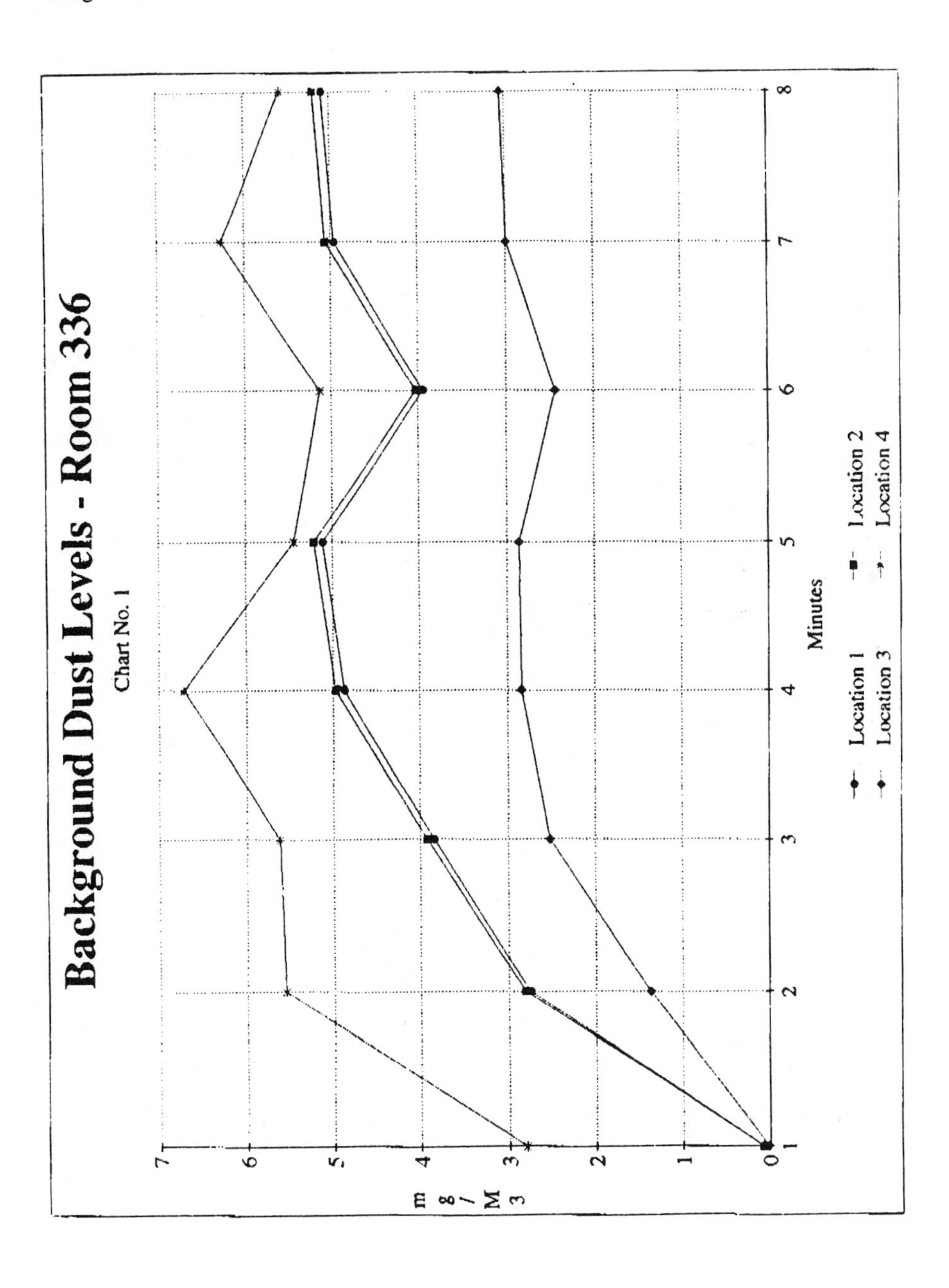

Settling Profile - Room 336

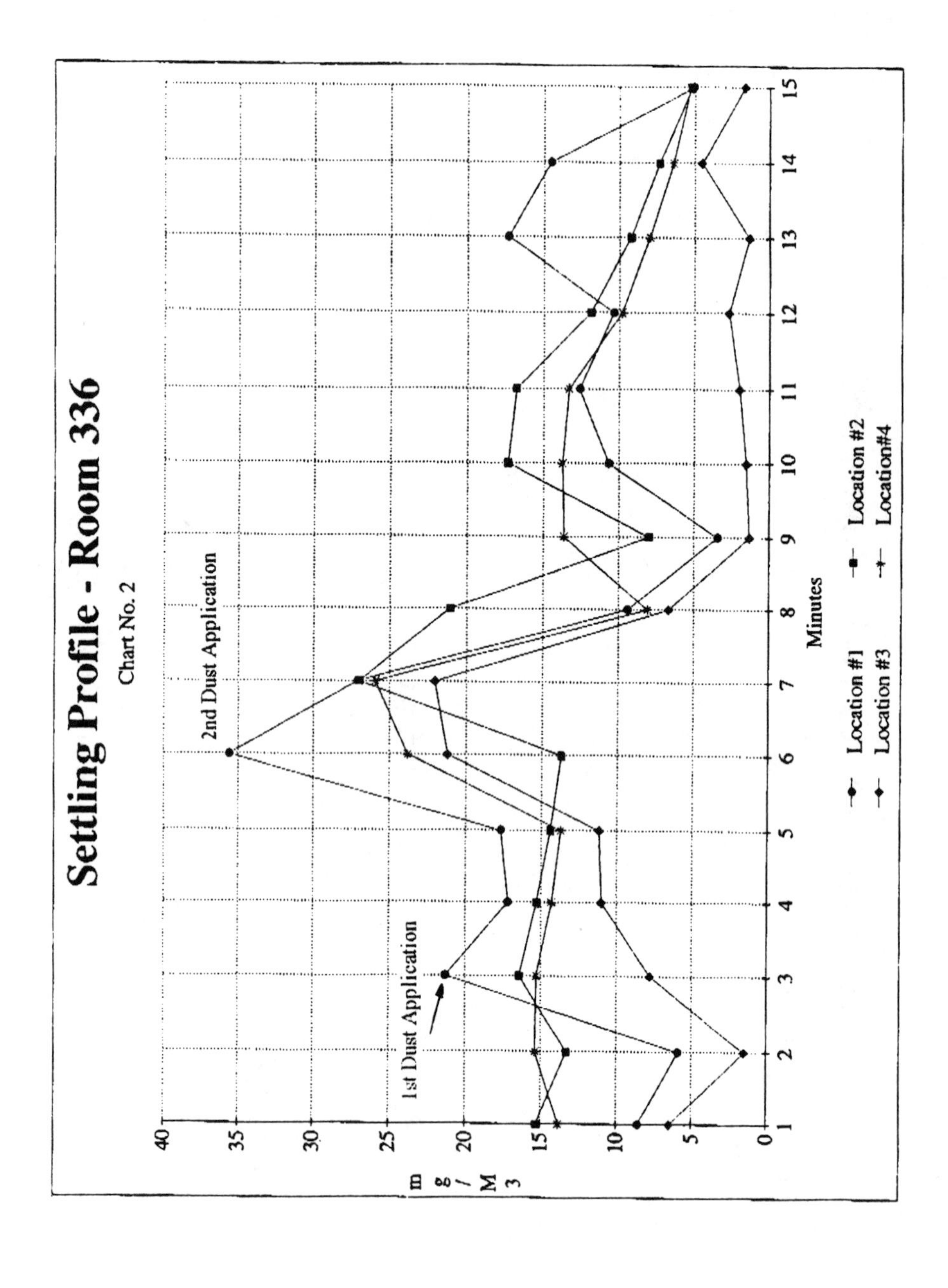

Test No. 1 - Room 336

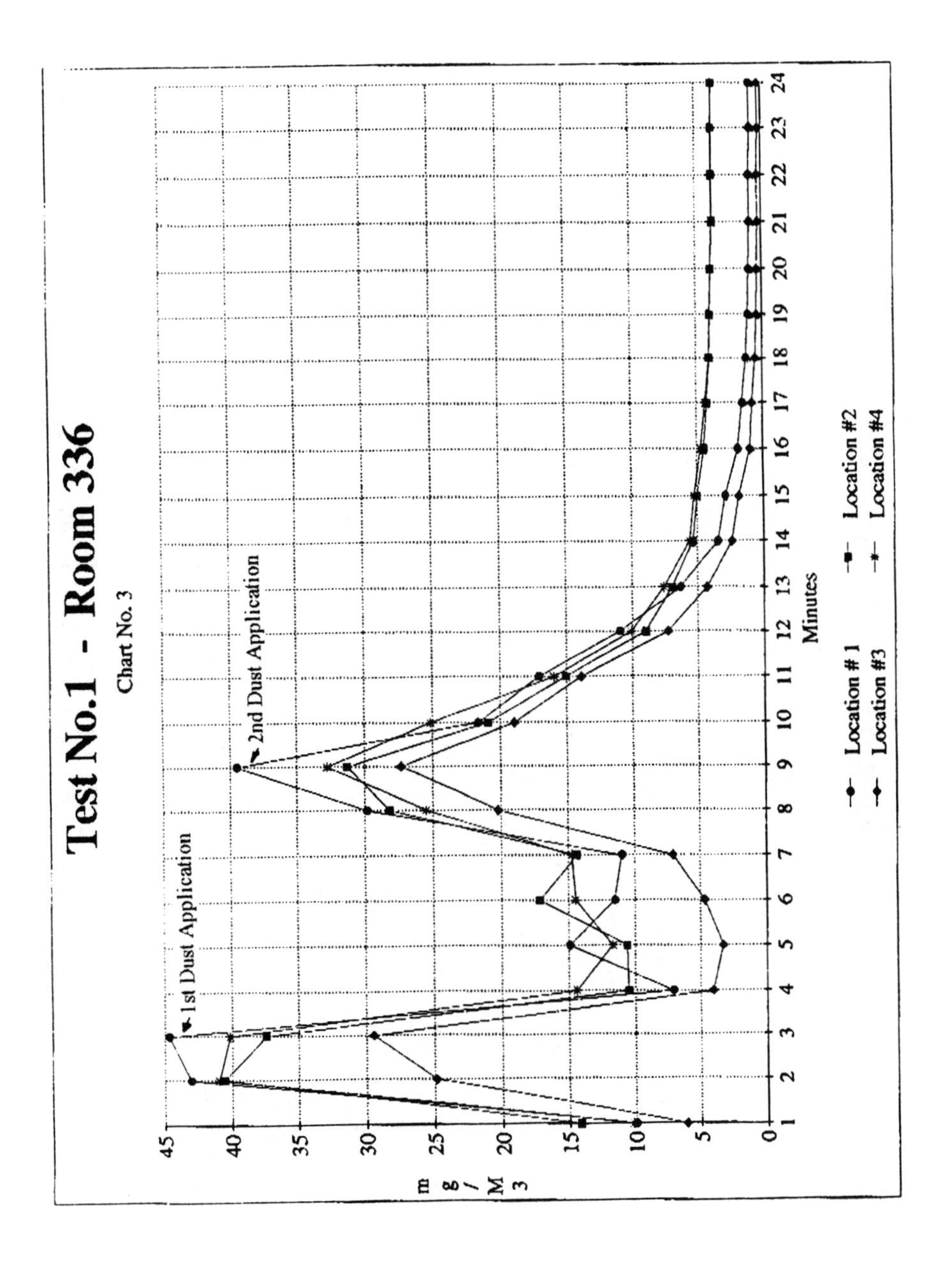

Test No. 2 - Room 336

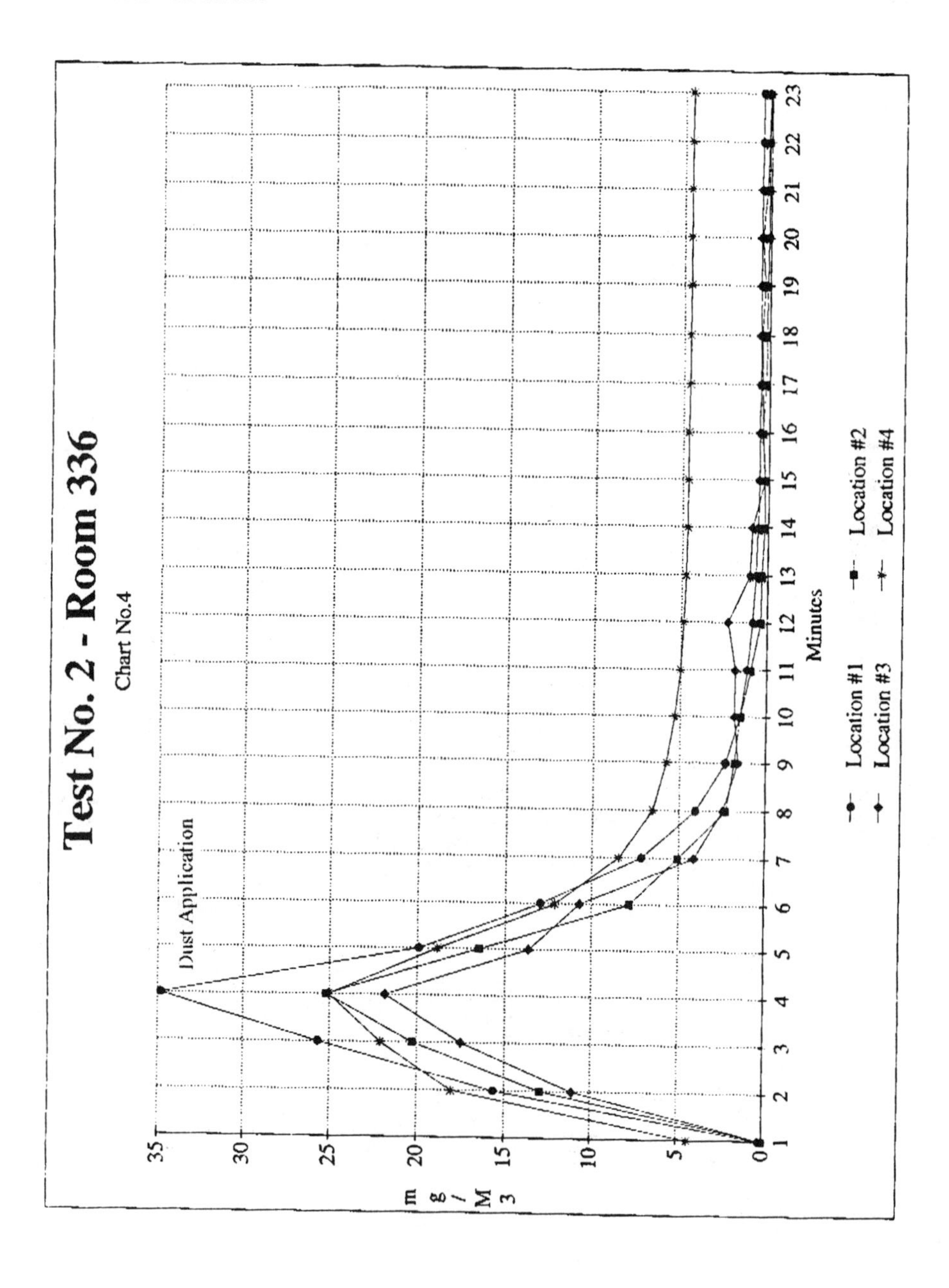

Test No. 3 - Room 336

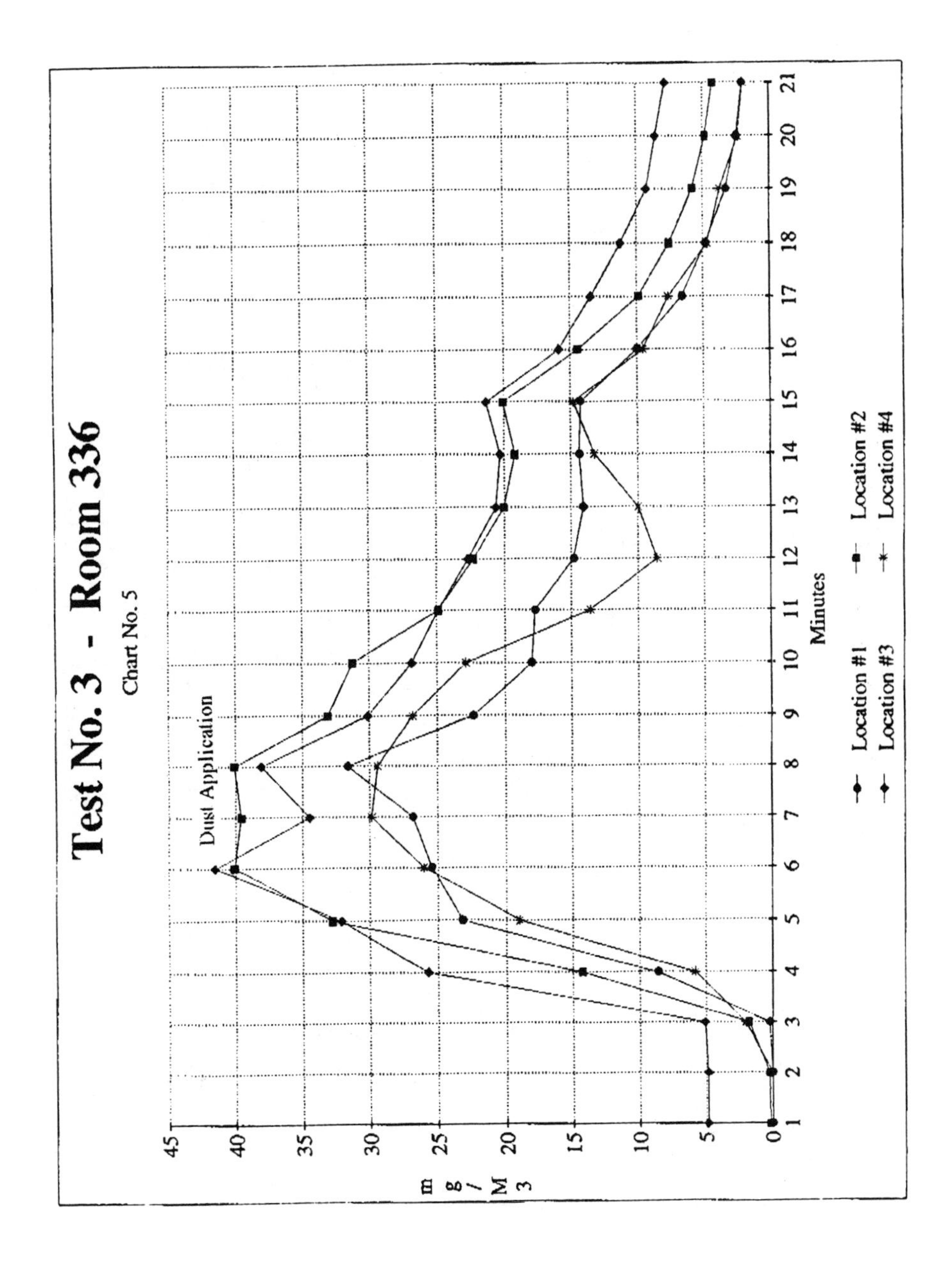

Test No. 4 - Room 335

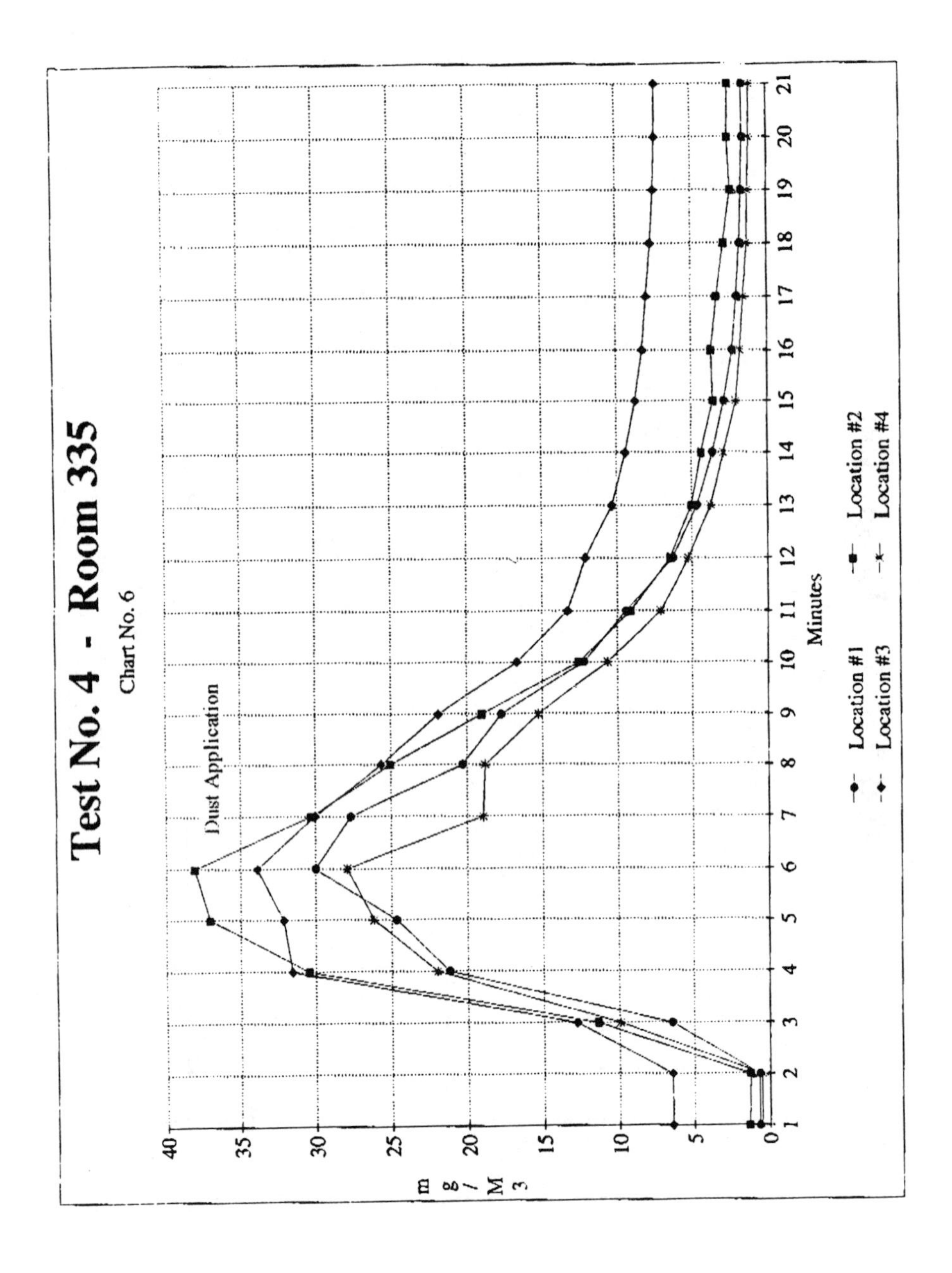

Test No. 5 - Room 335

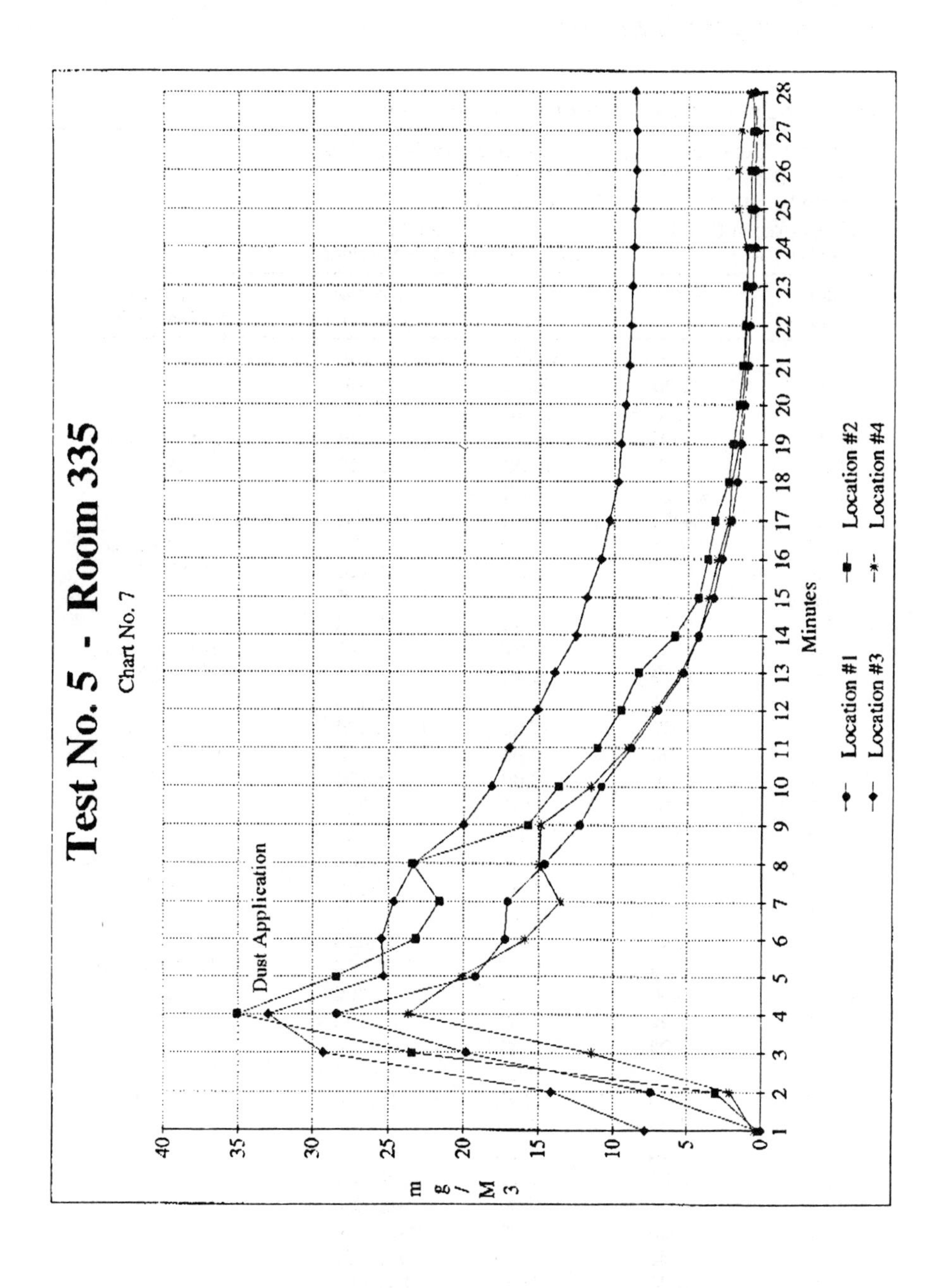

Appendix 4-2 Part II
Particulate (Dust) Size Data

Coulter Counter Model TAII Particle Size Analysis

Customer: General	Material: Arizona Road Dust
Date: 10-5-89	No. of Runs: 2
Sample No: 4281H	Aperture: 30
PTI Run No: 4281 F.F.	Cust. Lot No: 60029J
Operator: TAF	Mean Volume: 0.79 Microns

Total Data	Diff %	Cum %>	Diameter (Micron)	Active Channels
0	0.00	100.00	0.198	
0	0.00	100.00	0.250	
0	0.00	100.00	0.315	
0	0.00	100.00	0.397	0–2.5 microns
483	23.47	100.00	0.500	99.13
567	27.55	76.53	0.630	
513	24.93	48.98	0.794	
277	13.46	24.05	1.000	0–5 microns
135	6.56	10.59	1.260	100.00
48	2.33	4.03	1.590	
17	0.83	1.70	2.000	
8	0.39	0.87	2.520	5–10 microns
6	0.29	0.49	3.170	0.00
4	0.19	0.19	4.000	
0	0.00	0.00	5.040	
0	0.00	0.00	6.350	10–20 microns
0	0.00	0.00	8.000	0.00
0	0.00	0.00	10.080	
0	0.00	0.00	12.700	
0	0.00	0.00	16.000	20–40 microns
0	0.00	0.00	20.200	0.00
0	0.00	0.00	25.400	
0	0.00	0.00	32.000	
0	0.00	0.00	40.300	40–80 microns
0	0.00	0.00	50.800	0.00
0	0.00	0.00	64.000	
0	0.00	0.00	80.600	
0	0.00	0.00	101.600	80+ microns =
0	0.00	0.00	128.000	0.00
0	0.00	0.00	161.000	
0	0.00	0.00	203.000	
0	0.00	0.00	256.000	
0	0.00	0.00	322.000	
0	0.00	0.00	406.000	
0	0.00	0.00	512.000	
0	0.00	0.00	645.000	

Powder Technology Inc. (612) 894-8737

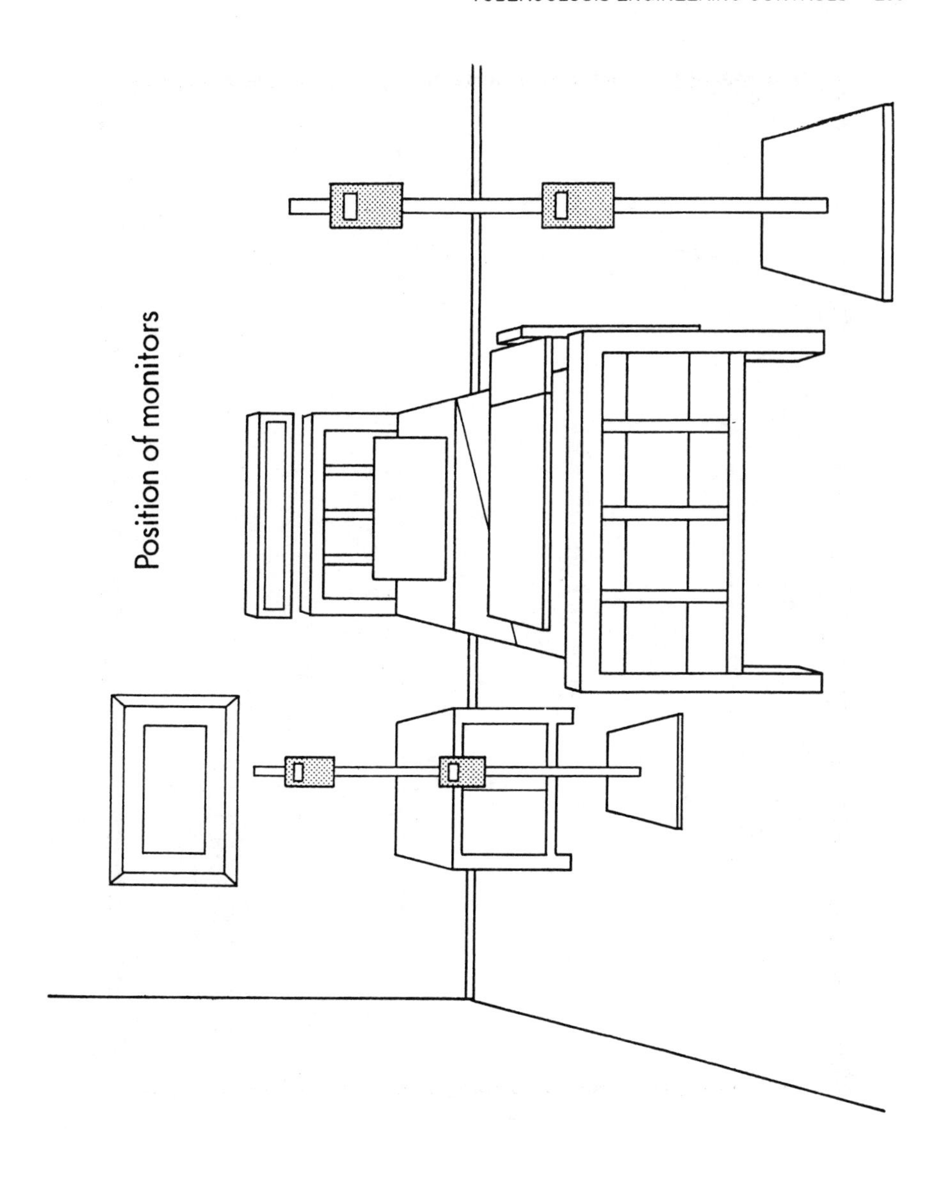

Position of monitors

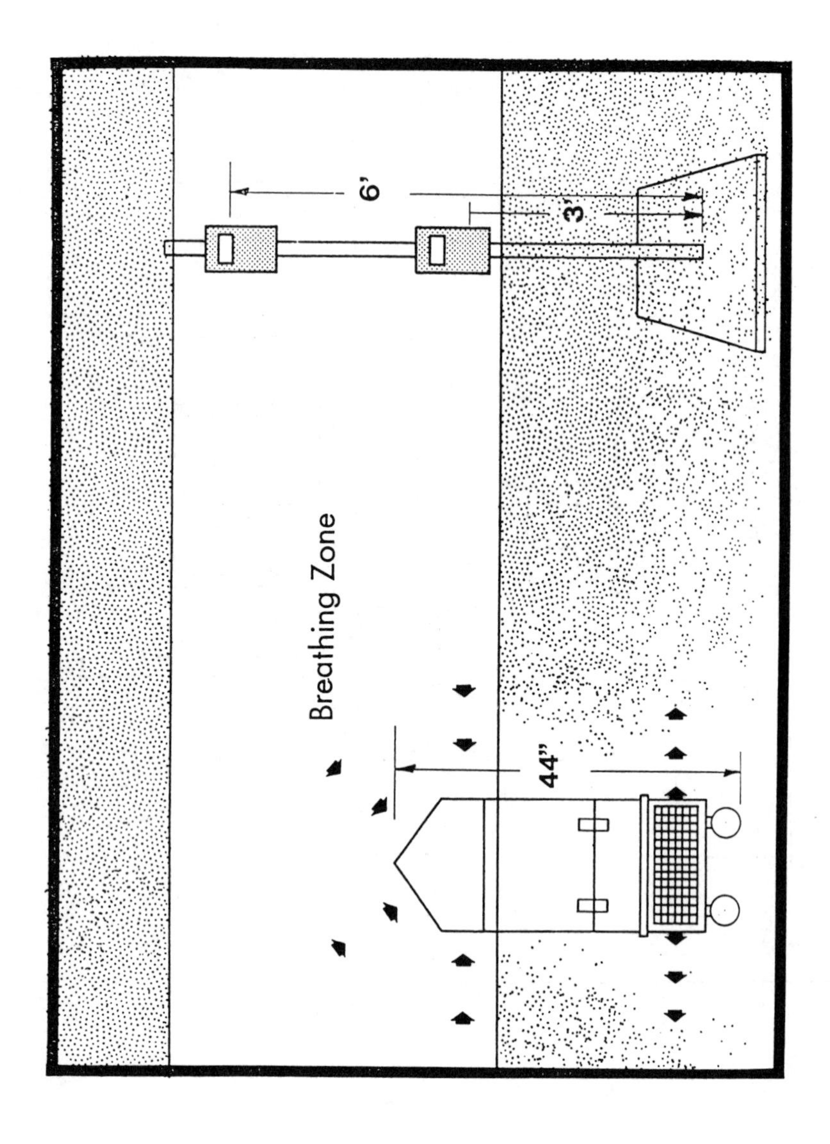

6'
3'
Breathing Zone
44"

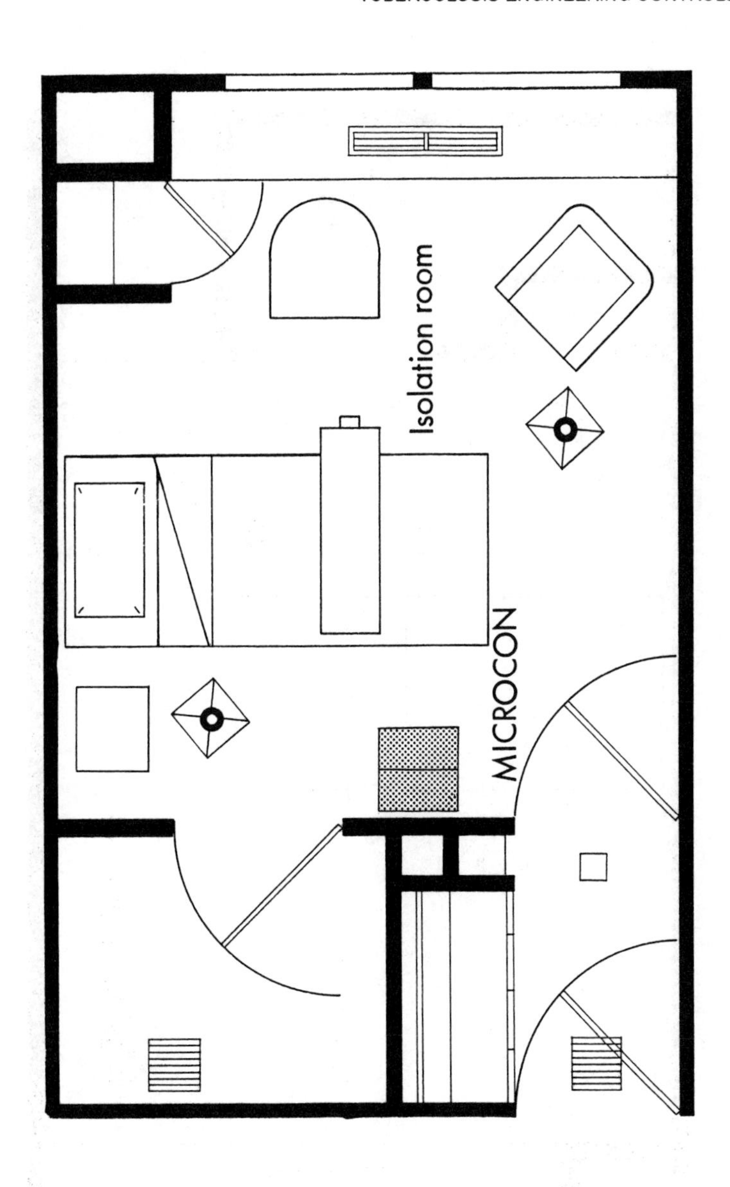
Isolation room
MICROCON

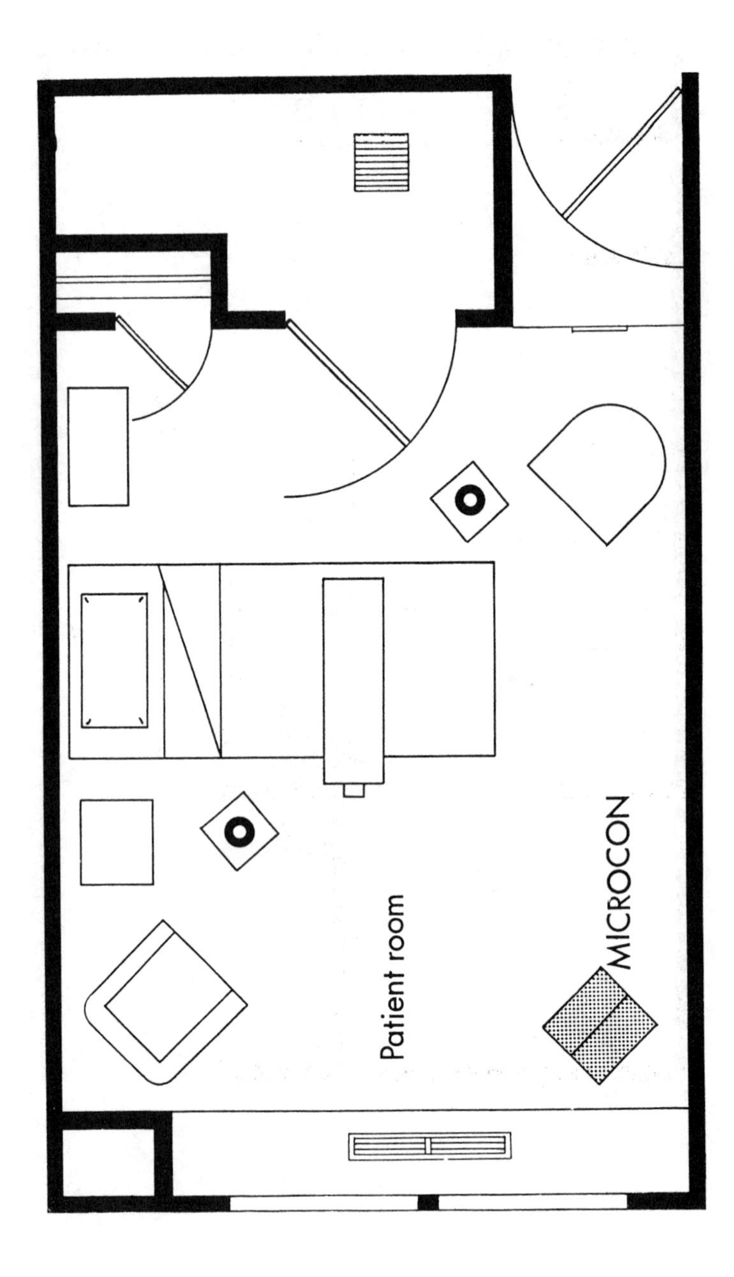

Source: Evaluation Report Purge Test of the MICROCON® in a Typical Hospital Environment Using a Simulated Human Bio-Aerosol. Prepared by: Hazardous Materials Technologies Corporation, Mount Laurel, NJ. Prepared for: Mr. Gary Messina, Biological Controls, Eatontown, NJ. Report date: January 27, 1993.

Appendix 4-2 Part III

Investigation of Isolation Room with the MICROCON® HEPA Air Filtration System

An investigation was made into the effectiveness of the MICROCON® HEPA Air Filtration System. A description of the equipment is included in this report. The survey involved sampling the air for microbiological viable organisms and total particulate levels. The samples were taken before running the MICROCON® and while the equipment was running. The results indicate that the MICROCON® HEPA Air Filtration System was effective at removing airborne particulate material and viable microbiological organisms. These results appear to support the manufacturer's claim that the filtration system will remove airborne droplet nuclei produced by persons actively infected with tuberculosis.

BACKGROUND

Over the past decade the incidence of respiratory tuberculosis (TB) infection has increased steadily in the New York Metropolitan area. Many hospitals in Newark have treated patients with actively infectious TB. Patients with active cases of TB can release infectious droplet nuclei containing TB into the air thereby exposing health care personnel in proximal areas. The risk of infection to the exposed persons is related to the quantities of TB bacillus in the air, the duration of exposure and the general health of the individual. The Centers for Disease Control (CDC) has a criteria document that suggests methods that may reduce the risk of TB exposure. These methods include:

1. Using isolation rooms for infectious persons. These rooms should be under negative pressure to the surrounding areas and should receive a minimum of 6 air changes per hour.
2. Providing adequate ventilation to reduce concentrations of TB in the air.
3. Using High Efficiency Particulate in Air (HEPA) filtration systems for recirculated air.
4. Using ultraviolet lamps to kill airborne TB.
5. Provide respiratory protection to exposed individuals in high risk areas such as isolation rooms.

This survey involved the evaluation of the MICROCON® Air Filtration System. This unit consists of a fan that pulls air into the top of the equipment through a HEPA filter and discharges the air back into the room through the base of the equipment. The unit operates at three fan settings:

1. 725 cubic feet per minute (cfm)
2. 675 cfm
3. 400 cfm

For a patient room that is 12 foot square and with an 8 foot ceiling this unit would provide over 20 air changes per hour at the lowest setting.

METHODS

Air sampling was conducted following NIOSH method 500 for total particulate in air. The 5.0 micron poly vinyl chloride (PVC) filters were preweighted and analyzed by the ITT Hartford Environmental Lab in Hartford, CT. This is an AIHA accredited laboratory. See Appendix 2 in this section for the lab report from the Hartford lab.

Fungi samples were collected on 3% malt extract agar using an Anderson N6 impaction sampler. Sample Petri dishes were prepared and analyzed by P&K Microbiology Services of Cherry Hill, NJ.

RESULTS

Both the dust samples and the fungi samples indicate a reduction of airborne particulate levels. Table 1 summarizes the dust level sampling results. Levels prior to operating the air cleaner were 180 micrograms per cubic meter of air. After running the air cleaner levels were below the limit of detection of 10 mcg/m^3.

Table 2 summarizes the bioaerosol monitoring the fungi levels in air. Levels were reduced from 40 cfu/m^3 prior to running the air cleaner to between 5 and 7 cfu/m^3 after running at low speed for approximately 1 hour.

DISCUSSION

The room tested is designed to operate as an isolation for TB patients. The room is designed to have 6 air changes per hour and is under negative pressure to the hallway. This design is adequate for preventing the spread of airborne TB to the corridor, but it does not appreciably reduce the level of airborne contaminants inside the room. Operating the air cleaner in the room reduced levels of particulate and bioaerosols.

Table 1. Airborne particulate sampling room tested.

Sample No.	Description	Result (μg/m^3)*
14024	Pretest - in room B521 prior to operating the MICROCON® Air Filtration System.	180
14023	Pretest - in the hallway outside room prior to operating the MICROCON® Air Filtration System.	20
14021	After operating the MICROCON® Air Filtration System inside the room.	Below 10
14016	Field Blank	<25 μg per filter

* μg/m^3 Indicates micrograms per cubic meter.

Table 2. Microbial evaluation room tested.

Sample No.	Location - Description	Species Identified	Result (CFU/M^3)
9306281	Inside room, pretest; before running the MICROCON® air cleaner 141.5 liters	Cladosporium, penicillium, yeasts	42.4
9306282	Outside room, pretest; before running the MICROCON® air cleaner 141.5 liters	Cladosporium	14.1
9306283	Field blank	No growth	
9306284	Inside room after turning on the MICROCON® air cleaner 141.5 liters	No growth	<5
9306285	Inside room after turning on the MICROCON® air cleaner 198.1 liters	Cladosporium	5
9306286	Inside room after turning on the MICROCON® air cleaner 424.5 liters	Cladosporium, sterile fungi	7.1

The results of this monitoring survey indicate the ability for HEPA filtration devices to reduce overall bioaerosol levels inside the isolation room. These results are consistent with our findings in a similar test conducted in May of 1993. Although we have not used the system on TB, we believe the results support the contention that these air filtration devices are effective at reducing airborne droplet nuclei.

There is no single way to protect persons from TB transmission and it is recommended that a combination of methods be used. These sample results indicate that HEPA filtration in combination with negative pressure in isolation rooms will provide an improved level of protection to the medical staff.

It should be noted that the MICROCON® HEPA Air Filtration System is an effective means of lowering the level of bioaerosols in the air. There is no engineering control that will eliminate TB exposure in the isolation room.

Source: Edward Olmsted, CIH, CSP, Olmsted Environmental Services, Inc. RR 1 Box 480, Garrison, NY 10524

Appendix 4-2 Part IV

Evaluation of MICROCON® HEPA Air Filtration System in an Emergency Room for Controlling Bioaerosols

An investigation was made into the effectiveness of the MICROCON® HEPA Air Filtration System. The survey was conducted on October 13, 1993, by Edward Olmsted, CIH, CSP. A description of the air filtration equipment is included in this report. The survey involved sampling the air for microbiological viable organisms and total particulate levels. The samples were taken before running the MICROCON® and while the equipment was running. The results indicate that the MICROCON® HEPA Air Filtration System was effective at reducing viable microbiological organisms. The particulate levels were not appreciably reduced by the MICROCON® HEPA Air Filtration System. These results appear to support the manufacturer's claim that the filtration system will remove airborne droplet nuclei produced by persons actively infected with tuberculosis; and thereby, reduce the number of airborne infectious organisms.

BACKGROUND

Over the past decade, the incidence of respiratory tuberculosis (TB) infection has increased steadily in the New York Metropolitan area. Many hospitals in Newark have treated patients with actively infectious TB. Patients with active cases of TB can release infectious droplet nuclei containing TB into the air, thereby, exposing healthcare personnel in proximal areas. The risk of infection to the exposed persons is related to the quantities of TB bacillus in the air, the duration of exposure and the general health of the individual. The Centers for Disease Control (CDC) has a criteria document that suggests methods that may reduce the risk of TB exposure. These methods include:

1. Using isolation rooms for infectious persons. These rooms should be under negative pressure to the surrounding areas and should receive a minimum of six (6) air changes per hour.
2. Providing adequate ventilation to reduce concentrations of TB in the air.
3. Using High Efficient Particulate in Air (HEPA) filtration systems for recirculated air.
4. Using ultraviolet lamps to kill airborne TB.
5. Provide respiratory protection to exposed individuals in high risk areas; such as isolation rooms.

This survey involved the evaluation of the MICROCON® Air Filtration System. This unit consists of a fan that pulls air into the top of the equipment through a HEPA filter and discharges the air back into the room through the base of the equipment. The unit operates at three fan settings:

1. 725 cubic feet per minute (cfm)
2. 675 cfm
3. 400 cfm

Previous tests of the MICROCON® have indicated that the system is effective at reducing airborne particulate and viable organisms in an AFB isolation room and in a general use patient room. In this case, the system was tested in an emergency room (ER) waiting area. The emergency room was under normal use by the hospital during all sampling. Figure 1 gives a layout of the ER.

Emergency rooms provide unique problems for hospitals attempting to develop infection control programs for TB. Large numbers of persons visit the emergency room each day; and in urban areas such as Newark, NJ, there is an increased potential for actively TB infected parsons to come to the ER. Controlling TB in emergency rooms has mostly consisted of the following:

1. Isolating persons that show signs and symptoms of TB
2. Installing UV lamps
3. Installing HEPA filtration systems

Very little is known about the effectiveness of these methods at controlling TB. The results of this survey indicate that the MICROCON® HEPA Air Filtration System does reduce airborne bioaerosols and may reduce infectious TB as well.

METHODS

Air sampling was conducted following NIOSH method 500 for total particulate in air. The 5.0 micron poly vinyl chloride (PVC) filters were preweighted and analyzed by the ITT Hartford Environmental Lab in Hartford, CT. This is an AIHA accredited laboratory. The samples were collected using a volume flow rate of 15 liters per minute calibrated using a rotameter. The lab report from the Hartford Lab is attached to this report.

Bacteria in air samples were collected using Tryptose soy agar and were collected using the Anderson N6 single stage impaction sampler operated at a flow rate of 28.3 lpm. The sampler was calibrated using an in-line rotameter. Sample Petri dishes were prepared and analyzed by P&K Microbiology Services of Cherry Hill, NJ.

Samples were collected before operating any air cleaning equipment in the waiting area. After the pretest, two MICROCONs® were operated at the medium speed. Figure 1 indicates the locations of the MICROCON® air cleaners. Samples of the air were taken after 90 minutes of air cleaning.

RESULTS

Bioaerosol Sampling

Attached to this report are the analytical results from P&K Microbiological Services. Table 1 gives a summary of the results.

PARTICULATE SAMPLING

Attached to this report is the analytical results for particulate sampling. Table 2 summarizes the results.

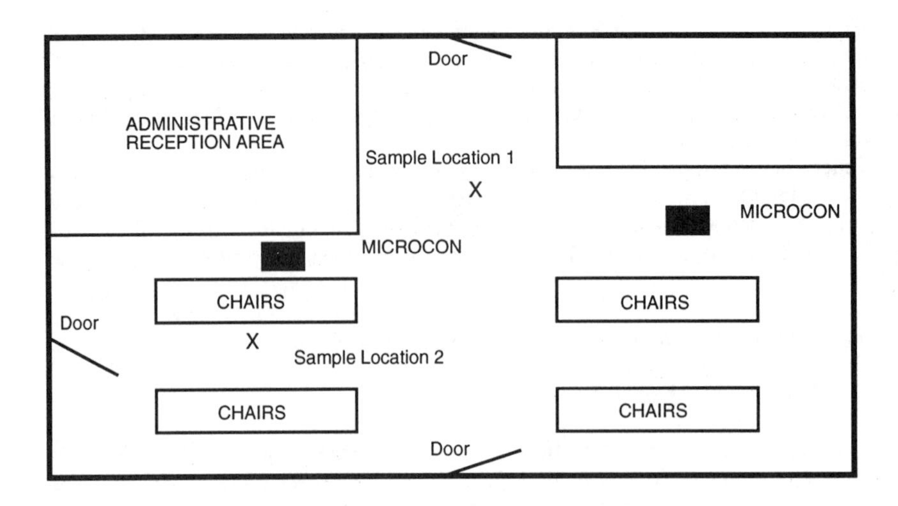

Figure 1. Emergency room.

Table 1. Bacteria in air sampling in emergency room waiting room.

Sample No.	Description	Result CFU/M^3
1	Location #1 - Pretest; before operating the MICROCON®	388.7
2	Location #2 - Pretest; before operating the MICROCON®	388.7
3	Location #1 - After running the MICROCON® 90 minutes	88.3
4	Location #2 - After running the MICROCON® 90 minutes	150.2
5	Location #1 - After running the MICROCON® 3 hours	91.9
6	Location #2 - After running the MICROCON® 3 hours	134.3

Average Bioaerosol Measurements (CFU/M^3)

Before the MICROCON® was run: 388.7
After 90 minutes of air cleaning via MICROCON®: 119.2
After 180 minutes of air cleaning via MICROCON®: 113.1

Table 2. Particulate sampling in emergency room waiting area.

Sample No.	Location	Result µg/m³
13969	Location #1; Pretest; before operating the MICROCON®	30
13965	Location #2; Pretest; before operating the MICROCON®	20
13966	Location #1; after running the MICROCON® 90 minutes	<20
13968	Location #2; after running the MICROCON® 90 minutes	50

Average Levels
Before running the MICROCON® HEPA Air Filtration System: 25 µg/m³
After 90 minutes of running the MICROCON®: 35 µg/m³

DISCUSSION

These results indicate that the MICROCON® HEPA Air Filtration System is effective at reducing airborne levels of bacteria. The results for particulate sampling did not show a reduction in airborne particulate levels. We believe that the level of particulate in the air was too low for statistically significant results. Nonetheless, the bioaerosol sample results clearly support the contention by the manufacturer that airborne viable bacteria are reduced by the system.

It should be noted that these units cannot be expected to eliminate exposure to TB in any setting. Other factors that are important include the rate of generation of TB in the room. This is dependent upon the degree of infectiousness of the patient. Our sampling results indicate that there is a fourfold reduction in airborne bacteria after running the equipment for 90 minutes, but no additional reduction is achieved with continual operation of the MICROCON® HEPA Air Filtration System.

Source: Edward Olmsted, CIH, CSP, Olmsted Environmental Services, Inc., RR 1 Box 480, Garrison, NY 10524

APPENDIX 4-3

ClestraClean Room Air Sterilizer Performance Testing

On March 12, 1993 and April 12, 1993 several performance tests were conducted on the new ClestraClean Room Air Sterilizer. Listed below are the tests that were performed and the results of those tests.

1) AIR FLOW MEASUREMENTS

Description of Testing Method:

All air flow measurements were accomplished by taking duct traverses either at the temporary inlet duct or at the 4″ø flex duct attached to the side of the unit.

Equipment Used:

Alnor Velometer, Model 6000 P

Results:

Mode: High speed, full recirculation, 525 CFM

Mode: Low speed, full recirculation, 230 CFM

Mode: High speed, portion of total (4″ round discharge) to ambient, 430 CFM discharge to room, 135 CFM (4″ round) to ambient

Mode: Low speed, portion of total (4″ round discharge) to ambient, 140 CFM discharge to room, 70 CFM (4″ round) to ambient

Mode: High speed, portion of total (4″ round suction) from ambient, 470 CFM from room, 75 CFM (4″ round) from ambient

Mode: Low speed, portion of total (4″ round suction) from ambient, 200 CFM from room, 40 CFM (4″ round) from ambient

2) VOLTS AND AMP DRAW MEASUREMENTS

Equipment Used:

AMP Probe, Model RS-3

Results:

High speed in full recirculation mode; 110 Volts and 3.0 AMPS

Low speed in full recirculation mode; 110 Volts and 6.0 AMPS

3) SOUND LEVEL MEASUREMENTS

Equipment Used:

Extech, Model 407703

Results:

High speed in full recirculation mode; 62 dBA (background with unit off was 48 dBA)

Low speed in full recirculation mode; 51 dBA (background with unit off was 48 dBA)

4) PRESSURIZATION MEASUREMENTS

Description of Testing Method:

In a reasonably tight room (approximately 1355 CU. FT.) the pressurization testing was performed. Negative pressurization was obtained by discharging a portion of the supply air to the ambient through the 4″ø flex duct connected to the 4″ø discharge port. Positive pressurization was obtained by drawing a portion of the return air from the ambient through the 4″ø flex duct connected to the 4″ø suction port.

Equipment Used:

Shortridge Air Data Meter, Model ADM 860

Results:

Room was -0.03″ W.C. to ambient with unit on high speed.

Duct discharging a portion of supply air to ambient.

Room was -0.01″ W.C. to ambient with unit on low speed and 4″ø duct discharging a portion of supply air to ambient.

Room was +0.002″ W.C. with unit on high speed and 4″ø duct drawing a portion of return air from ambient.

Room was +0.001″ W.C. with unit on low speed and 4″ø duct drawing a portion of return air from ambient.

5) PARTICLE COUNT MEASUREMENTS

Description and Testing Method:

Individual particle count measurements were sampled for 1 minute with a volume of 1 CU. FT. of air.

Baseline particle counts were measured with the unit off. With the unit on, the particle counter sensing probe was placed approximately 8′ from unit at 30″ above the floor. Several measurements were taken with the unit on high speed, low speed, and high speed with the 4″ø duct discharging a portion of the supply to simulate a room under negative pressurization.

Equipment Used:

Met One Particle Counter, Model 200

Results:

Baseline Particle Counts (µm)

0.3	0.5	1.0	2.0	5.0	10.0
256,252	149,066	19,341	4,548	372	86

With the unit on high speed, full recirculation, particle counts were taken again after 15 minutes:

Results:

0.3	0.5	1.0	2.0	5.0	10.0
10,965	4,670	481	98	11	3

With the unit on low speed, full recirculation:

Results:

0.3	0.5	1.0	2.0	5.0	10.0
11,403	5,099	657	195	21	8

With the unit on high speed, 4″ø duct discharging a portion of supply to maintain a slight negative room pressurization:

Results:

0.3	0.5	1.0	2.0	5.0	10.0
30,325	13,488	1,641	548	124	57

APPENDIX 4-4

The following is a partial list of commercially available HEPA filters. There are ceiling mount designs, roll-in floor designs. There are designs that offer air scrubbing and negative pressure in one module. Prices range from one hundred dollars to twenty-thousand dollars depending on type and power and add-ons.

Each facility needs to make individual assessments of needed capacity and function them order accordingly.

Model RS1000 Portable Air Scrubber/Negative Pressure Unit. Available from BioSafety Systems, Inc., San Diego, CA.

MICROCON®. Available from Biological Controls. One Industrial Way, Building D, Unit F, Eatontown, NJ 07724-3319, Telephone: 908-389-3319, Fax: 908-389-8821.

MICROCON® WallMAP™. Available from Biological Controls. One Industrial Way, Building D, Unit F, Eatontown, NJ 07724-3319, Telephone: 908-389-3319, Fax: 908-389-8821

Enviro™ Airlock Isolation Chamber. Available from Medical Safety Products, Inc., Marietta, GA.

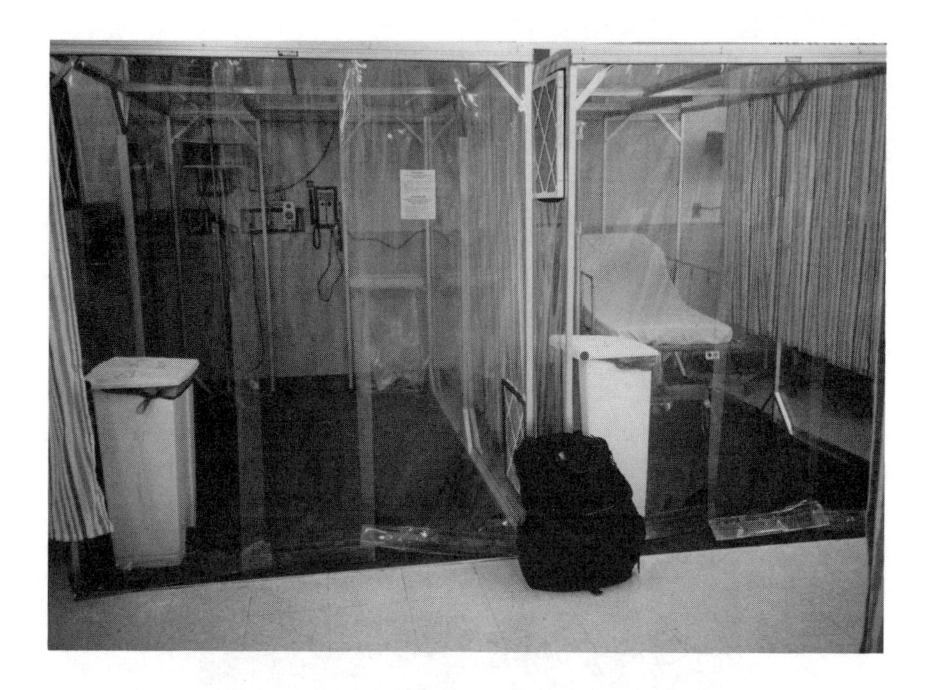

Enviro™ Airlock Isolation Chamber in Emergency Room. Available from Medical Safety Products, Inc., Marietta, GA.

Enviro™ Sputum Induction Chamber. Available from Medical Safety Products, Inc., Marietta, GA.

HEPAPORT. HEPA Fan/Exhaust. To create a negative pressure room as recommended by the Center for Disease Control (CDC), a filtered room exhaust may be installed in a window or wall powerful enough to overcome under-door or under-curtain leaks. Available from Modern Medical Systems Company, A Division of A. Kingsbury Co., Inc., 1655 Jericho Turnpike, New Hyde Park, NY 11040, 1-800-426-5304.

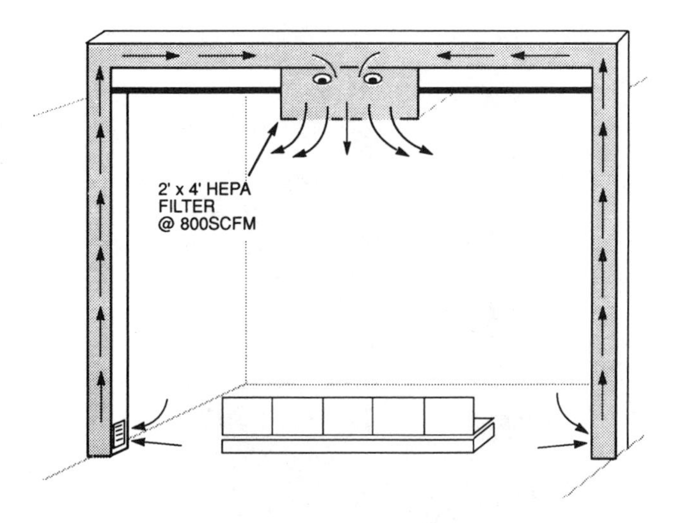

THE CRAB, Ceiling **R**ecirculating **A**ir **B**lower. For HEPA Air Filtration of waiting rooms, patient rooms, and treatment rooms. Available from Modern Medical Systems Company, A Division of A. Kingsbury Co., Inc., 1655 Jericho Turnpike, New Hyde Park, NY 11040, 1-800-426-5304.

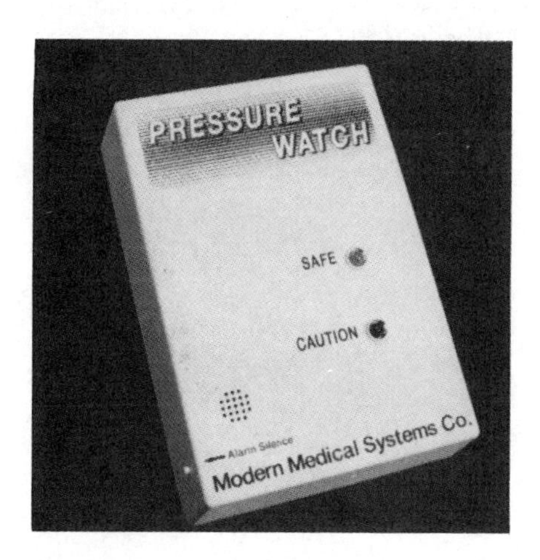

PRESSURE WATCH. For continuous monitoring of negative or positive pressure rooms. Available from Modern Medical Systems Company, A Division of A. Kingsbury Co., Inc., 1655 Jericho Turnpike, New Hyde Park, NY 11040, 1-800-426-5304.

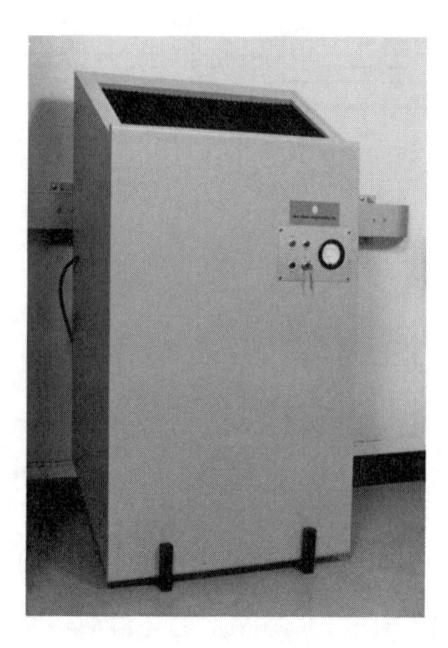

BioShield®. High efficiency hospital and medical grade air filtration unit. Available from Airo Clean, Inc., Pickering Creek Industrial Park, 212 Philips Road, Exton, PA 19341, Telephone: 610-524-8100, Fax: 610-524-8135.

CHAPTER 5

Needlesticks

PART 1

EPINet: A Tool for Surveillance and Prevention of Blood Exposures in Health Care Settings

Janine Jagger, Murray Cohen, and Beth Blackwell

BACKGROUND

Five million health care workers in the United States are routinely exposed to the blood and body fluids of patients; some while caring for patients, some while handling laboratory specimens, some while handling trash.[1] Although the transmission of the hepatitis B virus (HBV) to health care workers has long been recognized as a problem for health care institutions, it was not until the occupational transmission of human immunodeficiency virus (HIV) had been documented that a dramatic increase occurred in efforts to prevent blood exposures in the workplace. In response to this demand the Centers for Disease Control and Prevention (CDC) published Guidelines for the Prevention of HIV in Health Care Settings and recommended the infection control strategy of universal protection in 1987 as a measure to prevent or reduce contact with patient blood and body fluids.[2] These recommendations were based on the assumption that any patient could potentially harbor bloodborne pathogens and that a reduction in blood contacts would consequently reduce the likelihood of transmission of HIV, HBV, and any other bloodborne pathogens.

The CDC guidelines were embodied in the final rule promulgated by the Occupational Safety and Health Administration (OSHA) on the Prevention of Occupational Exposure to Bloodborne Pathogens.[1] The rule mandates risk reduction measure for two classes of blood and body fluid exposures: mucocutaneous (splashes, spray, direct contact) and percutaneous (needlestick and sharp object injuries). The distinction between these types of contacts or exposures is significant because they differ markedly in the mechanisms of exposure, in pathogen transmission rates and in prevention methods required.[1,3–5]

1-56670-083-3/94/$0.00+$.50

For the prevention of percutaneous injuries, OSHA mandates safe handling practices for sharp medical devices. Requirements include the immediate disposal of used needles and sharps in conveniently placed puncture-resistant containers and prohibition of recapping used needles except under specified circumstances. The OSHA rule goes beyond Universal Precautions in stating that, when available, engineering controls such as self-resheathing needles or other preventive technology be employed as the preferred method of exposure reduction. Furthermore, record keeping requirements for adverse blood and body fluid contacts are specified. These are intended to provide OSHA with a compliance evaluation tool, but they also afford hospitals an opportunity to implement an exposure surveillance system that can be used to evaluate and improve prevention programs.[1]

Further reduction of blood and body fluid exposures to health care workers requires broader implementation of engineering controls, which are especially important in the prevention of percutaneous exposures.[7–8] Studies at the University of Virginia suggest that the redesign of hollow-bore needle devices incorporating safety features that eliminate unnecessary needles, or shield hands from used needles, has the potential to prevent over 85% of needlesticks from those devices. This estimate is consistent with data reported in the New York State trial of needlestick prevention technology which showed a 93% decrease in needlesticks from needles used in intravenous systems when needleless or recessed needle systems replaced conventional, exposed-needle systems.[9] Another study of a similar intravenous safety system showed an 86% reduction in injuries.[10] A recent trial of a resheathing safety syringe resulted in an 86% reduction in needlesticks from 3-cc syringes.[11] A study of phlebotomy needles demonstrated an 82% reduction in needlesticks after the implementation of a blood collection tube holder that incorporated an after-use needle shield.[12]

These observations have important implications for reducing disease transmission, since hollow-bore needles have been shown to be the most common vehicle for transmitting HIV. Specifically, the needle devices known to have been involved in HIV transmission include 1) blood-drawing devices (vacuum tube blood collection needles, syringes and winged steel-needle intravenous sets), 2) intravenous catheter stylets and 3) intermittent I.V. needles (specifically, after blood had backed into the needles).[1,13–14] Engineering controls have been designed, and are currently available, that have the potential of preventing a high proportion of needlesticks from these devices.

However, there is evidence that this new technology can vary considerably in effectiveness, and not all purported safety designs actually reduce needlestick risk.[15] Therefore, comparative studies of conventional needles versus newer, protective designs are still needed to fully document the safety performance and user acceptance of the new devices. The rapid integration of effective preventive technology into the health care workplace depends upon carefully considered risk assessment and well-planned evaluations or trials in each hospital setting. Each hospital can become its own evaluation unit and can determine which devices and strategies can best reduce exposures to bloodborne pathogens.

The EPINet™ Surveillance System

The Exposure Prevention Information Network (**EPINet**) was developed to provide standardization of information on percutaneous injuries (and also blood and body fluid contacts) that all hospitals must collect to comply with OSHA recordkeeping requirements related to the Bloodborne Pathogen Standard of December 1991. Hospitals can use the **EPINet** system to compare and share information and identify successful prevention measures. An essential feature of the **EPINet** system is that it provides very specific identification of the devices causing injuries and the mechanisms by which the injuries occurred. This information allows hospitals to target high-risk devices and to evaluate the efficacy of new devices designed to prevent injuries. In short, the **EPINet** system is designed to overcome the lack of device-performance data that has been a major barrier to progress in reducing the risk of handling sharp equipment in the health care work place.

The **EPINet** system includes software for data entry and automatic report generation. It is a very simple and complete system that helps hospitals access their exposure data in a way that provides direction to prevention initiatives and enables hospitals to evaluate degrees of success in reducing the frequency of device-specific injuries. The **EPINet** system has been available since September 1992. In its first year, more than 1000 U.S. hospitals acquired it for use, and the Canadian government officially adopted the **EPINet** system as their national surveillance system. Each hospital using the **EPINet** system has the capacity to quickly retrieve device-specific data. Furthermore, many hospitals have joined data-sharing networks to conduct joint device trials to obtain efficacy results quickly, and communicate results efficiently.

The **EPINet** system includes a Uniform Needlestick and Sharp Object Injury Report and a Uniform Blood and Body Fluid Exposure Report (appended) to be filled out by employees reporting sharp object injuries or blood and body fluid exposures. The **EPINet** system also includes software programmed for entering, accessing and analyzing the data from the report. The data entry and analysis program, Epi Info, is a public domain (it can be copied freely) computer program for MS-DOS (IBM) compatible computers developed by the Centers for Disease Control. It is ideal for use with the **EPINet** system because it was designed for similar disease surveillance programs, it is easy to use, flexible and available at nominal cost.

DATA

Data from a network of nine hospitals follow. Participating hospitals and their coordinators are:

Florida Hospital, Orlando, FL: Sharon Kohler, Terri Shannon
Martha Jefferson Hospital, Charlottesville, VA: Pam Jones, Edwina Juillet
North Broward Hospital District, Ft. Lauderdale, FL: Marc Gomez, Lana Keough

Saint Joseph Hospital, Omaha, NE: Ann Lorenzen
Saint Vincent Health Center, Erie, PA: Mary Jo Dolecki, Diane Dougan
Saint Vincent Hospital, Indianapolis, IN: Dianne Spiller
Shands Hospital, Gainesville, FL: Deborah Boeff, Suzanne Hench
University Hospitals of Cleveland, Cleveland, OH: Pamela Parker Julia Taoras
University of Virginia Hospitals, Charlottesville, VA: Betty Joe Coyner, Vickie Pugh

Only injuries resulting from conventional devices (not safety designs) are presented, which include 1016 out of 1024 injuries in the database. The data were collected over a six-month period from 9/1/92 through 2/28/93. They are presented in order to demonstrate how systematic, standardized surveillance can be conducted in hospitals and to illustrate the baseline injury profile for the ten conventional devices most frequently associated with percutaneous injuries.

The data presented include: general frequency of all devices causing injury (Table 1), job classification of health care workers reporting injuries (Table 2) and device-specific mechanism of injury for selected devices most frequently causing injuries (Figures 1-10).

CONCLUSIONS

Device-specific surveillance of percutaneous injuries is an essential tracking and prevention tool for hospitals. The **EPINet** surveillance program was designed to meet these needs and also to assist hospitals in complying with OSHA recordkeeping requirements. Hospitals can use **EPINet** data in a variety of ways:

- Monitor reductions or increases in percutaneous injuries over time.
- Identify the devices causing percutaneous injuries.
- Track device-specific injury rates.
- Identify alternative devices to those causing injuries. Conduct trials of devices designed to prevent percutaneous injuries.
- Enhance ease of recordkeeping for monitoring compliance with OSHA regulations.
- Identify new device hazards.
- Quick retrieval of information and data to respond to in-house inquiries on topics such as cost of followup, products involved in exposures, etc. from the safety committee, product evaluation, risk management, infection control, employee health, environmental services and hospital administration.
- Provide user feedback to device manufacturers and product representatives.
- Compare exposure and injury rates to other hospitals using the **EPINet** system.

EPINet is a trademark of the University of Virginia. The development of the EPINet system was supported in part by Becton Dickinson and Company.

Table 1. Frequency of devices causing 1016 percutaneous injuries. Nine hospitals, six months. (Conventional, non-safety, devices.)

Hollow-Bore Needles*	**No.**	**%**
Disposable syringe/other syringe	302	29.8
Prefilled cartridge syringe	35	3.4
Blood gas syringe	18	1.8
Needle on I.V. line	100	9.8
Butterfly	43	4.2
I.V. catheter stylet	38	3.7
Vacuum tube phlebotomy needle	47	4.6
Spinal or epidural needle	4	0.4
Unattached hypodermic needle	22	2.2
Needle, not sure what kind	38	3.7
Other needle	34	3.3
Surgical Instrument or Other Sharp Item		
Lancet	37	3.6
Suture needle	108	10.6
Scalpel blade	63	6.2
Razor	6	0.6
Plastic pipette	1	0.1
Scissors	4	0.4
Bovie electrocautery	1	0.1
Towel clip	4	0.4
Microtome blade	1	0.1
Trocar	2	0.2
Plastic vacuum tube	1	0.1
Fingernails/teeth	8	0.8
Not sure what sharp item	17	1.7
Other item	55	5.5
Glass		
Medication ampule	1	0.1
Glass pipette	2	0.2
Glass vacuum tube	2	0.2
Glass specimen/test tube	9	0.9
Glass capillary tube	5	0.5
Other glass item	8	0.8

* Of 681 injuries with hollow-bore needles, 226 or 33%, were caused by *unnecessary needles,* that is, needles not used to pierce the skin.

Table 2. Job classification of 1024 hospital workers reporting percutaneous injuries during six months, nine hospitals.

	No.	%
M.D., attending, staff	48	4.7
M.D., resident, fellow	109	10.6
Medical student	15	1.5
Nurse	525	51.3
Nursing student	13	1.3
Respiratory therapist	23	2.2
Surgery attendant	25	2.4
Other attendant	41	4.0
Phlebotomist, venipuncture, I.V. team	40	3.9
Clinical laboratory worker	30	2.9
Technologist, non-lab	68	6.6
Dentist	5	0.5
Dental hygienist	8	0.8
Housekeeper/laundry worker	34	3.3
Other	39	3.8

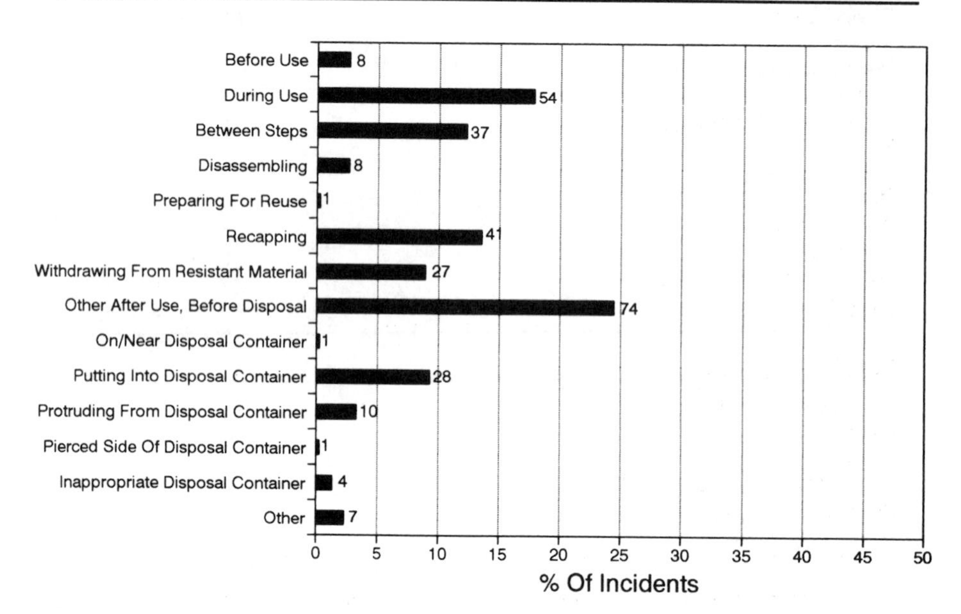

Figure 1. Injuries from disposable/other syringes. Six months, nine hospitals; N=302.

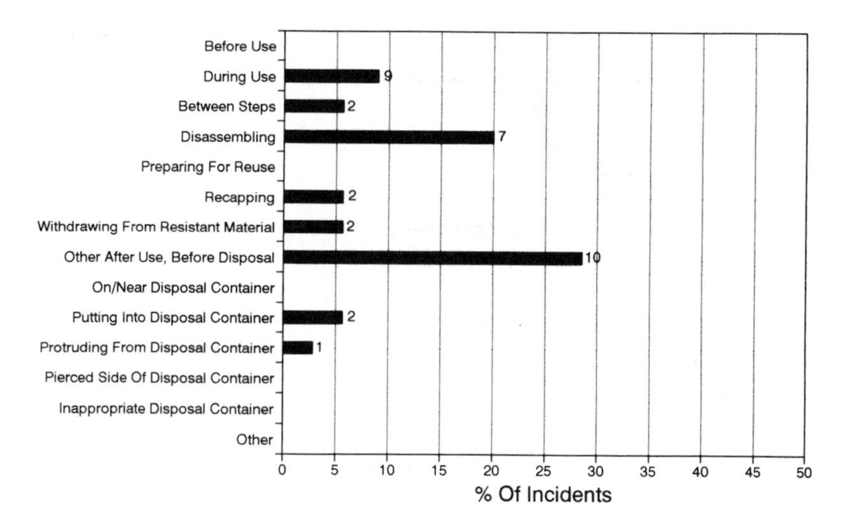

Figure 2. Injuries from prefilled cartridge syringes. Six months, nine hospitals; N=35.

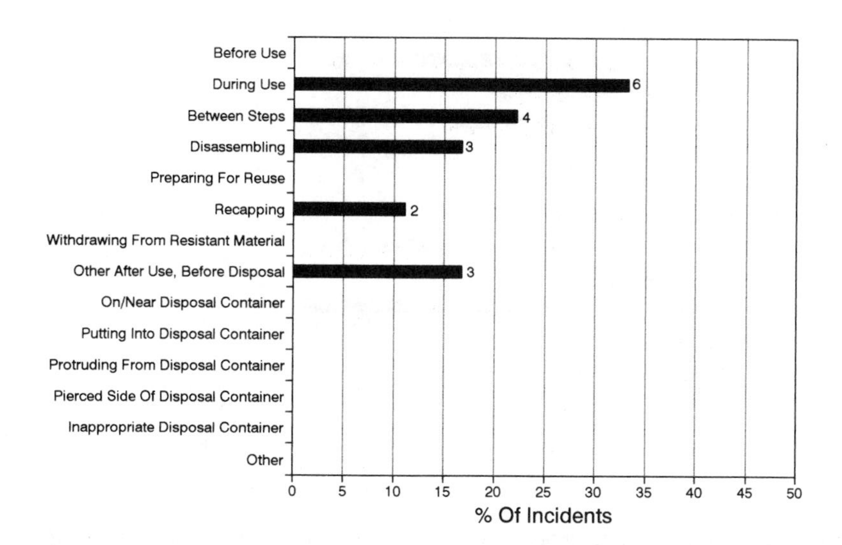

Figure 3. Injuries from arterial blood gas syringes. Six months, nine hospitals; N=18.

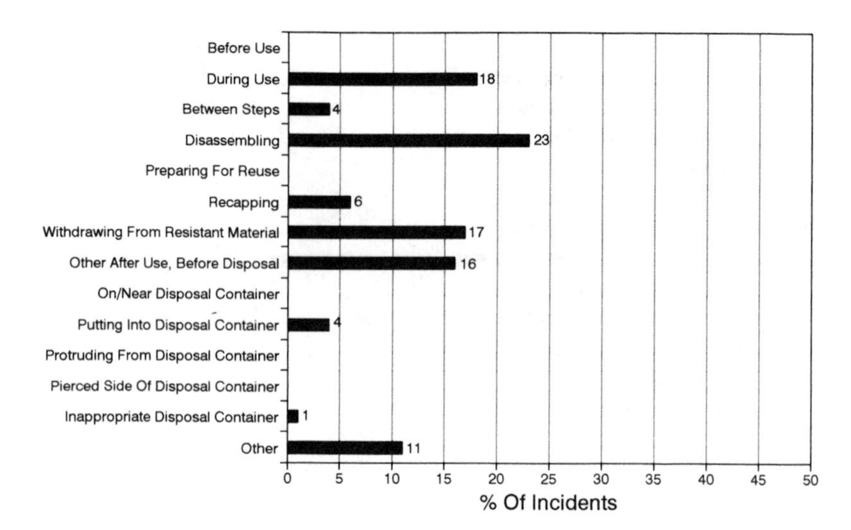

Figure 4. Injuries from needles on I.V. lines. Six months, nine hospitals; N=100.

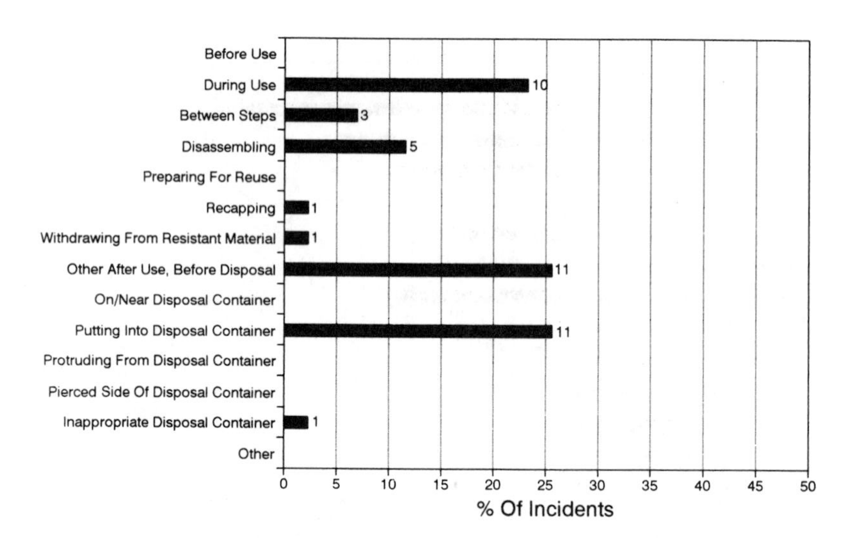

Figure 5. Injuries from winged steel needles. Six months, nine hospitals; N=43.

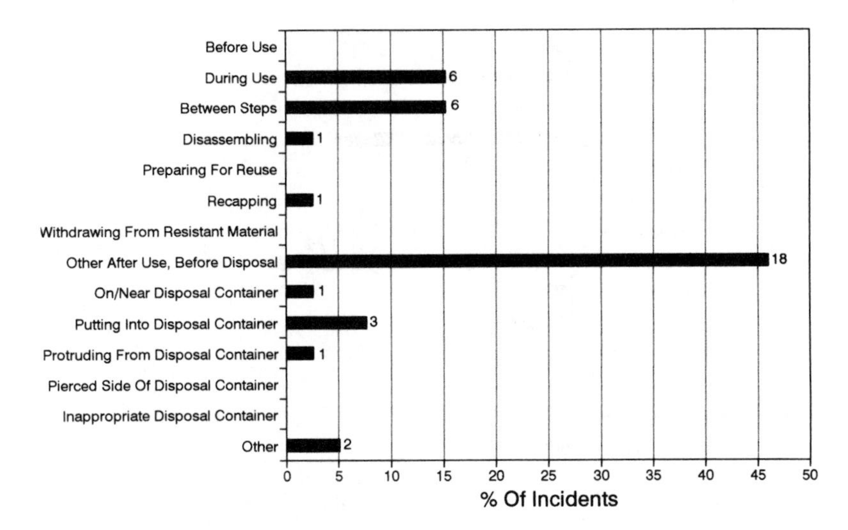

Figure 6. Injuries from I.V. catheter stylets. Six months, nine hospitals; N=39.

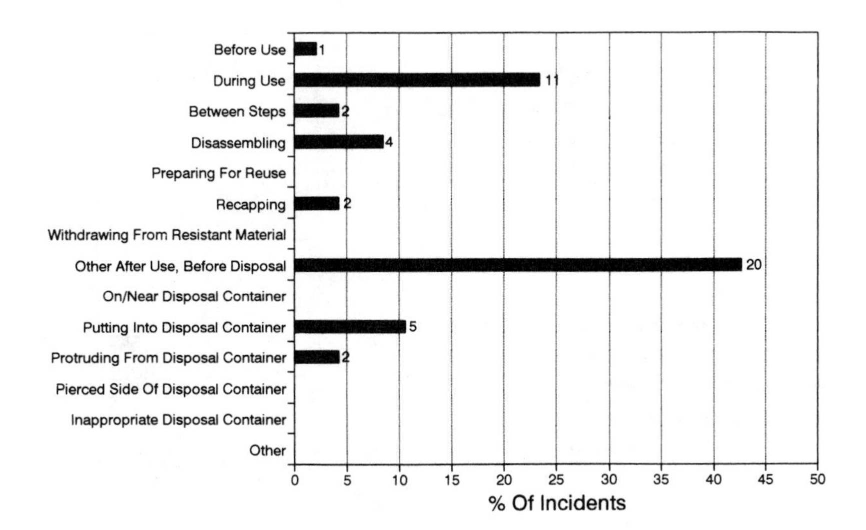

Figure 7. Injuries from vacuum tube/phlebotomy needles. Six months, nine hospitals; N=47.

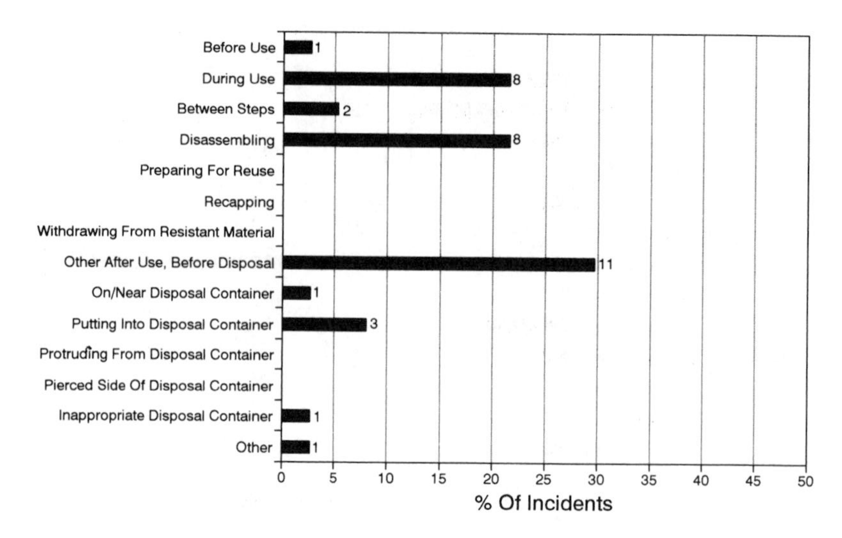

Figure 8. Injuries from lancets. Six months, nine hospitals; N=37.

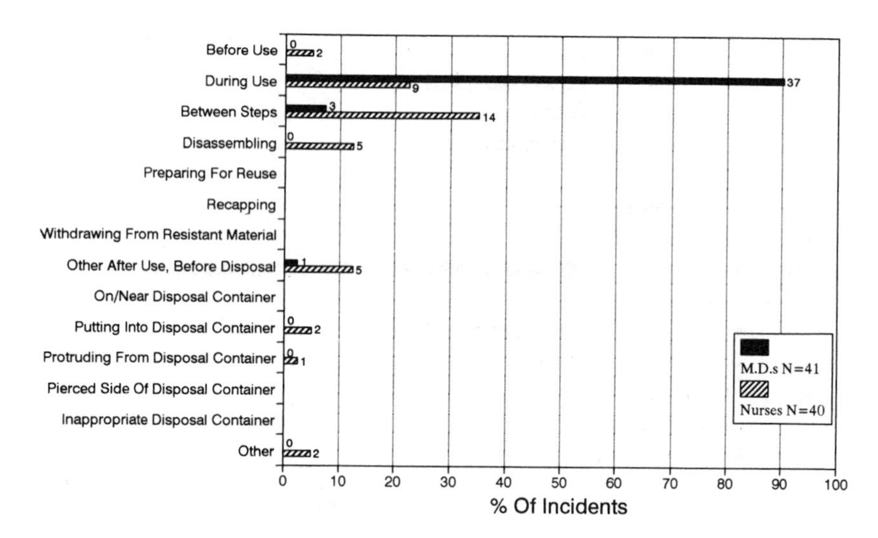

Figure 9. Injuries from suture needles. Six months, nine hospitals; M.D.s and nurses.

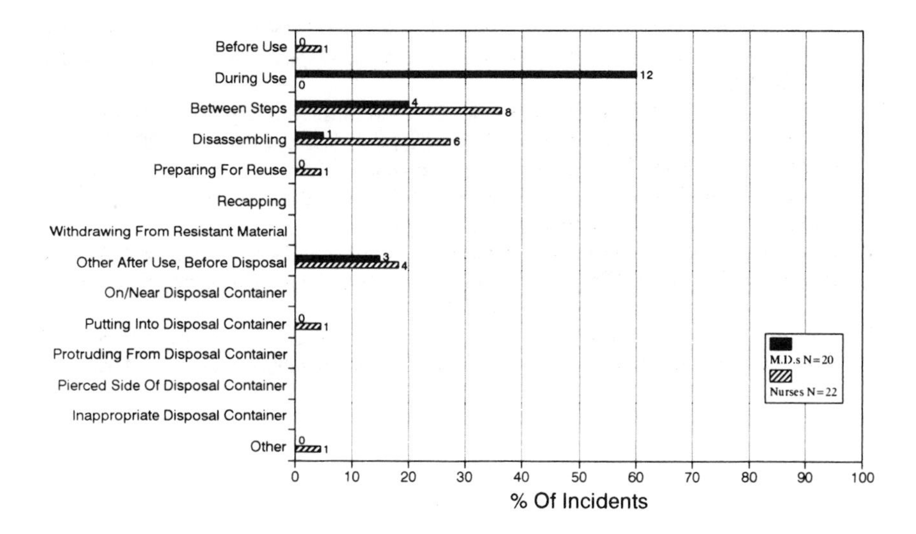

Figure 10. Injuries from scalpel blades. Six months, nine hospitals; M.D.s and nurses.

REFERENCES

1. Occupational Safety and Health Administration: 29 CFR Part 1910.1030: Occupational exposure to bloodborne pathogens. *Federal Register,* 56(235): 64004–64182, 1991.
2. Centers for Disease Control: Recommendations for prevention of HIV transmission in health care settings. *MMWR* 36(suppl 2S), 1987.
3. Marcus R and the CDC Cooperative Needlestick Surveillance Group: Surveillance of health care workers exposed to blood from patients infected with the human immunodeficiency virus. *N Engl J Med* 319:1118–23, 1988.
4. Henderson DK, Fahey BJ, Willy M, et al.: Risk for occupational transmission of HIV associated with clinical exposure. *Ann Intern Med* 113:740–6, 1990.
5. Bell DM: Human immunodeficiency virus transmission in health care settings: Risk and risk reduction. *Am J Med* 91:3B-294S-300S, 1991.
6. Wong ES, Stotka JL, Chinchilli VN, et al.: Are Universal Precautions effective in reducing the number of occupational exposures among health care workers? A prospective study of physicians on a medical service. *JAMA* 265:1123–1128, 1991.
7. Jagger J, Hunt EH, Brand-Elnaggar J, Pearson RD: Rates of needle-stick injury caused by various devices in a university hospital. *N Engl J Med* 319:284–288, 1988.
8. Jagger J, Hunt EH, Pearson RD: Sharp object injuries in the hospital; Causes and strategies for prevention. *Am J Infect Control* 4:227–231, 1990.
9. Chiarello L: Needlestick prevention technology. Testimony presented before the U.S. Congress House Subcommittee on Small Business Opportunities, and Energy. Washington, DC, February 7, 1992.
10. Gartner K: Impact of a needleless intravenous system in a university hospital. *Am J Infect Control* 20:75–79, 1992.

11. Younger B, Hunt EH, Robinson C, McLemore C: Impact of a shielded safety syringe on needlestick injuries among healthcare workers. *Infect Control Hosp Epidemiol* 13:349–353, 1992.
12. Billiet LS, Parker CR, Tanley PC, Wallas CW: Needlestick injury rate reduction during phlebotomy: A comparative study of two safety devices. *Lab Med* 22: 120–123, 1991.
13. Metler R, Ciesielski C, Marcus R: HIV seroconversions in nurses following blood contact. (poster) 5th National Forum on AIDS, Hepatitis, and Other Blood-Borne Diseases. NFID, CDC, Atlanta, March 31, 1992.
14. Testimony of Jane Doe Presented before the Occupational Safety and Health Administration Hearing on the Proposed Rule on the Occupational Exposure to Bloodborne Pathogens, San Francisco, January 16, 1990.
15. Chiarello L, Nagin D, Laufer F: Final Draft: New York State Department of Health Report to the Legislature, Pilot Study of Needlestick Prevention Devices. March 1992.

Appendix 1

4-92

EXPOSURE PREVENTION INFORMATION NETWORK

Uniform Needlestick and Sharp Object Injury Report

Name: ______________________ Hospital ID: [ ][ ][ ][ ]

1. ID: [ ][ ][ ][ ]

2. Date of injury: month [ ][ ] day [ ][ ] year [ ][ ]

3. Job Category: *(check one)*

- [1] M.D. *(attending/staff)*
- [2] M.D. *(intern/resident/fellow)*
- [3] medical student
- [4] nurse RN/LPN
- [5] nursing student
- [6] respiratory therapist
- [7] surgery attendant
- [8] other attendant
- [9] phlebotomist/venipuncture/I.V. team
- [10] clinical laboratory worker
- [11] technologist *(non-lab)*
- [12] dentist
- [13] dental hygienist
- [14] housekeeper/laundry worker
- [15] other, describe ______________________

4. Where did the injury occur? *(check one)*

- [1] patient room
- [2] outside patient room *(hallway, nurses' station, etc.)*
- [3] emergency department
- [4] intensive/critical care unit
- [5] operating room
- [6] outpatient clinic/office
- [7] blood bank
- [8] venipuncture
- [9] dialysis facility
- [10] procedure room *(X-ray, EMG, etc.)*
- [11] clinical laboratories
- [12] autopsy/pathology
- [13] service/utility area *(laundry, central supply, loading dock, etc.)*
- [14] other, describe ______________________

5. Was the source patient known?

yes [1] no [2] unknown [3] not applicable [4]

6. Was the injured worker the original user of the sharp item?

yes [1] no [2] unknown [3] not applicable [4]

7. Was the sharp item: *(check one)*

- [1] contaminated *(known exposure to patient or contaminated equipment)*
- [2] uncontaminated *(no known exposure to pt. or contaminated equipment)*
- [3] unknown

8. For what purpose was the sharp item originally used: *(check one)*

- [1] unknown/ not applicable
- [2] injection, intramuscular/subcutaneous , or other injection through the skin *(syringe)*
- [3] heparin or saline flush *(syringe)*
- [4] other injection into (or aspiration from) I.V. injection site or I.V. port *(syringe)*
- [5] to connect I.V. line *(intermittent I.V./ piggyback/ I.V. infusion/ other I.V. line connection)*
- [6] to start I.V. or set up heparin lock *(I.V. catheter or Butterfly™ -type needle)*
- [7] to draw a venous blood sample
- [8] to draw an arterial blood sample *(ABG)*
- [9] to obtain a body fluid or tissue sample *(urine/ CSF/ amniotic fluid/ other fluid, biopsy)*
- [10] fingerstick/ heel stick
- [11] suturing
- [12] cutting *(surgery)*
- [13] electrocautery
- [14] to contain a specimen or pharmaceutical *(glass items)*
- [15] other, describe ______________________

over⟶

4-92

9. Did the injury occur: *(check one)*

- [1] before use of item *(item broke or slipped, assembling device, etc.)*
- [2] during use of item *(item slipped, patient jarred item, etc.)*
- [3] between steps of a multistep procedure *(between incremental injections, passing instruments, etc.)*
- [4] disassembling device or equipment
- [5] in preparation for reuse of reusable instrument *(sorting, disinfecting, sterilizing, etc.)*
- [6] while recapping a used needle
- [7] withdrawing a needle from rubber or other resistant material *(rubber stopper, I.V. port, etc.)*
- [8] other after use, before disposal *(in transit to trash, cleaning up, left on bed, table, floor, or other inappropriate place, etc.)*
- [9] from item left on or near disposal container
- [10] while putting the item into the disposal container
- [11] after disposal, stuck by item protruding from opening of disposal container
- [12] item pierced side of disposal container
- [13] after disposal, item protruded from **trash bag** or **inappropriate** waste container
- [14] other, describe ____________________

10. What device or item caused the injury?
(refer to the list of items and enter the item code number here): [ ][ ]
If the item is coded as "other" (29, 59, 79), then please describe the item:
____

11. If the item causing the injury was a needle, was it a "safety design" with a shielded, recessed, or retractable needle?

yes [1] no / not applicable [2]

12. Mark the location of the injury:

13. Was the injury: *(check one)*

- [1] superficial *(little or no bleeding)*
- [2] moderate *(skin punctured, some bleeding)*
- [3] severe *(deep stick/cut, or profuse bleeding)*

14. Describe the circumstances leading to this injury:

____
____
____
____
____
____
____

Costs: for office use only *(round to nearest dollar)*

				laboratory charges for employee and source *(Hb, HIV, other tests)*
				treatment, prophylaxis *(HbIG, Hb vaccine, tetanus, AZT, other treatments)*
				service charges *(Emergency Department, Employee Health, other services)*
				other costs *(Worker's Compensation, surgery, other)*
				total

Is this incident OSHA reportable?
- *Medical treatment (H-BIG, Hepatitis vaccine, gamma globulin, AZT, etc.; Not first aid, not tetanus)*
- *Restricted/lost work time; transferred*
- *Illness/death*

[1] yes [2] no

If yes, enter: days away from work [ ][ ][ ]
days restricted work activity [ ][ ][ ]

4-92

Items Causing Needlestick and Sharp Object Injuries

NEEDLE

(for suture needle see "surgical instruments")

Item Codes

(1) disposable syringe *(includes standard syringes, insulin, tuberculin syringes)*
(2) prefilled cartridge syringe *(includes Tubex TM*/Carpuject TM* -type syringes)*
(3) blood gas syringe *(ABG)*
(4) syringe, other type or not sure what kind
(5) needle on I.V. line *(includes piggybacks and I.V. line connections)*
(6) winged steel needle I.V. set *(includes Butterfly TM* -type devices)*
(7) I.V. catheter *(stylet)*
(8) vacuum tube blood collection holder/needle *(includes Vacutainer TM* -type devices)*
(9) spinal or epidural needle
(10) unattached hypodermic needle

(28) needle, not sure what kind
(29) other needle *(please describe device on the report form)*

SURGICAL INSTRUMENT OR OTHER SHARP ITEM

(For glass items see below)

(30) lancet *(finger or heel sticks)*
(31) suture needle
(32) scalpel blade
(33) razor
(34) pipette *(plastic)*
(35) scissors
(36) bovie electrocautery device
(37) bone cutter
(38) bone chip
(39) towel clip
(40) microtome blade
(41) trocar
(42) vacuum tube *(plastic)*
(43) specimen/test tube *(plastic)*

(58) sharp item, not sure what kind
(59) other item *(please describe item on the report form)*

GLASS

(60) medication ampule
(61) medication vial *(small volume with rubber stopper)*
(62) medication/I.V. bottle *(large volume)*
(63) pipette *(glass)*
(64) vacuum tube *(glass)*
(65) specimen/test tube *(glass)*
(66) capillary tube

(78) glass item, not sure what kind
(79) other glass item *(please describe item on the report form)*

*Tubex TM is a trademark of Wyeth Ayerst; Carpuject TM is a trademark of Sanofi Winthrop; Butterfly TM is a trademark of Abbott Laboratories; Vacutainer TM is a trademark of Becton Dickinson. Identification of these product categories does not imply involvement of these specific brands.

492

EXPOSURE PREVENTION INFORMATION NETWORK

Uniform Blood and Body Fluid Exposure Report

Name: ______________________ Hospital ID: [ ][ ][ ][ ]

1. ID: [ ][ ][ ][ ]   **2. Date of injury:** month [ ][ ] day [ ][ ] year [ ][ ]

3. Job Category: *(check one)*

- [1] M.D. *(attending/staff)*
- [2] M.D. *(intern/resident/fellow)*
- [3] medical student
- [4] nurse RN/LPN
- [5] nursing student
- [6] respiratory therapist
- [7] surgery attendant
- [8] other attendant
- [9] phlebotomist/venipuncture/I.V. team
- [10] clinical laboratory worker
- [11] technologist *(non-lab)*
- [12] dentist
- [13] dental hygienist
- [14] housekeeper/laundry worker
- [15] other, describe ______________

4. Where did the exposure occur? *(check one)*

- [1] patient room
- [2] outside patient room *(hallway, nurses' station, etc.)*
- [3] emergency department
- [4] intensive/critical care unit
- [5] operating room
- [6] outpatient clinic/office
- [7] blood bank
- [8] venipuncture
- [9] dialysis facility
- [10] procedure room *(X-ray, EMG, etc.)*
- [11] clinical laboratories
- [12] autopsy/pathology
- [13] service/utility area *(laundry, central supply, loading dock, etc.)*
- [14] other, describe ______________

5. Was the source patient known? *(check one)*

yes [1]   no [2]   unknown [3]   not applicable [4]

6. Which body fluids were involved in the exposure? *(check all that apply)*

- [ ] blood or blood product
- [ ] vomit
- [ ] CSF
- [ ] peritoneal fluid
- [ ] pleural fluid
- [ ] amniotic fluid
- [ ] urine
- [ ] other, describe ______________

7. Was the exposed part: *(check all that apply)*

- [ ] intact skin
- [ ] non-intact skin
- [ ] eye(s)
- [ ] nose
- [ ] mouth
- [ ] other, describe ______________

8. Did the blood or body fluid: *(check all that apply)*

- [ ] touch unprotected skin
- [ ] touch skin through gap between protective garments
- [ ] soak through protective garment
- [ ] soak through clothing

9. Which protective items were worn at the time of the exposure? *(check all that apply)*

- [ ] single pair latex/vinyl gloves
- [ ] double pair latex/vinyl gloves
- [ ] goggles
- [ ] eyeglasses
- [ ] faceshield
- [ ] surgical mask
- [ ] surgical gown
- [ ] plastic apron
- [ ] lab coat, cloth
- [ ] lab coat, other, ______________
- [ ] other, describe ______________

over→

4-92

10. Was the exposure the result of: *(check one)*

- 1 direct patient exposure
- 2 specimen container leaked/spilled
- 3 specimen container broke
- 4 IV tubing/bag/pump leaked
- 5 other body fluid container spilled/leaked
- 6 touched contaminated equipment
- 7 touched contaminated drapes/sheets/gowns, etc.
- 8 unknown
- 9 other, describe ______________________

11. For how long was the blood or body fluid in contact with your skin or mucous membranes? *(check one)*

- 1 less than 5 minutes
- 2 5-14 minutes
- 3 15 minutes to 1 hour
- 4 more than 1 hour

12. Estimate the quantity of blood/body fluid that came in contact with your skin or mucous membranes: *(check one)*

- 1 small amount *(up to 5 cc, or up to a teaspoon)*
- 2 moderate amount *(up to 50 cc, or up to a quarter cup)*
- 3 large amount *(more than 50 cc)*

13. Mark the size and location of the exposure:

Right 1 2
Left 3 4

Front: 33 39 34 40 32 31 30 45 46 47 35 41 36 42 37 43 38 44

Back: 51 57 52 58 50 63 49 64 48 65 53 59 54 60 55 61 56 62

14. Describe the circumstances leading to this exposure:

____

____

____

____

____

____

____

Costs: for office use only *(round to nearest dollar)*

				laboratory charges for employee and source *(Hb, HIV, other tests)*
				treatment, prophylaxis *(HbIG, Hb vaccine, tetanus, AZT, other treatments)*
				service charges *(Emergency Department, Employee Health, other services)*
				other costs *(Worker's Compensation, surgery, other)*
				total

Is this incident OSHA reportable?

- *Medical treatment (H-BIG, Hepatitis vaccine, gamma globulin, AZT, etc.; Not first aid, not tetanus)*
- *Restricted/lost work time; transferred*
- *Illness/death*

1 yes
2 no

If yes, enter: days away from work ☐☐☐
days restricted work activity ☐☐☐

PART 2

Strategies for Integrating Health Care Workers Into the Process of Design, Selection and Use of Control Technology

June Fisher

OVERVIEW

With the advent of AIDS and, in particular, with the adoption of OSHA's Bloodborne Pathogens Standard, there has been a major focus in diverse sectors on prevention of needlestick and sharps injuries in the health care setting.

Historically, occupational health and safety for health care workers has received little attention. Though increasing attention is being given to the health care sector in this regard, health care institutions and organizations have had little experience in, and few guidelines for, developing effective occupational health and safety prevention programs.

With the recognition that serum hepatitis was caused by a virus, and with the development of techniques to identify serum markers of the hepatitis B virus, it became evident in the 1970s that needlesticks and sharps injuries were serious occupational hazards.[1–2] Early epidemiology studies suggested that recapping might be linked with such injuries.[3] What little interest there was in preventing such injuries focused on behavioral strategies to reduce recapping. With the advent of AIDS, such efforts were dramatically increased[4] and currently are the most popular strategies in the health care industry to reduce such injuries. However, these efforts have not only been nonproductive but are also perceived by health care workers as "blaming the victim."

Engineering Controls

An occupational health and safety approach to reduction of needlesticks and sharps injuries would give priority to the use of engineering controls. In the mid-1980s this concept began to be introduced within some health care sectors, even though the availability of such controls was extremely limited.[5]

Jagger et al.'s landmark article, which analyzed needlestick injuries (NSI) from an injury prevention epidemiology perspective, clearly indicated that such injuries could be largely eliminated by better designed technology; that is, by emphasizing engineering controls rather than behavior as a primary prevention strategy.[6]

Over the six years since publication of this paper, there has been slow but increasing interest in the development and use of needles and sharps with safety

1-56670-083-3/94/$0.00+$.50

features. This interest has grown dramatically in the past two years. Some of the major factors are:

- A small but increasing number of health care workers have occupationally acquired HIV infections.
- There is increasing awareness that a puncture that does not result in HIV conversion nevertheless has a tremendous negative impact on a health care worker's well-being. Such injuries occur in the hundreds of thousands annually.
- There is increasing recognition that behavioral strategies alone have had little impact on NSI.
- There is increasing availability of equipment with safety features.
- The promotion of use of engineering controls has become a national priority for most health care workers' unions.

However, the single most important factor is that the use of engineering controls has been mandated as the first priority for compliance with the OSHA Bloodborne Pathogens Standard.

The present wide interest in the use of engineering controls was clearly evidenced by the over-subscribed conference sponsored by OSHA, FDA and CDC in August 1992.

Participation of Front-Line Health Care Workers

The two major topics addressed at this meeting were the primary role of engineering controls in prevention of NSI and the role of the front-line health care worker in the process of design, development, selection and use of control technology.

In his keynote address, Norman Estrin, former Associate Director for Scientific Affairs of the Health Industry Manufacturing Association, outlined the areas in which health care workers' participation was essential. Defining both health care workers' needs and what industry needs from health care workers, he enumerated health care workers' needs as:

1. Safer devices
2. More opportunities to look at products
3. Input into device design
4. Information about what safer devices are available
5. Greater involvement in purchasing decisions
6. Time to adapt to changes
7. Support by their employers
8. More research devoted to methods for preventing needlesticks, not just on what to do after the injury occurs
9. Rigorous in-service training at all levels and opportunities to interact with manufacturer representatives on new products

He delineated what industry needs from health care workers as:

1. Being open to new technologies and breaking old habits
2. Maximizing input on developing technologies and practices
3. Bringing their vast experience to bear on dealing with this problem

His concluding remarks succinctly stated the importance of involving health care workers in technology development and transfer:

> Let's face it, it is the health care workers who are at the front lines. They have the experience to know what works and what doesn't work. They know which products present more risks than others. We need from them to communicate their views to help develop better products and practices and keep the pressure on all groups to keep this problem as a top priority.[7]

Dr. Estrin's analysis of the need for the health care worker's involvement in the process reinforces the author's own experiences as (1) the director of an employee health service for a major public hospital renowned for its AIDS activities; (2) chair of a pioneer hospital needlestick prevention committee; (3) co-producer of the AHA training module for preventing needlesticks and (4) Director of the Training for Development of Innovative Development Technology, a user-driven collaborative project bringing together product design and industrial hygiene disciplines with front-line health care workers.

A participatory and activated workforce approach to reduction of occupational injury and illness has been promoted since the 1970s. The evidence accumulated over the past several years has demonstrated the effectiveness and vitality of this approach.[8] While it has been limited in the health care sector, an increasing literature addresses the necessity of the employee participatory approach.[9–10] In Washington and Oregon it is mandated that all workplaces, including hospitals, must have joint labor-management health and safety committees. California is moving in this direction.

In our own programmatic activities directed toward the evaluation and recommendations for improved control technology, we have found that the involvement of health care workers has been essential. Their participation in focus groups and as mentors for product design and industrial hygiene students has been essential in defining health care workers' needs, developing criteria for equipment to be used and developing evaluation forms that could be used by health care workers to assess current technology.[11]

Barriers to Prevention of Needlestick Injuries

Over the course of work for the past three years, we have identified a series of barriers to the use of control technology to prevent NSI.

- Manufacturers originally had not been at the forefront of developing new designs. Most recently, more safety products have appeared on the market.
- To a large degree, existent controls are still in early stages of development, and the failure of certain products to perform adequately has discouraged exploration on the part of health care institutions for other equipment.
- At times, key safety features of medical devices have been misapplied in product development.
- Almost all products reviewed could use better design.
- There is restricted user input during the development of new products.
- Manufacturers' methods for testing their equipment are varied and unspecified.
- Performance standards for such equipment currently do not exist.
- Cost of new products with safety features is generally higher than standard products, therefore discouraging widespread use.
- New equipment evaluation is costly for the individual institutions and creates a barrier to the introduction of new types of equipment.
- The process for selection of equipment varies from institution to institution. Some have product evaluation committees; sometimes the materials managers make the decision; increasingly the decisions are made by a buying consortium. The role of line users in the selection process is generally very limited.
- Costs of pilot testing and training mitigate against the introduction of new equipment.
- Quality assurance in regard to patient outcomes is of the highest concern, yet there are rarely evaluation tools for this use.

These areas are specific to the role of control technology in the prevention of NSI. To have an effective prevention program, it is important that the use of engineering controls (safer devices) is viewed to be part of a comprehensive NSI prevention program. It is also essential that this program, in turn, be part of a comprehensive occupational health and safety program so that recommendations for NSI prevention are not at cross-purposes with other required OHS measures. An example of this would be the failure to acknowledge the specialized handling of chemotherapy agents when recommending sharp boxes for needle disposal, or the failure to recognize the need of respiratory equipment when recommending certain procedures for spills that involve blood or body substances.

RECOMMENDATIONS FOR A COMPREHENSIVE PARTICIPATORY PROGRAM

Involving line health care workers in all phases of control technology to prevent exposure to bloodborne pathogens provides the means of:

- Tapping the expertise of these workers
- Assuring that product design is user-friendly and truly effective
- Developing systems that improve compliance
- Improving employee morale
- Improving patient care

The following pages outline a series of recommendations to accomplish this objective. Group I, which can be seen as structural, involves creation of committees that either (A) specifically mandate NSI prevention or (B) impact on prevention strategies. Group II deals with the process of implementing specific actions or programs.

At the end of these is a list of resources that should assist in implementing recommended actions. In addition, there is an appendix of additional materials.

GROUP I RECOMMENDATIONS

Needlestick Prevention Committee

Increasingly, health care institutions are developing specific committees or task forces to develop programs to reduce needlestick injuries. Such groups are crucial in focusing on this problem and integrating activities that must be instituted system wide in order to be effective.

The needlestick prevention committee:

- May or may not be a subcommittee of the health and safety committee
- Should be composed of 50 percent direct care health care workers and 50 percent supervisors and/or administrators
- Should have a charge from the CEO and should have access to administrative authority
- Should be guaranteed that all reports and recommendations will have a prompt administrative response
- Degree of authority should be clearly defined; should not be just an advisory committee
- Should review all exposure control plans relating to needlestick prevention
- Should review all data on NSI on a frequent and regular basis
- Should work closely with the
  — Infection Control Committee
  — Health and Safety Committee
  — Product Evaluation Committee
  — Materials Management
  — Employee Health Service
- Should develop a mechanism to involve broad groups of health care workers in NSI prevention programs
- Should have oversight responsibility for a comprehensive program, including:
  — defining of problems
  — ongoing surveillance at the worksites
  — developing strategies for improved NSI reporting
  — developing surveillance systems to monitor needlestick injuries
  — obtaining and disseminating information on new devices
  — assuring health care workers' input into product selection
  — assuring appropriate pilot testing of new devices
  — in-service training for introduction of new products
  — exposure treatment program
  — exposure control plan as mandated by OSHA

Health and Safety Committee

Ideally the needlestick prevention committee should be a subcommittee of a joint labor-management health and safety committee. If this is not the process in an individual hospital, there should be representation from the needlestick prevention committee on the health and safety committee. Representatives should include line health care workers.

The health and safety committee should:

- Review the needlestick prevention committee's activities on a regular basis
- Forward recommendations to appropriate administrators
- Assure that there is prompt response to proposed activities
- Monitor consistency of NSI prevention activities with other recommended OHS programs
- Assist the needlestick prevention committee in developing broad-based focus groups
- Have input and review of exposure control plan as mandated by OSHA

Infection Control Committee

The infection control committee, in all institutions, has a critical role to play in a needlestick prevention program. In many institutions the infection control practitioner will be charged with developing the OSHA-mandated exposure control plan. The infection control committee should:

- Work closely with the needlestick prevention committee
- Review all devices being considered for purchase for potential of nosocomial infections
- Develop a plan for monitoring patient outcomes when control technologies are introduced
- Seek input via the worksite and needlestick prevention committees to the development and implementation of the exposure control plan
- Review on a regular basis all data on NSI

Product Evaluation Committee

Many institutions have product evaluation committees. Where they exist, they play a key role in identifying available medical devices and making recommendations for equipment purchase. For advancing the use of control technologies for prevention of NSI, product evaluation committees should:

- Actively identify all products available for purchase
- Develop methods for informing all the health care workers in the institution about these products
- Develop systematic methods for evaluating these products

- Develop a program and schedule so that line health care workers are involved in all evaluation processes
- Develop a written plan for the above activities
- Report regularly to the needlestick prevention committee

Materials Management and Purchasing Departments

It is crucial that the materials managers are aware of the need for safety devices to prevent NSI and are committed to identifying and purchasing such equipment. Selection of such products should be predicated on health care worker selection and preference.

In many instances, purchasing may be done by a consortium outside the institution. Such arrangements may result in cost savings; however, it is crucial that such purchases do not provide inferior or inappropriate products for a specific area.

Public institutions may be obligated to use a bidding process. Though this occasionally has been an excuse for not providing control technology, development of appropriate specificities should assure purchase of appropriate equipment.

Materials Management and Purchasing departments should:

- Develop educational programs on control technology for their staff
- Interact with the needlestick prevention committee
- Develop specificities for bids
- Assure that safety devices and protective equipment are available at all times

Employee Health Service

The Employee Health Service provides both preventative and post-exposure services. While the services they render must guarantee confidentiality, it is critical that health care workers be involved in the development of these services and the oversight of activities to assure that (1) they are client-oriented; (2) all employees are aware of and have confidence in them and (3) services are fully utilized when appropriate.

Where a designated employee health service does not exist, the providers of these services should be clearly identified.

All providers of clinical services for hepatitis B vaccination and post-exposure treatment and counseling for NSI should:

- Interact with the needlestick prevention committee
- Understand the principles of control technology
- Develop an injury reporting form that provides broad insights to NSI
- Review monitoring programs for NSI with the needlestick prevention committee
- Provide regular updates on monitoring programs to appropriate committees and worksites

- Assure that all such data be in formats that are understandable to all workers
- Develop a program for improving NSI reporting
- Develop a vaccination program that assures broad utilization
- Develop a post-vaccination program to assure that boosters are given in accordance with appropriate recommendations
- Assure that services are available at all times to injured employees

Administration

The Chief Executive Officer is responsible for the facility's compliance with the institution's exposure control plan mandated by OSHA's Bloodborne Pathogens Standard. Full commitment to a comprehensive needlestick prevention program should also include:

- Regular review of all programs related to NSI prevention and treatment
- Prompt response to all recommendations and proposals from the appropriate committee
- Assurance that broad groups of health care workers are involved in these activities
- Assurance that participation in these activities is part of health care workers' regular activities and that they are compensated accordingly
- Resources for assuring that control technology and other protective equipment are available at all times

GROUP II RECOMMENDATIONS

Group I recommendations have described structures and activities within those structures that would promote a comprehensive needlesticks and sharps injury prevention program. The following section outlines formats and activities that are necessary for allowing maximum worker participation in a comprehensive needlestick prevention program that emphasizes the use of control technology.

Worksite Committee

Each worksite should have a worksite committee. The committee should:

- Represent all classifications of workers in that site
- Allow for participation as part of a work assignment, with allotted paid time
- Meet on a regular basis
- Keep records of time and content of meetings
- Develop an action plan for the specific worksite
- Review on a regular basis all exposures and near misses in the worksite (confidentiality must be assured)
- Report to the needlestick prevention committee on a regular basis

Focus Groups

Focus groups are an excellent means of tapping worker expertise and identifying problems related to worker health and safety. Focus groups in specific worksites provide a means for involving a broad group of workers in the development of a worksite action plan. A successful focus group:

- Allots paid time for participation
- Guarantees release time for participants (This is very important in an environment where short staffing is frequently the norm.)
- Has a facilitator
- Encourages "brainstorming"
- Identifies the characteristics of the ideal device

The facilitator is crucial for the success of focus group activities. The facilitator should:

- Not be a member of the worksite
- Have the respect of the workers
- Be non-judgmental
- Be familiar with the work and occupational health and safety hazards in the specific worksite. Prior to the focus group, the facilitator should:
  - Do a walkaround of the site with members of the site staff and hospital health and safety staff
  - Review all health and safety regulations pertinent to that worksite
  - Review all accidents and injuries in that site for the preceding two years

Needlestick Injury Statistics

It is essential that statistical data be kept in order to identify problems and monitor programs established to eliminate these problems. Activities to enhance reporting should include:

- A firm commitment to confidentiality in all reporting and review by oversight committees
- Developing a survey, in coordination with the needlestick prevention and worksite committees, to ascertain the percentage of unreported injuries and barriers to reporting
- Developing a plan for improving reporting
- Doing follow-up surveys to ascertain whether this plan is effective
- Review of data, done in such a way as to assure confidentiality
- Development of a comprehensive reporting form that includes:
  - Job classification
  - Nature of accident
  - Devices used
  - Equipment availability
  - Staffing
  - Shift/hours worked

  - — Workload
  - — Development of method for reporting near misses
- Compilation of data on an ongoing statistical basis in a format that can be easily understood by most line health care workers
- Regular report-back to the needlestick prevention and worksite committees

Product Identification

Identification of safety feature products available on the market has proven to be difficult. At the present time there is no centralized resource for this. The Service Employees International Union (SEIU) has, at the present time, the most comprehensive list of products available. This list is in their updated fact pack.

Once products are identified, several samples should be requested. The Training for Development of Innovative Control Technology (TDICT) product and Kaiser Permanente Medical Center, San Francisco have developed boards for displaying all products in various categories. These boards are easily transported to the worksite. They are extremely useful in informing line health care workers about what equipment is available. It is important that *all products*, regardless of efficacy, are displayed so that health care workers can review the entire range of products.

Product Evaluation

Product evaluation should involve health care workers in targeted areas where the product will be used.

At the present time, the Trainer for Development of Innovative Control Technology Project has published product safety evaluation sheets for five categories of equipment.[11]

These evaluation criteria pertain only to the safety features. Other criteria should be developed by the worksite and needlestick prevention committees in conjunction with other appropriate committees. These may pertain to clinical appropriateness, storage, etc.

Product evaluation is a time-consuming effort. Release work time should be guaranteed for participants. A product should be evaluated in a setting where clinical procedures can be evaluated using appropriate models. A resource coordinator who is familiar with the products and the evaluations forms should be involved throughout the process.

Manufacturer's representatives *should not be present* during these evaluations.

Pilot Testing of Products

Prior to pilot testing, a written evaluation should be developed cooperatively by the designated coordinator and the users. The form should be at the language level of the majority of the users. Goals should be set regarding the percent of responders and worksite strategies developed to assure that goal.

The results should be tabulated and reviewed with the target group prior to review by the needlestick prevention committee and other appropriate groups.

Device inadequacies should be recorded, even if the device is accepted, and should be shared with the manufacturer as well as with other hospital groups.

In-Service Education

Introduction and implementation of control technology facility-wide involves:

- Adequate in-service training
  - Coordination with manufacturer and education staff of facility
  - Review of manufacturer's written program
  - Assurance that manufacturer's representatives have clinical experience
  - Development of performance testing
  - Allowance for extended training for workers who fail initial test or are uncomfortable with the device
  - Followup within a specified time
- Monitoring of surveillance data regarding needlestick injuries before and after introduction of control technology.

Product Failure and Inadequacies

Current available control technologies to prevent NSI are in early stages of development. Rarely do the ideal devices exist. However, the use of appropriate control technology should not be deferred until the ideal device is available.

It is imperative, however, that the manufacturers understand the inadequacies of their products. At the present time, it is not clear how widely products are pretested before they reach the market. Therefore, the findings of worksite and needlestick prevention committees become important resources for better understanding the inadequacies of current designs and the needs of the health care workers.

Currently there is no centralized resource for such findings. At the present time we recommend that the findings be sent to the individual manufacturer and to the FDA. The Training for Development of Innovative Control Technology would welcome these findings to assist in updating and developing new product evaluation forms.

GOAL SETTING

It is important to establish goals for all the groups involved in the NSI Prevention Program. In keeping with the concept of the participatory model, each committee or activity should establish its own goals. However, it is imperative to have an overall annual goal for NSI reduction. Setting this goal should be a priority for the Needlestick Prevention Committee, once it has reviewed the baseline injury data. Recognizing that NSI reporting should initially increase as a result of strategies to encourage reporting, the first year's goal might be set to begin several months after the program is initiated. The goal should be realistic but high enough to stimulate concerted efforts for significant reduction. As devices are introduced, individual goals for each device can be set based on other hospitals' experiences. EPINet might prove to be a valuable resource for this.[12]

The CEO should review accomplishments with the Needlestick Prevention Committee on a bi-annual basis. Failure to meet the overall goal for NSI reduction should be regarded as grave, and all activities in this regard should receive in-depth scrutiny. A written plan for remedial action should be implemented promptly.

ACCOUNTABILITY

The success of a comprehensive NSI Prevention Program is dependent on the efforts of a broad group of people, involving both labor and management from many different departments. For both groups there needs to be a commitment to the project and a willingness to work cooperatively. The former may be easier to achieve. To work cooperatively within a participatory model means that the joint committees need to be empowered to make changes and be given the resources to do this. They need true cooperation from management groups, who may feel that their authority is being usurped, and from labor representatives, who may be more used to working in an adversarial mode.

All hospital groups involved must understand that they are not only accountable to the CEO but to the entire population at risk. While it is imperative that the CEO receive regular reports from all sectors of the NSI Prevention Program, it is as imperative that all hospital employees receive information about the program on an ongoing basis and that each and every employee have a means for providing input into the program.

To accomplish this, every group involved in the NSI Prevention Program should:

(1) Have written goals
(2) Review these goals at regular intervals and provide remedial plans when necessary
(3) Maintain written records of activities
(4) Provide all written and statistical material in a manner that is understandable by most employees
(5) Make all records available to employees when requested

The above should be provided to the NSI Prevention Committee on a regular basis. The reports should then be edited for clarity, reviewed by the committee and distributed to all employees. This comprehensive review should not be a public relations document. It should include problems and proposed remedial actions, as well as accomplishments.

Such recordkeeping should not be viewed as an additional burden, but as an important means for providing accountability to the population at risk—the front line health care workers.

REFERENCES

1. Pattison, C.P. et al. 1975. Epidemiology of hepatitis B in hospital personnel. *Am. J. Epidemiol.* 101:59–64.
2. Dienstag, J.L., Ryan, D.M. 1982. Occupational exposure to hepatitis B virus in hospital personnel: Infection or immunization? *Am. J. Epidemiol.* 115:26–39.
3. McCormick, R.D., Maki, D.G. 1981. Epidemiology of needle stick injuries in hospital personnel. *Am. J. Med.* 70:928–932.
4. *Mortality and Morbidity Weekly Report* 1987. Recommendations for Prevention of HIV Transmission in Health Care Settings. *MMWR* 36:2S:1S–18S.
5. San Francisco General Hospital Health and Safety Committee 1987. Recommendations for Reducing Needlestick Incidents/Body Fluid Exposures at San Francisco General Hospital. Needlestick Committee Report.
6. Jagger, J. et al. 1988. Rates of needle stick injuries caused by various devices in a university hospital. *New Engl. J. Med.* 319:284–288.
7. Estrin, N.F. 1992. Market changes that affect changes and introduction of new technology. Keynote address, at Frontline Healthcare Workers: A National Conference on Prevention of Device-Mediated Bloodborne Infections. Washington, DC. Aug. 17–19, 1992.
8. Deutsch, Steven 1992. New Technologies and the Work Environment: Challenge for Hospital and Health Care Industry, in *Technical Change in the Workplace: Health Impacts for Workers.* Edited by M. Brown and J. Froines. UCLA: Center for Occupational and Environmental Health. In press.
9. Buback, K.A. and Grant, M.K. eds. 1985. *Quality of Work Life: Health Care Applications.* St. Louis: Catholic Health Association of the U.S.
10. American Hospital Association 1989. *Working Together: Needlestick Prevention.* Chicago, AHA.
11. Fisher, J.M., and Schmitz, J. 1992. *Safety Evaluation Forms: Criteria for Evaluating Medical Devices Safety Features to Prevent Exposure to Bloodborne Pathogens. Implementing Safety Medical Devices.* American Hospital Association, Chicago, 1992.
12. Jagger, J., Cohen, M., and Blackwell, Beth 1994. EPINet: A Tool for Surveillance and Prevention of Blood Exposures in Health Care Settings. *Essentials of Modern Hospital Safety,* Volume III. Edited by William Charney. Lewis Publishers, Boca Raton, FL, 1994.

SELECTED RESOURCES

General Hospital Health and Safety

1. Charney, W. and Schirmer, J., eds. 1992. *Essentials of Modern Hospital Safety,* Vols. 1 and 3. Michigan: Lewis Publishers.
2. National Institute for Occupational Safety and Health 1988. *Guidelines for Protecting the Safety and Health of Health Care Workers.* DHHS (NIOSH) Publication No. 88-119.

Health and Safety Inspection of Hospital Worksites (Walkarounds)

1. Charney, William. Checklists for Walk Through Surveys. Prepared for the American Hospital Association.
2. Bracker, A.L. 1991. *Hospital Hazards: A Union Guide to Inspecting the Workplace.* Berkeley: Labor Occupational Health Program, University of California, Berkeley.

Needlestick Prevention and Engineering Controls

1. American Hospital Association 1989. *Working Together: Needlestick Prevention.* Chicago: AHA. (This is a training kit that includes two videos and a training manual.)
2. Service Employees International Union 1992. Needlestick Prevention Fact Pack (2nd Edition). Washington: SEIU.
3. Needlestick Prevention Devices: Special Report and Product Review 1991. In *Health Devices* 20:154–180.

Worker Participation Programs

1. Karasek, R. and Theorell, T. 1990. *Healthy Work: Stress, Productivity and the Reconstructing of Working Life.* New York: Basic Books.
2. Labor Studies Center 1991. Participatory Approaches to Improving Workplace Health Presentation Summaries to an International Conference. Ann Arbor: University of Michigan Labor Studies Center.

Focus Groups and Worker Participatory Training

1. Basch, C.E. 1987. Focus Group Interviews: Underutilized Research Techniques for Improving Theory and Practice in Health Education. *Health Education Quarterly* 14:411–448.
2. Mergler, D. 1987. Worker Participation in Occupational Health Research: Theory and Practice. *International Journal of Health Service* 17:151–167.
3. American Hospital Association 1989. *Working Together: Needlestick Prevention Training Manual.* (See Appendix.)
4. Weinger, M. and Wallerstein, N. 1992. Education for Auction: An Innovative Approach to Training Hospital Employees. In *Essentials of Modern Hospital Safety, Vol. 1.* Michigan: Lewis Publishers.

Supported in part by CDC contract 9186796, Dr. Murray L. Cohen, Project Officer

CHAPTER 6

Electrocautery Smoke: Reasons for Scavenging

This section concentrates on the health and safety criteria for plume control during laser and electrocautery surgery. Many articles recommend scavenging during these procedures.[1-5] Presented in this section are: 1. The Importance of Understanding Surgical Smoke; 2. Electrocautery Health and Safety; 3. Health Hazard Evaluation Report; 4. Laser Hazards in the Health Care Industry; 5. Proposed Recommended Practices; and 6. Laser Plume Quantification. For recommended standards and regulations the reader is referred to the American National Standards Institute (ANSI Z136.3) and OSHA, respectively.

PART 1

The Importance of Understanding Surgical Smoke

Stackhouse, Inc.

The advent of laser technology in surgery raised concerns over the impact of smoke on the health of patients and the surgical staff.

In the early days of laser surgery, the primary motivation for evacuating smoke was the acrid odor that was present during most procedures.[1]

Over time, other concerns emerged regarding the inhalation of vaporized human tissue. These concerns were shared by leading safety organizations and health care workers alike and resulted in a great deal of research into the hazards of all forms of surgical smoke, including what is generated during electrosurgery.

Understanding the dangers of surgical smoke begins with an overview of the gases and particulate matter that are generated when human tissue comes in contact with laser and electrosurgery heat sources.

THE KNOWN CONTENT OF SURGICAL SMOKE

New evidence suggests there is little difference between the smoke generated from electrosurgery and smoke from lasers. Each contains carbonized tissue,

1-56670-083-3/94/$0.00+$.50

blood and virus.[1–10] In addition, surgical smoke contains the gases benzene, toluene, formaldehyde and polycyclic aromatic hydrocarbons; these are known carcinogens.[11–14] These gases create the acrid smell, but the real danger from smoke comes from the particle content of the smoke.

Knowledge that each form of particulate matter and gases present in surgical smoke are known to pose health risks, a number or researchers set out to gain a better understanding of the specific health risks to surgical teams.

CURRENT RESEARCH ON SURGICAL SMOKE

One of the first and most comprehensive animal studies on the physiological effects of surgical smoke was conducted by Dr. Michael Baggish, 1992 president of the American Society of Lasers in Medicine and Surgery and currently the chairman of the Obstetrics and Gynecology Department at Ravenswood Hospital in Chicago, Illinois.[1] His objective was twofold:

- Identify any physiologic risks associated with the inhalation of surgical smoke.
- Determine the effectiveness of commercially available evacuation equipment in protecting rats from these effects.

Dr. Baggish vaporized pig skin with a CO_2 laser to create surgical smoke. He found that accumulated particulate matter on the rat's lung tissue caused interstitial pneumonia, bronchitis and emphysema. The severity of disease increased in proportion to increased exposure to smoke.

Dr. Baggish verified the value of protection from smoke one year later by studying the value of smoke evacuation.[5] Again, vaporizing pig skin, this time Dr. Baggish divided the rats into two groups. The control group experienced the same accumulated particulate matter on their lungs as in the first study. However, rats receiving air filtered through a smoke evacuator equipped with an ULPA filter experienced no pathological changes over the course of the study.

In the 1987 Garden and O'Banion found living virus (human papilloma) in a CO_2 laser smoke.[3] In 1989 Sawchuk and colleagues found viral DNA in the vapor of warts.[10] These findings have led to current concerns that the thermal effects of laser and electrosurgery may not be counted on to destroy all viable DNA from viruses.

In 1988, Dr. Barry Wenig found the conclusions of Baggish regarding the damaging effects of smoke on lung tissue applied also to the use of electrosurgical units and the Nd:YAG laser. Wenig also recommended the use of an efficient smoke evacuator in all procedures that generate surgical smoke.

Further studies by Baggish have shown the presence of HIV DNA captive in the suction tubing of the evacuator used to suction the surgical smoke while lasing 10 mL of HIV-infected cells.[7] This further substantiates the importance of proper smoke evacuation.

There have been no documented cases of health care workers being infected with these viruses by means of surgical smoke. Considerably more research needs to be done in this area clearly defining the risk of airborne pathogens. However, preliminary indications have captured the attention of regulatory agencies charged with setting forth policies that assure health work environments.

STANDARDS AND RECOMMENDATIONS

The American National Standards Institute (ANSI) is responsible for setting safety standards for health care workers while on the job. ANSI has compiled evidence that airborne contaminants during class 4 laser surgery, (laser procedures that use a power setting of 0.5 watts or greater) can cause lacrimation, nausea, vomiting and abdominal cramping. As a result, ANSI has recommended that all class 4 laser procedures producing smoke must have it removed by localized exhaust ventilation.[16]

The National Institute of Occupational Safety and Health (NIOSH), has made recommendations at specific hospitals employing either electrosurgical equipment or lasers. Based on the mutagenicity of compounds NIOSH collected at these facilities they recommended that smoke evacuation units be used to reduce the potential for chronic health effects.[13–14]

The Emergency Care Research Institute (ECRI) and independent medical device test organization has issued statements that the treatment of electrosurgery smoke possesses the same danger as laser smoke.[15] The Food and Drug Administration (FDA) has actively challenged current acceptance of smoke in the surgical suite during any procedure.

The position of these regulatory agencies make it clear that surgical smoke should be evacuated, not only for the safety of the surgical staff, but for the safety of the patient as well.

Quoted from the “Hazards of Surgical Smoke” video study guide by Stackhouse, Inc.

BIBLIOGRAPHY

1. Baggish MS, Elbakry M: The effects of laser smoke on the lungs of rats. *Amer J Ob/Gyn* 5/87; Vol. 156, No. 5: 1260–1265.
2. Intact viruses in CO_2 laser plumes spur safety concern. *Clinical Laser Monthly* 9/87; Vol. 5, No. 9:101–103.
3. Garden JM, O'Banion MK, Shelnitz LS, Pinski KS, Bakus AD, Reichmann ME, Sundberg JP: Papillomavirus in the vapor of carbon dioxide laser-treated verrucae. *JAMA* 2/26/88; Vol. 259, No. 8:1199–1202.
4. Carbon dioxide laser irradiation of bacterial targets *in vitro*. *J Hosp Infection* 6/86.
5. Baggish MS, Baltoyannis P, Sze E: Protection of the rat lung from the harmful effects of laser smoke. *Laser Surg Med* 1988; 8:248–253.

6. Carlson SE, Schwartz DE, Wentzel JM: Properties of aerosolized blood produced during a dermabrasion medical procedure. *Amer Ind Hyg Conf,* St. Louis, MO; 5/19/89.
7. Baggish MS, Poiesz BJ, Joret D, Wiliamson P, Refai A: Presence of human immunodeficiency virus DNA in laser smoke. *Laser Surg Med* 1991; 11: 197–203.
8. Research confirms earlier study on plume hazard: viral contaminants. *Clinical Laser Monthly* 4/89.
9. Some problems about condensate induced by CO_2 laser irradiation. *Laser Safety* 6/85.
10. Sawchuk WS, Wedber PJ, Lowy Dr, Dzubow LM: Infectious papillomavirus in the vapor of warts treated with carbon dioxide laser or electrocoagulation: detection and protection. *J Amer Acad Derm* 7/89; Vol. 21, No 1:44–49.
11. Kokosa JM, Doyle DJ: Chemical composition of laser-tissue interaction smoke plume. *J Laser Appl* 7/89; Vol 1, No 3:59–63.
12. Kokosa JM, Doyle DJ: Chemical by-products produced by CO_2 and Nd:YAG laser interaction with tissue. SPIE 1988; 908:51–53.
13. Bryant C, Gorman R, Stewart J, Whong WZ: Health Hazard Report HETA 85-126-1932, Bryn Mawr Hosp, Bryn Mawr, Pennsylvania; 9/88.
14. Moss CE, Bryant C, Stewart J, Whong WZ, Fleeger A, Gunter BJ: Health Hazard Report HETA 88-101-2008, University of Utah Health Sciences Center, Salt Lake City, Utah; 1/90.
15. Health Devices, ECRI; Vol. 19, No. 1, 1/90.
16. American National Standard for the safe use of lasers in health care facilities; ANSI Z136.3-1988.
17. Baumann N: The plume hazard: ICALEO panel calls for more suction. *Laser Med & Surg News & Adv* 2/89; 16–18.
18. AORN Standards and Recommended Practices for Perioperative Nursing, 1990.
19. OSHA cites surgicenter for "serious" safety violations. *Clinical Laser Monthly* 5/90: 69–70.
20. Newman Dorland WA: The American Illustrated Medical Dictionary 21st Edition; 1947.
21. Neufeldt V, Guralnik DB: Webster's New World Dictionary of American English, 3rd College Edition; 1988.

PART 2

Electrocautery Health and Safety

Jacob D. Paz

HISTORY OF ELECTROCAUTERY

Bovie[1] in 1928 discovered that high-frequency alternating current in the range of 250,000 to 2,000,000 Hz could be used to incise coagulated tissue to obtain homeostasis. This technique was first popularized by Cushing[2] in neurosurgery and was subsequently used in other types of surgery. Heat achieves homeostasis by denaturation of protein, which results in coagulation of large areas of tissue. With actual cautery, heat is transmitted from the instrument by conduction directly to the tissue; with electrocautery, heating occurs by induction from an alternating current source.

When electrocautery is employed, the amplitude setting should be high enough to produce prompt coagulation, but not so high as to set up an arc between the tissue and the cautery tip. Strict control prevents burns outside the operative field and the exit of current through electrocardiographic leads and other monitoring devices.

A negative plate should be placed beneath the patient whenever cautery is used so that severe skin burns do not occur. The advantage of cautery is that it saves time; its disadvantage is that more tissue is killed than with precise ligature. In the past certain anesthetic agents, such as cyclopropane, could not be used with electrocautery because of the hazard of explosion.[3]

TYPES OF ELECTROCAUTERY

Currently, two types of electrocautery devices are in use, unipolar and bipolar.

The unipolar electrosurgical unit is used both for surgical dissection and homeostasis. When undampened high-frequency electrical current is passed through tissue, the active electrode functions as a bloodless knife, and the cells at the edges of the wound disintegrate. A mild thermal injury occurs away from the plane of cutting, and blood vessels thrombose. When the oscillations are dampened, homeostasis is accomplished without cutting. During this procedure, the cells rapidly dehydrate, the vessels within the tissue coagulate and damage to adjacent tissue may be extensive. The precise tip of the divided vessel is all that requires coagulation, however, and the power of the unit should be set at the lowest level possible.

The bipolar electrocautery unit confines the damage to tissues between the tips of the cauterizing forceps. Notably, the bipolar instrument can be used in a wet environment, and it is indicated to control bleeding in microvascular and microneural surgery.

1-56670-083-3/94/$0.00+$.50

ELECTROCAUTERY HEALTH HAZARDS

Explosion, burns, cardiac effects and muscle excitation are the main health hazards often associated with electrosurgical units (ESUs), are well-documented in the literature and are discussed elsewhere.[4] To this list we can now add electromagnetic radiation, smoke and bioaerosol hazards, which are discussed here.

Electromagnetic Radiation

Microwave (MW) and radiofrequency (RF) radiation are electromagnetic energies in the form of waves that travel at the speed of light; RF ranges from 100 KHz to 300 MHz and MW from 300 MHz to 300 GHz. They are classified as nonionizing radiation because the energy of these photons is relatively low. Only limited information on exposure to electric or magnetic fields generated by ESU has been reported in the literature.

Adverse biological effects may occur from the heating of deep body tissues by MW and RF radiation. Such heating may produce damaging cell alterations, and other non-thermal effects, such as neurological, behavioral and immunological changes resulting in leukemia, cataract, mood swings and dizziness.[5–6]

Effects due to heating are well-documented in animals, but the evidence is incomplete and disputable for those effects occurring without an accompanying increase in tissue temperature. Thermal effects occur in direct proportion to the field strength or power density. When the amount of heat generated from the absorbed energy is too great to be released into the surrounding environment, the temperature of the body gradually increases and can lead to heat stress.

Reports of human effects cover a series of clinical and epidemiological investigations into the association between RF radiation and damage to the eyes, central nervous system and reproductive capability. Only a few studies have assessed RF/MW exposure levels to medical personnel. The OSHA recommended standard for exposure to microwaves is 10 mW/cm^2. Both ANSI and ACGIH have published guidelines for occupational exposure to electromagnetic radiation.[7–8]

Any area where RF/MW radiation exposure exceeds permissible levels should be considered potentially hazardous. The area should be clearly identified, and warning signs posted. Interlocks may be used to prevent unauthorized entry. Basic protective measures include the provision of shields or absorbing enclosures for equipment. Personal protective equipment may be used (e.g., gonad shields, protective suits, and wire-netting helmets). Although protective goggles have been developed, they may not provide sufficient protection. Implementing such precautions, however, would be difficult in the operating room.

Carpenter,[9] Birenbaum[10] and Gay et al.,[11] have studied the relationships of MW to subsequent developments of ocular cataracts. The severity of ocular damage induced depends upon MW intensity, wavelength and duration of exposure. Cataractogenesis required a power density above 100 mW/cm^2 and localized ocular hypothermia of 41°C. In addition to power density, time and frequency are important factors with respect to lens changes. Not only the penetration depth of energy, but also the specific absorption rate is distribute frequency dependent. It appears that a critical range of frequency extended from 0.8 to 10 GHz. No lens changes were detected at frequencies of 386,486 MHz or at 35 and 107 GHz.

Fox et al.,[12] were one of the first groups to measure the frequency spectrum and power density of ESU. They reported that the spark gap ESU emission extended up to 1 GHz and maximum energies concentrated below 100 MHz with a peak of 2.4 MHz. They reported that the ESU power density was 150 mW/cm^2 (which exceeds the present OSHA standard of 10 mW/cm^2 for 6 minutes of exposure), and stated that "...the surgeon and patient are most exposed because the active electrode is manipulated by the surgeon and the radiating lead is draped over the patient." Urologists doing transurethral resections are heavily exposed since the radiating active lead enters near the eye. We do not have sufficient information about the new ESU.

A literature review indicates that ESUs are operated at a frequency of 650 Hz.[23] However, in many cases spectrum energy of ESU could not be found in the ESU manual as well as in the professional literature. Electric field (E) strength measurements of ESU by Ruggera and Segerson[14] in 1977 found that the H field was 1000 mW/cm^2 at 16 cm from the ESU source.

Paz et al.,[5] in 1987 conducted a simulation study, to evaluate electric and magnetic field strength generated by a monocular ESU. Test results are listed in Tables 1 and 2. Experimental data indicated that the surgeons may expose themselves to high levels of electric and magnetic fields exceeding ACGIH threshold values. Ocular exposures were especially high: 20 cm from the active lead at the eye/forehead position the E field ranged from 4.0×10^4 to 9.0×10^6 V^2/m^2 and the magnetic field was 3.5 A^2/m^2 in the coagulation mode, compared to 0.01 A^2/m^2 in the cutting mode.

A real time survey of E fields was conducted during surgery by Paz et al.[14] Experimental data illustrated in Table 3 showed that the electric field strength in the cutting mode ranged from 1.0×10^3 to 1.0×10^5 V^2/m^2. By comparison, a higher electric field value was noticed in the coagulation mode to the extent of 1.0×10^3 to 7.0×10^7 with a peak of $>10 \times 10^7$ V^2/m^2. Electric field strength (mean values) measurements both in cutting mode and coagulation mode exceeded both ACGIH and ANSI standards.

Electrocautery Smoke

In 1976 Goldstein and Paz[15] reported increasing levels of NO_2 where high energy devices were used. The average NO_2 level in the operating room during 11 procedures in which ESU and X-ray were used was 0.114 ppm, and in the control area was 0.57 ppm. Area monitoring for NO_2 by a chemiluminescent probe located about 3 m on a perpendicular line from the middle of the operating room table showed an NO concentration of about 0.6 ppm and NO_2 levels of 0.14 ppm. High levels of >5.0 ppm NO_2, NO and NO_x were recorded sparking ESU

Table 1. Electric field strength V^2/m^2: bipolar electrosurgical unit simulation studies.

Body Organ	Coagulation Mode V^2/m^2	Cutting Mode V^2/m^2
Eye/forehead	9.0×10^6	2.0×10^4
Neck	1.0×10^6	3.0×10^4
Chest	1.0×10^5	3.0×10^4
Upper arms	1.0×10^5	5.0×10^4
Lower arms	5.0×10^4	4.0×10^4
Waist	1.0×10^4	4.0×10^4
Gonads	1.0×10^4	4.0×10^4

Table 2. Magnetic field strength A^2/m^2 bipolar electrosurgical unit simulation studies.

Body Organ	Coagulation Mode A^2/m^2	Cutting Mode A^2/m^2
Eye/forehead	3.50	0.01
Neck	0.50	0.01
Chest	0.04	0.06
Upper arms	0.05	0.04
Lower arms	0.05	0.05
Waist	0.04	0.03
Gonads	0.01	0.02

Table 3. Real time monitoring of electric field strength V^2/m^2 bipolar electrosurgical unit.

Operation Type	Low Value V^2/m^2	Peak Value V^2/m^2	Mean Value A^2/m^2
Cutting mode	1.0×10^5	7.0×10^7	1.0×10^6
Cutting mode	1.0×10^3	5.0×10^5	2.9×10^5

in a stream of air, and N_2O was measured using a chemiluminescence monitor. They hypothesized that high-energy devices may modify operating room atmosphere and may lead to the formation of other new toxic byproducts.

Paz et al.[17] studied the effect of ESU sparking on N_2O, air and halothane in an environmental chamber. Experimental results showed an increase in values which reached levels of about 40 ppm of NO_2, measured by use of Miran 1A. Sparking halothane mixture in air caused an increase in 8.8 microns infrared (IR) wavelength, indicating that new material was being produced. One explanation is that the new IR peak on the spectrogram indicated the formation of halothane degradation products.

The experimental data of Bosterling and Trudell[18] using gas chromatography-mass spectrometry (GC-MS) demonstrated that ultraviolet (UV) irradiation of halothane in air is capable of producing free radicals to form new toxic byproducts. This work confirmed the earlier hypothesis of Paz et al.[16–17]

A recent study by Gatti et al.[18] revealed electrocautery smoke produced during breast surgery contained unidentified organic compounds undetected by current analytical techniques. Using the Ames test, Gatti reported that these compounds, found in air samples from an operating room, were mutagens.

Electrocautery and Bioaerosol Hazards

The AIDS epidemic has focused attention on the routes by which HIV virus may be transmitted. One potential exposure route is inhalation of blood-containing aerosols infected with the virus in the operating room. Early studies reported that common surgical power tools produced blood-containing aerosols composed of particles in the <5 mm size range.[18] Johnson and Robinson[19] reported that some of the aerosols found to be infected by HIV-infected blood, which also were generated by a surgical tool, had the ability to infect T-cell tissue culture.

In more recent studies, Jewett et al.[20] studied the potential hazard of blood aerosol generated by a variety of power tools, such as the Hall drill, Shea drill and ESU operating in cutting and coagulation modes during surgery. The experimental data showed that surgical tools capable of generating a wide distribution of particle sizes produced blood-containing particles in the respirable range. Surgical masks do not provide adequate respiratory protection against these aerosols. The surgical mask does not provide effective protection for removal of bioaerosol particles, the authors speculated that the heat produced by ESU may inactivate the virus.

CONTROL OF ESU SMOKE AND RF/MW RADIATION

Radiofrequency and Microwave

Any area where RF/MW radiation exposure exceeds permissible levels should be considered potentially hazardous. The area should be clearly identified and warning signs posted. Interlocks may be used to prevent authorized entry. Basic protective measures include the provision of shields or absorbing enclosures for equipment. Personal protective equipment may be used (e.g., gonad shields, protective suits, and wire-netting helmets). Implementing such precautions, however, would be difficult in the operating room.

Engineering Control of ESU Smoke

Both the National Institutes of Health and the Laser Institute have recommended smoke evacuators and filtration be used during Laser and ESU procedures to eliminate odor and reduce the health risks associated with generation of toxic byproducts during surgery. An earlier version of smoke evacuator was designed with 0.5-µ particle filtration with charcoal to evacuate noxious odors. Today, as more and more questions arise over the presence of airborne chemicals, microorganisms, and viable DNA in smoke, an evacuator is considered more than an instrument of convenience; it is an important part of protective equipment.

A typical smoke evacuator consists of a Vacuum pump, operated at 50 CFM, which provides high static suction; and three filters completely enclosed. First, the filter stage draws air throughout a pre-filter, to capture large particles, collecting up to 80 cc of fluid. At a second stage, Ultra efficiency filters capture potentially viable microorganisms and carbonized tissue as small as 0.01-µ with the efficiency of 99.9999%. At the third stage activated carbon adsorbs organic compounds and odors by products. As air is drawn into the final stage, flow velocity reduced by filtration allows the gases more time to be absorbed on activated carbon and thereby increasing adsorption efficiency. Figure 1 illustrates a smoke filtration system.

CONCLUSION

The author recommends that immediate research is needed to: 1. Identify ESU toxic byproducts, mutagen and potential carcinogen generated during ESU surgery; 2. Perform real time exposure and dosimetry to RF/MW on OR personnel; 3. Test and analyze all ESU for spectrum energy; 4. Conduct epidemiological studies on the potential health effects of ESU smoke; 5. Educate OR personnel on health hazards associated with the use of ESU.

REFERENCES

1. Textbook of Surgery. The biological basis of Modern Surgical Practice. Sabiston CD (Ed). W.B. Saunders Company. pp 135–136, 1987.
2. Textbook of Surgery. The biological basis of Modern Surgical Practice. Sabiston CD (Ed). W.B. Saunders Company. pp 214–215, 1993.
3. Bailey MK, Bromley HR, Allison JM, Conroy JM, Krzyzaniak W: Electrocautery-induced Airway Fire During Tracheostomy. *Anesth Analg* 84: 2376–2382, 1991.
4. Schellhammer FP: Electrocautery: Principle Hazards and Precaution. *Urology* 3:261–268, 1991.
5. Paz JD, Milliken R, Ingram WT, Frank A: Potential Ocular Damage from Microwave Exposure During Electrosurgery: Dosimetric Survey. *J Occup Med* 29:580–583, 1987.
6. Peterson RC: Bioeffects of Microwaves: A Review of Knowledge. *J Occup Med* 25: 103–111, 1993.

Figure 1. Stackhouse AirSafe® ES-2000 Electrosurgical Smoke Filtration System (Riverside, CA).

7. American Conference Governmental Industrial Hygiene, Threshold Limit Value for Chemical Substances and Physical Agent and Biological Exposure Indicts, 1993–1994.
8. ANSI C95. 1-1992/IEEEC95. 1-1991.
9. Carpenter RL, Van Ummersen CA: The Action of Microwave on the Eye. *J Microwave* 3:3–19, 1968.
10. Birenbaum L, Grosof GM, Rosental SW, Zaret MM: Effects of Microwave on the Eye, IEEE Trans, *Bio-Medical Eng* BME-161:7–14, 1969.
11. Guy AW, Lin JD, Kramer PO, Emary AF: Effects of 2450 MHz Radiation on Rabbit Eye IEEE Trans Microwave Techq. MTTT-23 6:492–498, 1975.
12. Fox J, Kendel RT, Brook HR: Radiofrequency in the Operating Room Theater, *The Lancet* 31:962, 1976.
13. Ruggera PS, Segerson DA: Quantitative Measurements Near Electrocautery Unit, AAMI 12th Ann Meeting, March, 1977.
14. Paz JD: Real Time Non-Ionized Radiation Survey in the Operating Room, Paper Presented at the Annual Meeting AIHA San Francisco, May, 1988.
15. Goldstein B, Paz JD, Giufreida JJ, Palmes ED, Ferrand FD: Atmospheric Derivatives of Anesthesia Gases as a Possible Hazard to Operating Room Personnel, *The Lancet* 31: 235–237, 1976.

16. Tomita A, Shigenobu M, Kazgata N, Setsuo U, Masakazu F, Minoru H, Tomio H: Mutagenicity of Smoke Condensate Included by CO_2 Laser Irradiation and Electrocauterization. *Mut Res* 89:145–149, 1981.
17. Paz JD, Milliken RA: Radiofrequency Degradation of Anesthetic Gases as a Possible Health Hazard. Paper Presented at the Annual Meeting of the American Chemical Society Kansas City, September, 1982.
18. Bosterling B, Trudell JR: Production of 5- and 15-Hydroperoxyeicosatetaenoic Acid from Arachidonic Acid by Halothane-free Radicals Generated by UV-Irradiation. *Anesthesiology* 60:209–213, 1984.
19. Gatti JE, Bryant CJ, Barrett M, Murphy JB: The Mutagenicity of Electrocautery Smoke. *Plast and Reconst Surgery* 89:781–785, 992.
20. Heinsohn DL, Jewett L, Bazer CH, Bennett, Seipel P, Rosen A: Aerosol Created by Some Surgical Power Tool; Particles Size Distribution, and Qualitative Hemoglobin Content. *Appl Occup Env Hyg* 6:773–776, 1991.
21. Johnson GK, Robinson WS: Human Immunodeficiency Virus Type 1 in the Vapor of Surgical Masks. *Am Ind Hyg Assoc J* 46:308–312, 1985.
22. Jewett DL, Heinsohn P, Bennett C, Rosen A, Neuilly: Blood-containing Aerosols Generated by Surgical Techniques: A Possible Infectious Hazard. *Am Ind Hyg Assoc J* 53:229–231, 1993.
23. Kopencky KK, Steidle CP, Eble JN, Birhrle R, Dreesen RG, Sutton GP, Becker GJ: Endoluminal Radio-Frequency Electrocautery for Permanent Urethral Occlusion in Swine. *Radiology* 170:1043–1046, 1989.

PART 3

Health Hazard Evaluation Report

National Institute for Occupational Safety and Health
HETA 85-126-1932, Bryn Mawr Hospital,
Bryn Mawr, Pennsylvania

PREFACE

The Hazard Evaluations and Technical Assistance Branch of NIOSH conducts field investigations of possible health hazards in the workplace. These investigations are conducted under the authority of Section 20(a)(6) of the Occupational Safety and Health Act of 1970, 29 U.S.C. 669(a)(6) which authorizes the Secretary of Health and Human Services, following a written request from any employer or authorized representative of employees, to determine whether any substance normally found in the place of employment has potentially toxic effects in such concentrations as used or found.

The Hazard Evaluations and Technical Assistance Branch also provides, upon request, medical, nursing and industrial hygiene technical and consultative assistance (TA) to federal, state and local agencies; labor; industry and other groups or individuals to control occupational health hazards and to prevent related trauma and disease.

I. SUMMARY

On January 7, 1985, the National Institute for Occupational Safety and Health (NIOSH) received a request from a group of plastic surgeons at the Bryn Mawr Hospital in Bryn Mawr, Pennsylvania, to evaluate exposure to emissions generated by the use of electrocautery knives during reduction mammoplasty surgical procedures. Numerous health effects (headache, nausea, upper respiratory and eye irritation) reported by operating room personnel were cited in the request.

An initial on-site survey was conducted on February 14, 1985, with follow-up surveys performed on December 12, 1985 (Pennsylvania Hospital), April 28, 1987 and August 26, 1987. Industrial hygiene sampling was conducted to evaluate exposure to hydrocarbons, nitrosamines, total particulates, benzene soluble fraction, polynuclear aromatic compounds (PNAs) and airborne mutagens. In addition, since very little data have previously been collected for this exposure situation, sampling was performed to obtain qualitative exposure data utilizing a variety of solid sorbent tubes (high volume sampling), fourier transform infrared spectroscopy (FTIR) and aldehyde screening sorbent tubes (Orbo-23).

Personal (breathing zone) and area samples collected for hydrocarbons contained isopropanol at concentrations well below all relevant criteria.

None of the seven nitrosamines or sixteen PNAs that were evaluated were found in detectable quantities. Since several nitrosamines and PNAs are carcinogenic, any detectable levels would have been considered potentially significant.

Concentrations of airborne particulates ranged from 0.4 to 9.4 milligrams per cubic meter of air (mg/m^3) with a mean of 2.75 mg/m^3. Although these levels all were below the OSHA PEL (15 mg/m^3) and ACGIH TLV (10 mg/m^3) for total nuisance particulates, it is not known at this time whether this particulate is biologically inert; comparison with the nuisance dust evaluation criteria may not be appropriate.

The benzene-soluble fraction of the particulate samples ranged from 0.5 to 7.4 mg/m^3, averaging 2.4 mg/m^3. Seven of the 11 samples exceeded the NIOSH recommended exposure limit of 0.1 mg/m^3 and OSHA PEL of 0.2 mg/m^3. The purpose of these exposure criteria are to minimize worker exposure to carcinogenic PNA compounds. However, this is based on industrial settings (coke ovens, asphalt, petroleum coke) and may not apply to a non-industrial hospital environment.

Sorbent tubes (charcoal, silica gel, Tenax-TA) that were utilized at high sampling volumes, qualitatively revealed a trace (between the limit of detection and quantitation) amount of hydrocarbons. All of the concentrations were far below evaluation criteria and would not be expected to cause any health effects. FTIR analysis identified a component of the smoke as a compound or compounds related to fatty acid esters. None of the aldehydes (C_1-C_8 aldehydes) evaluated were detected.

Solvent extracts of airborne particles were mutagenic (with microsomal (S9) activation, and slightly mutagenic without activation) to the *Salmonella typhimurium* TA 98 strain, clearly indicating operating room personnel exposures to potentially genotoxic agents. However, whether exposure of operating room personnel to agents that are mutagenic to bacteria or the level of these agents to which workers are exposed pose any genotoxic hazards is not known.

On the basis of the mutagenicity of the airborne compounds collected during this evaluation and the acute health effects reported by operating room personnel, NIOSH investigators determined that there is a potential hazard from exposure to smoke generated by electrocautery knives during reduction mammoplasty surgical procedures. Recommendations aimed at reducing exposure among operating room personnel are presented in Section VIII of this report.

Key words: SIC 8062 (General Medical and Surgical Hospitals), operating rooms, electrocautery knives, electrocautery smoke, reduction mammoplasty, mutagenicity assessment, polynuclear aromatics (PNAs), nitrosamines, benzene soluble fraction.

II. INTRODUCTION

On January 7, 1985, NIOSH received a request for a health hazard evaluation at the Bryn Mawr Hospital, Bryn Mawr, Pennsylvania. The request was submitted by a group of surgeons who were concerned about exposure to emissions generated by electrocautery knives when performing reduction mammoplasties. NIOSH investigators conducted environmental surveys at the Bryn Mawr and Pennsylvania Hospitals on February 14 and December 12, 1985, April 28, 1987 and August 26, 1987.

III. BACKGROUND

The surgical procedure known as breast reduction is one of the most common procedures where considerable smoke is produced. The plastic surgeons at Bryn Mawr Hospital became concerned about the chemical composition and toxicity of this smoke, after noticing that several operating room personnel were experiencing acute health effects during this procedure. Reported health effects included upper respiratory and eye irritation, headache and nausea (obnoxious odors).

The electrosurgical knife (ESK) is presently used for a wide variety of surgical procedures in many health care facilities throughout the United States. Currently there may be as many as 30–40 U.S. manufacturers of ESK devices. These devices cut or coagulate body tissues utilizing an electromagnetic (EM) field that is focused onto the body site. The presence of this EM field requires the use of a grounding pad to be placed on the opposite side of the body being cut in order to collect all fields produced by the ESK devices. The ESK units used in this evaluation were a Valley Laboratory (model SSE2L) and a Neo-Med (model 3000). The operating parameters used on these systems during surgical procedures were the same (i.e., mid-range cut and coagulate settings estimated to be 120 watts delivered to the cutting area).

On February 14, 1985, NIOSH personnel conducted an initial environmental survey at the Bryn Mawr Hospital. Personal (breathing zone) and area air samples were taken for hydrocarbons, nitrosamines, total particulates, benzene soluble fraction and polynuclear aromatic compounds (PNAs). Findings from this visit were presented in a letter dated May 7, 1985.

A follow-up visit was made on December 12, 1985, at the Pennsylvania Hospital in Philadelphia, Pennsylvania. Environmental samples were taken (at the suggestion of NIOSH chemists) for PNAs, total particulates, benzene soluble fraction, qualitative organic sorbent tube sampling (charcoal, silica gel, Tenax-TA) and FTIR for qualitative organic analysis. Results were reported on February 14, 1986.

On April 28, 1987, NIOSH conducted an additional follow-up study at the Bryn Mawr Hospital. Environmental air samples for qualitative aldehyde scans were obtained. Monitoring for airborne mutagens was performed on April 28 and August 26, 1987. Results were reported in a letter dated December 1, 1987.

Although the purpose of this evaluation was to determine the nature of the emissions produced by ESK devices in the operating room, it should be realized that there are other potential occupational health issues in addition to the chemical and environmental concerns. One issue is the production of EM radiation. Previous NIOSH research work on such systems has indicated that radiofrequency radiation at 0.5 megahertz is produced by these systems.[1] This finding has also been confirmed in another report.[2]

IV. METHODS AND MATERIALS

A. Total Particulates, Benzene Solubles and PNAs

Personal and area air samples were collected utilizing a sampling train consisting of a Zefluor 2-micron filter (Membrana Co.) and a cellulose acetate O-ring in a cassette, followed by a 7-mm O.D. glass tube containing two sections of prewashed XAD-2 resin (100 mg/50 mg) connected to a battery-operated sampling pump calibrated at a flowrate of 2.0 liters per minute (1pm).

Total particulate weights were determined by weighing the samples plus the filters on an electrobalance and subtracting the previously determined tare weight of the filters. The instrumental precision is 0.01 mg.

The benzene soluble fraction of the filter samples was determined by placing the filters in screw-cap vials with 5 mL of benzene and sonifying for 15 minutes. The extract was filtered through a Millex-HV 0.45 µm filter and collected in a screw-cap vial. Each sample was transferred into a tared Teflon® cup and evaporated to dryness in a vacuum oven at 40°C. The Teflon® cups were again weighed and the difference recorded. The analytical limit of detection is 0.05 mg/sample.

The filter and tube samples were analyzed for PNAs following NIOSH Technical Bulletin TB-001 issued December 1, 1982. The filters and tubes were desorbed in 5 mL of benzene and sonicated for 30 minutes. The resulting solution was filtered through a 0.45 µm nylon filter. The samples and standards desorbed in benzene were solvent exchanged to acetonitrile by alternate and multiple additions of acetonitrile and evaporation. The samples and standards were not allowed to go to dryness at any time during the exchange. Analysis was then performed by high performance liquid chromatography (HPLC) with a fluorescence/UV detector. The retention times of the analytes in the standards were compared to the retention times in the sample chromatograms for analyte identification. The standard analytes and their associated analytical limits of detection (LOD) are listed below:

Analyte	LOD nanograms/sample
Acenaphthene	100
Acenaphthylene	500
Anthracene	250
Benz(a)anthracene	25
Benzo(a)pyrene	25
Benzo(b)fluorathene	25
Benzo(e)pyrene	50
Benzo(k)fluoranthene	25
Benzo(g,h,l)perylene	50
Chrysene	25
Dibenz(a,h)anthracene	25
Fluoranthene	50
Fluorene	100
Indeno(1,2,3,c,d)pyrene	50
Phenanthrene	100
Pyrene	50

B. Nitrosamines

Personal and area air samples for nitrosamines were collected on Thermosorb/N tubes attached to battery-operated sampling pumps operating at a flowrate of 2.0 lpm. The tubes were desorbed with 2.0 mL of a solution of 75% methylene chloride and 25% methyl alcohol. The samples were then analyzed by gas chromatography with a thermal energy analyzer in the nitrosamine mode. The analytical limits of detection for this method ranged from 10–100 nanograms per sample (depending upon the particular nitrosamine that was to be identified).

C. Hydrocarbons

The air samples for hydrocarbons were collected by drawing air through a glass tube containing 150 milligrams of activated charcoal at a flowrate of 1.0 lpm (qualitative samples) and 0.2 lpm (quantitative samples) using calibrated, battery-operated sampling pumps. The samples were desorbed with 1 mL of carbon disulfide and analyzed by gas chromatography (GC) with a flame ionization detector (FID). Some of the samples were concentrated and analyzed by GC using a mass spectrometer for major compound identification.

D. Sorbent Tubes—Qualitative Organic Analysis

Personal and area air samples for qualitative organics analyses were collected on charcoal, silica gel and Tenax-TA tubes attached to battery-operated sampling pumps operating at a flowrate of 1.0 lpm (Tenax-TA, 0.5 lpm).

The high volume Tenax tubes were analyzed first by thermal desorption. A Tekmar model 4000 dynamic headspace concentrator equipped with a heated sampler module and capillary cryofocusing interface was used for this procedure. The concentrator unit was interfaced directly to a GC/MS system. Front 100 mg sections of the Tenax sample tubes were put into the sampler module heated to 200°C. The headspace was continually purged during this time and the effluent trapped on an internal Tenax trap. The trap was then thermally desorbed onto the front end of a 30-meter DB-1 capillary column, flash heated, and injected into the gas chromatograph and mass spectrometer for analysis.

Both the charcoal and silica gel sorbent tubes were screened by gas chromatography (FID) and GC/MS. Charcoal tubes were desorbed with 1 mL carbon disulfide and the silica gel tubes were desorbed with 1 mL ethanol. Both front sections and the front glass wool plugs were desorbed together. All analyses were performed using 30-meter DB-1 fused silica capillary columns (splitless mode).

E. Fourier Transform Infrared Spectroscopy Qualitative Organic Analysis

Personal and area samples for FTIR analysis were collected by drawing air through a Zefluor filter at a flowrate of 1.0 lpm using calibrated, battery-operated sampling pumps.

Of the six samples submitted, one area and one personal sample were selected for analysis after visually inspecting the filters. The initial analysis involved the analysis of the filters using attenuated total reflectance spectroscopy (ATR). The area sample filter was removed from its cassette and placed in a Barnes Model 305 Horizontal ATR cell. The crystal in the cell was KRS-5 (thallous bromide). Spectra were collected with a Nicolet 60SX Fourier Transform Infrared Spectrometer using a combined Indium Actinimide/Mercury Cadmium Telluride detector at 0.5 cm^{-1} spectral resolution. After correcting the recorded sample spectra for the background effects of the filter, there appeared to be some compound present. When the filter was removed from the ATR cell, an oily residue was noted to remain on the ATR cell crystal. A spectra of this material was recorded and corrected for the background of the ATR cell. The corrected spectra was compared with the Aldrich FTIR spectral search library using Nicolet searching software. This filter was then desorbed with 1,1,2-trichlorotrifluoroethane. This solution was then evaporated onto the ATR crystal. The spectra was similar to that obtained from the residue of the filter. This desorption procedure was also used for the personal sample.

F. Qualitative Aldehyde Screen

Samples for airborne aldehydes were collected by drawing air through Orbo-23 tubes at a flowrate of 0.08 lpm. Samples were desorbed with 1 mL of toluene in an ultrasonic bath for 60 minutes. Aliquots of the sample extracts were then screened by gas chromatography (FID) twice; first with a 30-meter DB-1 GC column, and second with a 30-meter DB-was fused silica capillary column (split-less mode).

G. Airborne Mutagens

Airborne particles were collected on glass-fiber filters (type A/E, 4″ diameter) using Hi-Vol pumps (General Metal Works) at flow rates between 17 and 24 cfm. Filters were changed if the flow rate dropped below 17 cfm.

Samples from the first survey (April 28, 1987) were first extracted with 150 mL of methylene chloride (DCM) then with 150 mL of acetone plus methanol (A+M). Samples from the second survey (August 25, 1987) were divided, because of the quantity of particles. One half was extracted as that in the first survey, the other half was extracted with an XAD-2 resin column. Each extract was filtered and concentrated to a final volume of 0.45 and 0.3 mL in dimethyl sulfoxide for the first and second surveys, respectively.

The same sampling sites were used for both surveys. In the operating room (OR), air was sampled 3 ft directly above the operation. As a control (CR), samplers were placed 1/2 ft above the floor in the anteroom.

All extracts were tested for the mutagenic activity in both tester strains TA98 and TA100 of *Salmonella typhimurium* using the Salmonella/microsomal micro-suspension test.[3] The system is characterized by adding increased numbers of bacterial cells (approx. 10^9) which are exposed to airborne particle extracts with or without S9 in a concentrated treatment mixture. For metabolic activation, 0.065 mL of S9 mix (10% S9) was also added to each treatment tube. The S9 was prepared from the livers of male Fischer rats pretreated with Aroclor 1254 (500 mg/kg body wt). The micro-suspension test is a suitable assay system for limited quantities of test materials. After 90 minute pre-incubation at 37°C, the mixture is processed according to the standard Ames test protocol.[4] The mutagenic activity was scored in tester cells from histidine-dependence to histidine-independence.

In the *in situ* assay, samples were taken at intervals of 2, 4 and 6 hours post-operation from the trapping media and were plated on the appropriate agar plates to determine survival and mutation frequencies.[5] Plates were scored after incubation at 37°C for 2 days.

V. EVALUATION CRITERIA

As a guide to the evaluation of the hazards posed by workplace exposures, NIOSH field staff employs environmental evaluation criteria for assessment of a number of chemical and physical agents. These criteria are intended to suggest levels of exposure to which most workers may be exposed up to 10 hours per day, 40 hours per week, for a working lifetime, without experiencing adverse health effects. It is, however, important that not all exposures are maintained below these levels. A small percentage may experience adverse health effects because of individual susceptibility, a pre-existing medical condition, and/or hypersensitivity (allergy).

In addition, some hazardous substances may act in combination with other workplace exposures, the general environment or with medications or personal habits of the worker to produce health effects, even if the occupational exposures are controlled at the level set by the evaluation criteria. Also, some substances are absorbed by direct contact with the skin and mucous membranes, and thus, potentially increase the overall exposure. Finally, evaluation criteria may change over the years as new information on the toxic effects of an agent become available.

The primary sources of environmental evaluation criteria for the workplace are: 1) NIOSH criteria documents and recommendations, 2) the ACGIH TLVS and 3) the U.S. Department of Labor (OSHA) occupational health standards. Often, the NIOSH recommendations and ACGIH TLVs are lower than the corresponding OSHA standards. Both NIOSH recommendations and ACGIH TLVs usually are based on more recent information than are the OSHA standards. The OSHA standards also may be required to take into account the feasibility of controlling exposures in various industries where the agents are used; the NIOSH recommended standards, by contrast, are based primarily on concerns relating to the prevention of occupational disease. In evaluating the exposure levels and the recommendations for reducing these levels found in the report, it should be noted that industry is legally required to meet those levels specified by an OSHA standard.

A time-weighted average (TWA) exposure refers to the average airborne concentration of a substance during a normal 8- to 10-workday. Some substances have recommended short-term exposure limits or ceiling values which are intended to supplement the TWA, where there are recognized toxic effects from high short-term exposures.

A. Isopropanol

Isopropyl alcohol causes mild irritation of the eyes, nose and throat. High vapor concentrations may cause drowsiness, dizziness and headache. Repeated skin exposure may cause drying and cracking. NIOSH recommends an exposure limit of 980 mg/m^3.[6] The OSHA standard and ACGIH TLV are the same.[7–8]

B. Nitrosamines

Nitrosamines are a class of compounds which are readily formed by the interaction of secondary amines and nitrites or oxides of nitrogen. Because these precursors are ubiquitous, nitrosamines have been found in air, water, tobacco smoke, cured meats, cosmetics and in many industrial processes, including leather tanneries, pesticide formulations and tire and rubber manufacturing facilities.[9]

Nitrosamines are considered to be among the most potent of animal carcinogens. Of more than 150 nitrosamine compounds tested approximately 80% have been found to be carcinogenic in at least one species of animal. To date, there are no standards for employee exposure to airborne nitrosamines. OSHA has a regulation regarding work practices and handling of liquid and solid *N*-nitrosodimethylamine in concentrations greater then 1%.[8] In addition, the Food and Drug Administration has limited the amount of nitrosamines allowed in beer to 5 parts per billion (ppb) and the United States Department of Agriculture has limited nitrosamine concentration in cooked bacon to 10 ppb. The International Agency for Research on Cancer recommends that *N*-nitrosodimethylamine, *N*-nitrosodiethylamine, *N*-nitrosodibutylamine, and *N*-nitrosomorpholine be regarded for practical purposes as if they were carcinogenic to humans.[10] NIOSH policy on human exposure to known or suspected carcinogens is to reduce exposure to the lowest feasible level.[11]

C. PNAs and Benzene of Cyclohexane Solubles

PNAs are condensed ring aromatic hydrocarbons normally arising from the combustion of organic matter. They are commonly emitted into the air when coal tar, coal tar pitch or their products are heated, but can result from burning the heavy petroleum fraction used in petroleum coke.[12] A number of PNAs, including benzo(a)pyrene and anthracene are carcinogenic (lung and skin). There are no federal standards pertaining to airborne concentrations of individual PNAs. In 1967, the ACGIH adopted a TLV of 0.2 mg/m^3 for coal tar pitch volatiles (CTPV), described as a "benzene-soluble" fraction, and listed certain carcinogenic components of CTPV. The TLV was established to minimize exposure to the listed substances believed to be carcinogens, viz, anthracene, BaP [benzo(a)pyrent], phenanthrene, acridine, chrysene and pyrene. CTPVs are among the seven substances listed as "Human Carcinogens" in Appendix A of the current ACGIH TLVs. This group consists of "a substance, or substances, associated with industrial processes, recognized to have carcinogenic or cocarcinogenic potential with an assigned TLV." The TLV was promulgated as a federal standard under the Occupational Safety and Health Act of 1970 (29 CFR 1910.1000).[13] In 1972, the Federal Register (37:24749, November 21, 1972) contained an interpretative rule of the term "coal tar pitch volatiles:" "...coal tar pitch volatiles include the fused polycyclic hydrocarbons which volatilize from the distillation residues of coal, petroleum, wood, and other organic matter." This has been reprinted as 29 CFR 1910.1002. The general philosophy behind this interpretation was that "all of these volatiles have the same basic composition and...present the same dangers to a person's health."[14]

In the development of the NIOSH recommended standard, it was concluded that CTPVs are carcinogenic and can increase the risk of lung and skin cancer in workers. Since no absolutely safe concentration can be established for a carcinogen, NIOSH recommended the exposure limit be the lowest concentration that can be reliably detected by the recommended method of environmental monitoring. At that time (September, 1977) the lowest detectable concentration for CPTVs was 0.1 mg/m^3 for the recommended sampling method.

Although the benzene or cyclohexane extractable fraction offers an easier, less expensive method of analysis than PNA quantitation, there is no certainty that there is a correlation between the two. The analytical method for measuring the benzene soluble fraction is not limited to PNAs but will include all other organic compounds collected on the filter and soluble in benzene.[15]

D. Mutagenicity Assay

All the overlayed plates were scored for *his+* revertants after 2 days of incubations. An extract was considered mutagenic if the number of revertants in any of the four concentrations tested (undiluted, 1 to 2, 1 to 4, 1 to 8) was twofold or greater than the control, and showed a dose-related response.

VI. RESULTS

A. Hydrocarbons

Table 1 presents the results of the air samples taken for hydrocarbons. Four of the five samples taken contained isopropanol in concentrations ranging from 1.4 to 4.9 mg/m^3. All of the samples were well below the evaluation criteria for 980 mg/m^3. The likely source was not the emissions from surgery, but the isopropanol used as a sanitizing agent in the operating room.

B. Nitrosamines

The results of the air samples taken for nitrosamines are presented in Table 2. None of the seven nitrosamines evaluated were detected. The specific compounds evaluated included the nitrosamines of dimethyl, diethyl, dipropyl and dibutylamine, plus those of pyrrolidine, piperidine and morpholine. The limit of detection for these compounds ranged from 25–150 nanograms/sample. Since several nitrosamines are carcinogenic, any detectable levels would have been considered significant.

C. Particulates, Benzene Solubles and PNAs

Tables 3, 5 and 6 contain the data from the analysis of the Zenfluor filters and the Orbo-43 tubes. Each filter and tube sampling train provided the following three types of data:

Table 1. Isopropanol, Bryn Mawr Hospital, Bryn Mawr, Pennsylvania, HETA 85-126, February 14, 1985.

Sample Location/Job	Sample Type	Sampling	Isopropanol (mg/m^3)*
Assistant Surgeon	Personal	13:30–15:20	1.4
Surgeon	Personal	13:30–15:20	2.1
Anesthesia area or operating room lights	Area	13:30–15:20	1.4
Surgical nurse	Personal	13:30–15:20	4.9
Laser used on breast tissues	Area	15:30–15:41	ND**
Evaluation Criteria: (8 hr. time-weighted average)		NIOSH	980
		ACGIH	980
		OSHA	980

* = All air concentrations are reported as time-weighted averages for the time period sampled.

** = Non-detectable. Limit of detection is 0.01 mg/sample, which would correspond to an atmospheric concentration of 0.48 mg/m^3 when the average sample air volume (21 liters) is considered.

Table 2. Nitrosamines, Bryn Mawr Hospital, Bryn Mawr, Pennsylvania, HETA 85-126, February 14, 1985.

Sample Location/Job	Sample Type	Sampling	Nitrosamines*
Assistant surgeon	Personal	13:30–15:20	ND**
Surgeon	Personal	13:30–15:20	ND**
Anesthesia area or operating room lights	Area	13:30–15:20	ND**
Surgical nurse	Personal	13:30–15:20	ND**
Laser used on breast tissues	Area	15:30–15:41	ND**
Evaluation criteria			LFL***

* = NIOSH currently uses a seven standard mixture to calibrate and identify specific nitrosamines. The mixture contains the nitrosamines of dimethyl, diethyl, dipropyl and dibutylamine plus those of pyrrolidine, piperidine and morpholine.

** = Non-Detectable. Limits of detection range from 25–150 ng/sample (aid adjusted concentrations would range from 114–682 $\mu g/m^3$).

*** = No evaluation criteria has been established for nitrosamines. Exposure should be reduced to lowest feasible level (LFL).

Table 3. Particulates, benzene solubles and PNAs, Bryn Mawr Hospital, Bryn Mawr, Pennsylvania, HETA 85-126, February 14, 1985.

Sample Location/Job	Sample Type	Sampling Period	Exposure Concentrations*		
			Total Particulates (mg/m^3)	Benzene Soluble (mg/m^3)	PNAs**
Assistant surgeon	Personal	13:30–15:20	1.6	ND	ND
Surgeon	Personal	13:30–15:20	ND***	ND	ND
Anesthesia area of operating room lights	Area	13:30–15:20	1.2	ND	ND
Surgical nurse	Personal	13:30–15:20	0.4	ND	ND
Laser used on breast tissue	Area	15:30–15:41	9.4	7.4	ND
Evaluation criteria (8 hr. time-weighted average)		NIOSH	-	0.1	****
		ACGIH	10	0.2	****
		OSHA	15	0.2	****

* = All air samples are reported as time-weighted averages for the time period sampled.

** = Represents the following EPA priority PNAs: acenapthene, acenaphthylene, anthracene, benz(a)anthracene, benzo(a)pyrene, benzo(a)fluorathene, benzo(e)pyrene, benzo(k)fluoranthene, benzo(g,h,i)perylene, chrysene, dibenz(a,h)anthracene, fluoranthene, fluorene, ideno(1,2,3,c,d)pyrene, phenanathrene, pyrene.

*** = Non-Detectable. Limits of detection are 0.01 mg/sample for total particulates, 0.05 mg/sample for benzene soluble fraction, and 25–500 ng/sample for the various PNAs analyzed.

**** = No criteria currently exists for total PNAs, however, a number of individual PNAs are carcinogenic (benzo(a)pyrene, anthracene, chrysene) and exposures should be controlled to the lowest feasible level.

Table 4. Sorbent Tube Sampling, Pennsylvania Hospital, Philadelphia, Pennsylvania, HETA 85-126, December 12, 1985.

Sorbent Tube	Substance Identified
Tenax-TA*	Isoflurane (anesthetic gas) Halothane (anesthetic gas)
Charcoal**	Isoflurane Isopropanol 1,1,1-Trichloroethane Trichloroethylene Toluene Perchloroethylene Xylene Several aliphatic hydrocarbons
Silica Gel	None detected

* = Thermally desorbed tubes cannot be quantitated.
** = Substances were present in trace quantities, between the limit of detection (1–5 μg/sample) and limit of quantitation (5–10 μg/sample).

1. Total Particulates - represents the total weight of the smoke per cubic meter of sampled air. Five personal breathing-zone air samples for total particulate ranged from 0.4 to 2.0 mg/m^3 with a mean of 1.0 mg/m^3. Six area samples ranged from 0.7 to 9.4 mg/m^3, with a mean of 4.2 mg/m^3.

2. Benzene Soluble Fraction - represents the total weight of the smoke that is benzene soluble per cubic meter of sampled air. No benzene soluble fraction was detected in any of the particulate samples taken on February 14, 1985. All of the samples taken on December 12, 1985, had a benzene soluble fraction ranging from 0.7 to 6.7 mg/m^3, well above the NIOSH evaluation criteria of 0.1 mg/m^3.

3. PNAs - represents the analysis for 17 polynuclear aromatic hydrocarbons. None of the 17 PNAs which are monitored in the NIOSH standard method were detected in any of the samples.

D. Sorbent Tubes - Qualitative Organic Analysis (Table 4)

Table 4 lists the substances that were identified by sorbent tube sampling during the survey on December 12, 1985. The air samples indicate that trace amounts (less than 10 micrograms) of hydrocarbons were present within the operating room. The substances identified would not be expected to cause ill health effects in most people at the levels detected.

Table 5. Zefluor filter sample/Fourier Transform infrared spectroscopy analysis, Pennsylvania Hospital, Philadelphia, Pennsylvania, HETA 85-126, December 12, 1985.

Sample Type	Substance Identified
Zefluor (area)	Compound or compounds related to fatty acid esters
Zefluor (personal)	Compound or compounds related to fatty acid esters

E. Fourier Transform Infrared Spectroscopy Qualitative Organic Analysis

The search area of the spectra indicated that the compounds found on both the area and personal filters were related to fatty acid esters, based on spectral similarities. A search of the library using the absolute derivative search algorithm indicated that olive oil, cottonseed oil, methyl stearate and castor oil were all close matches to the sample spectra for the bulk sample. The goodness-of-fit values associated with each of these matches were all equivalent and higher than what would be expected for an exact match. A low goodness-of-fit value indicates a good match of library spectra. For the qualitative sample, matches included ethyl stearate, castor oil and other straight chain hydrocarbon compounds. Based on the source of the sample, the identification of the sample as fatty acid esters is logical.

Attempts made to analyze these samples by gas chromatography were unsuccessful. The samples were not volatile enough to allow chromatography.

As a final attempt at characterization of the samples, the two desorbed filter solutions were submitted for direct probe mass spectral analysis. The solutions were evaporated onto the direct probe of the mass spectrometer and heated under vacuum. Results from this analysis indicated that the samples gave mass spectral detail related to straight chain hydrocarbons, i.e., methylene group (CH_2) fragments. Since fatty acid esters contain long chain alkyl groups, this fragmentation pattern was not unexpected. The mass spectra were compared to methyl stearate and stearyl palmitate and were found to be similar to the sample spectra but were not an exact match, indicating that there was probably a mixture of fatty acid esters in the samples.

F. Qualitative Aldehyde Screen

Table 7 presents the results of the area and personal samples taken for qualitative aldehyde scan analysis. No aldehydes were detected. It should be noted that the aldehyde scan has only been tried on the low molecular weight (C_1 - C_8) aliphatic aldehydes.

Table 6. Particulates, benzene solubles and PNAs, Pennsylvania Hospital, Philadelphia, Pennsylvania, HETA 85-126, December 12, 1985.

Sample Location/Job	Sample Type	Sampling Period	Exposure Concentrations*		
			Total Particulates (mg/m^3)	Benzene Soluble (mg/m^3)	PNAs**
Surgeon	Personal	11:30–12:40	2.0	1.4	ND
Assistant surgeon	Personal	11:30–12:40	0.9	0.7	ND
Hand-held/operative site	Area	11:35–12:35	0.7	1.3	ND
Hand-held/operative site	Area	11:35–12:35	0.7	1.7	ND
Hand-held/breast tissue/post-surgery	Area	13:00–13:15	8.7	6.7	ND
Hand-held/breast tissue	Area	13:00–13:15	4.7	6.7	ND
Evaluation criteria (8 hr. time-weighted average)		NIOSH	-	0.1	***
		ACGIH	10	0.2	***
		OSHA	15	0.2	***

* = All air samples are reported as time-weighted averages for the time period sampled.

** = Represents the following EPA priority PNAs: acenapthene, acenaphthylene, anthracene, benz(a)anthracene, benzo(a)pyrene, benzo(a)fluorathene, benzo(e)pyrene, benzo(k)fluoranthene, benzo(g,h,i)perylene, chrysene, dibenz(a,h)anthracene, fluoranthene, fluorene, ideno(1,2,3,c,d)pyrene, phenanathrene, pyrene.

*** = No criteria currently exists for total PNAs, however, a number of individual PNAs are carcinogenic (benzo(a)pyrene, anthracene, chrysene) and exposures should be controlled to the lowest feasible level.

ND = Non-Detectable. Limits of detection are 0.01 mg/sample for total particulates, 0.05 mg/sample for benzene soluble fraction and 25–500 ng/sample for the various PNAs analyzed.

Table 7. Qualitative aldehyde scan, Bryn Mawr Hospital, Bryn Mawr, Pennsylvania, HETA 85-126, April 28, 1987.

Sample Location	Type Sample	Aldehydes Identified*
Scrub nurse	Personal	None
Surgeon	Personal	None
Assistant surgeon	Personal	None
Anesthetist	Area	None
Operating room	Area	None

* = Aldehyde scan has only been tried on the low molecular weight (C_1-C_8) aliphatic aldehydes.

Table 8. Mutagenicity of airborne particle extracts, Bryn Mawr Hospital, Bryn Mawr, Pennsylvania, HETA-85-126, April 28, 1987.

			His Rev/Plate			
			TA98		TA100	
Sample Location	Particles µg/Plate	Air Vol. m³/Plate	-S9	+S9	-S9	+S9
DCM extraction						
Operating room	78	0.25	4	6	54	55
	155	0.49	6	8	64	52
	310	0.98	10	19	62	50
	620	1.95	10	31	62	55
Control room	9	0.25	4	8	51	43
	18	0.49	7	9	60	47
	37	0.98	8	8	48	45
	73	1.95	8	11	53	48
Filter control			5	8	54	45
Negative control			7	5	44	46
Positive control*			1608			1926

* = 2-Aminoanthracine; 2.5 µg/plate
His Rev = Histidine revertants

Table 9. Mutagenicity of airborne particle extracts, Bryn Mawr Hospital, Bryn Mawr, Pennsylvania, HETA 85-126, August 26, 1987.

Sample Location	Particles µg/Plate	Air Vol m^3/Plate	His Rev/Plate		
			TA98		TA100
			-S9	+S9*	+S9
DCM extraction					
Operating room	265	0.35	5	37	54
	530	0.70	6	69	67
	1060	1.41	12	92	79
Control room	17	0.73	4	8	52
	33	1.46	2	7	73
Filter control			3	6	64
XAD2 Column Extract					
Operating room	265	0.35	4	24	57
	530	0.70	6	45	45
	1060	1.41	10	57	57
Control room	17	0.73	2	8	55
	33	1.46	5	8	49
Filter control			4	7	59
Negative control			4	7	68
Positive control**				1551	1610

* = Average of two experiments
** = 2-Aminoanthracine; 2.5 µg/plate
His Rev = Histidine revertants

G. Airborne Mutagens

Airborne particles collected on glass-fiber filters from both surveys were found to be mutagenic (TA98). Samples from the first survey on April 28, 1987 (Table 8), showed a positive response only with S9, but the second survey (August 26, 1987) samples also showed a slight response without activation. The mutagenic response of extracts from the organic solvent extraction of the second survey was higher than those from the XAD-2 column extraction (Table 9). No significant mutagenic response was found with the *in situ* assay system in either survey (Table 10). The population of TA98W for the *in situ* testing was too low and gave sporadic results. This was not reported.

Table 10. ***In situ*** **testing of airborne particle extracts*, Bryn Mawr Hospital, Bryn Mawr, Pennsylvania, HETA 85-126.**

	Control Room				Operating Room			
	C		T		C		T	
Hour	% Sur	Rev/ 10^7	% Sur	Rev/ 10^7	% Sur	Rev/ 10^7	% Sur	Rev/ 10^7
			April 28, 1987					
2	100	8	104	7	100	7	97	8
4	100	5	111	7	110	6	100	7
6	104	8	114	7	124	6	107	8
2S	100	7	79	10	100	11	123	10
4S	74	11	58	12	85	14	108	10
6S	40	17	33	21	52	20	92	13
			August 25, 1987					
2	100	14	164	12	100	12	131	12
4	118	13	182	10	108	8	123	11
6	109	12	200	13	108	10	131	12
2S	100	21	127	16	100	26	112	23
4S	97	21	116	21	102	24	138	19
6S	64	33	80	30	57	37	75	35

* = TA100 only
S = With S-9 activation
C = Recirculating closed system (control)
T = Ambient room air
% Sur = Percent survival
Rev/10^7 = Number of revertants/10^7 living cells

An extract was considered mutagenic if the number of revertants in any of the four concentrations tested (undiluted, 1 to 2, 1 to 4, 1 to 8) was twofold or greater than the control and showed a dose-related response.

VII. CONCLUSIONS

A. No specific organic vapors, other than isopropanol, were quantitatively identified during the surgical procedures.
B. There were no PNAs or nitrosamines detected during the procedures.
C. The exposures to particulates, which ranged from 0.4 to 2.0 mg/m^3, confirm that the visible emissions are more than just water vapor. However, there are no exposure criteria with which to compare this exposure. Exposure criteria

of 10–15 mg/m^3 has been established for nuisance dust; however, to apply this criteria, the particulate would need to be biologically inert, which in this case, is not known. Airborne particulate can contribute to the eye irritation which has been reported during these procedures. We would have expected the samples from the surgeons and some of the area samples (hand-held approximating breathing-zone) to be similar. The "not-detected" on one surgeon suggests that we encountered a flow problem with that sample train even though the flow looked fine at the end of the sampling period.

D. The sample (Table 3) taken during a brief demonstration of the laser-cutting technique measured 9.4 mg/m^3 of total particulate of which 7.4 mg/m^3 was found to be benzene soluble. It is difficult to make conclusions based on one sample, but this data suggest that there is a tendency for the laser method to produce more particulate that is soluble in benzene. Whether this produces more PNAs is unknown.

E. All of the particulate samples taken on December 12, 1985, contained benzene soluble fractions above the evaluation criteria. However, it should be noted that the following three issues complicate the interpretation of the benzene soluble fraction data:

1. Three of the four area samples contained benzene soluble fractions that were higher than the corresponding total particulate values. This is improbable; the benzene soluble fraction can be equal to, but never exceed the total particulate value. A review of the blank values revealed no discrepancies in the analytical procedures. No explanation for this anomaly can be offered at this time.
2. None of the electrocautery samples taken during the February 14 survey contained a benzene soluble fraction, while all the samples taken on December 12 had a benzene soluble fraction above the NIOSH evaluation criteria. Sampling conditions were almost identical. Again, no explanation for the difference in concentrations can be given.
3. PNAs are benzene soluble and would therefore be contained in the benzene soluble fraction of the total particulate samples. It follows that the higher the value for benzene solubles, the more potential for PNAs and, therefore, the greater the risk from the exposure. However, it should be noted that this concept is based on industrial exposures (coal tar pitch volatiles, petroleum coke, asphalt fumes, etc.). Although the benzene extractable fraction offers an easier, less expensive method of analysis than PNA quantitation, there is no certainty that there is a correlation between the two (especially in a non-industrial setting).

F. Trace amounts of hydrocarbons were identified, (utilizing high volume sorbent tube sampling) and would not be expected to cause ill health effects in most people at the levels detected.

G. Based on the FRIT (qualitative organic) analysis, the major component of the samples is a compound or compounds related to fatty acid esters.

H. Aldehydes (C_1 - C_8) aliphatic) were not present in quantifiable levels.

I. The results of the studies for airborne mutagens indicate that the solvent extracts of airborne particles collected from the hospital operating room using cauterization were mutagenic. The mutagenic activity varies from patient to patient; age, fat content and size. The patient in the first survey was older with more fat in the tissue than the patient in the second survey. By comparison, samples from the second survey showed at least double the mutagenic activity than those of the first. Whether exposure of operating room personnel to agents that are mutagenic to bacteria or the level and condition of these agents to which workers are exposed pose any genotoxic hazards is not known. Limited information suggests that there is a correlation between the bacterial mutagenicity level of airborne particles and lung cancer incidence.[16–17] Index of the mutagenicity of air particles has been considered to be a more powerful measure of the human health hazard of air pollution than the traditional indices of particulate concentration.[17] This information is yet to be validated by further epidemiological studies where the mutagenic activity of collected air samples is known. In the meantime, it may be prudent to monitor operating room personnel for any adverse health effects and to reduce mutagenically active contaminants whenever possible in the operating room.

J. Operating room staff experience acute health effects (upper respiratory and eye irritation, headache, nausea (obnoxious odors)) during this type of surgery where electrocautery techniques are used for a substantial part of the total operative procedure.

VIII. RECOMMENDATIONS

1. Engineering ventilation controls (smoke evacuation units) should be utilized to minimize the acute health effects and further reduce the potential for any chronic health effects. The smoke evacuation units will also eliminate the emissions that can impair the surgeon's vision.
2. Any further acute or chronic health effects experienced by the operating rom staff should be evaluated and documented.
3. Exposure to electrocautery smoke should be reevaluated if other techniques for identifying and quantitating the smoke emissions can be found or developed.

IX. REFERENCES

1. Moss CE: Evaluation of Body Currents from Exposure to Radiofrequency Fields. Paper Presented at the American Industrial Hygiene Conference, San Francisco, CA, May 15–20, 1988.
2. Paz J, Ingram TW, Milliken R, Hartstein G: Real Time Dosimetric Survey During Electrocautery Surgery. Paper Presented at the American Industrial Hygiene Conference, San Francisco, CA, May 15–20, 1988.
3. Kado et al.: A simple modification of the Salmonella liquid-incubative assay. *Mutation Res.* 121:25–32, 1983.

4. Ames et al.: Methods for detecting carcinogens and mutagens with Salmonella/mammalian microsome mutagenicity test. *Mutation Res.* 31:347–363, 1976.
5. Whong W-Z et al.: Development of an *in situ* microbial mutagenicity test system for airborne workplace mutagens: laboratory evaluation. *Mutation Res.* 130:45–51, 1984.
6. National Institute for Occupational Safety and Health. Current Intelligence Bulletin 48: Organic Solvent Neurotoxicity. Cincinnati, Ohio: National Institute for Occupational Safety and Health, 1987. (DHHS Publication No. (NIOSH) 87-104.)
7. American Conference of Governmental Industrial Hygienists. Threshold Limit Values and Biological Exposure Indices for 1987–1988. Cincinnati, Ohio: ACGIH, 1987.
8. Occupational Safety and Health Administration. OSHA Safety and Health Standards. 29 CFR 1910.1000. Occupational Safety and Health Administration, revised 1987.
9. Frank CW and Berry CM: N-nitrosamines. In: Patty's Industrial Hygiene and Toxicology Vol IIB, Chapter 43, 3rd revised edition, New York, John Wiley and Sons, 1981.
10. Bogovski R, Preussman EA, Walker EA, and Davis W: Evaluation of Carcinogenic Risk of Chemicals to Man. IARC Monographs, Vol. 1, International Agency for Research on Cancer, World Health Organization, Lyon, France, 1972.
11. National Institute for Occupational Safety and Health. Working with Carcinogens. Cincinnati, Ohio: National Institute for Occupational Safety and Health, 1977. (DHEW publication no. (NIOSH) 77-206.)
12. Scala RA: "Toxicology of PPOM" *J. Occupational Medicine,* 17:784–8, 1985.
13. National Institute for Occupational Safety and Health. Criteria for a recommended standard: occupational exposure to coal tar products. Cincinnati, Ohio: National Institute for Occupational Safety and Health, 1978. (DHEW publication no. (NIOSH) 78-107.)
14. National Institute for Occupational Safety and Health. Criteria for a recommended standard: occupational exposure to asphalt fumes. Cincinnati, Ohio: National Institute for Occupational Safety and Health, 1978. (DHEW publication no. (NIOSH) 78-106.)
15. National Institute for Occupational Safety and Health. Petroleum Refinery Workers Exposure to PAHs at Fluid Catalytic Cracker, Coker, and Asphalt Processing Units. Cincinnati, Ohio: National Institute for Occupational Safety and Health, 1981. (NIOSH Contract No. 210-78-0082.)
16. Kaiser C, Keer A, McCalla DR, Lockington JN, Gibson ES: Use of bacterial mutagenicity assays to probe steel foundry lung cancer hazard. Polynucl. Aromat. Hydrocarbons int. Symp. Chem. Biol. Effects, 5th, Ohio 583–592, 1981.
17. Walker RD, Connor TH, MacDonald EJ, Trieff NM, Legator MS, MacKenzie KW, Dobbins JG: Correlation of mutagenic assessment of Houston air particulate extracts in relation to lung cancer mortality rates. *Environ. Res.* 28: 303–312, 1982.

X. AUTHORSHIP AND ACKNOWLEDGMENTS

Report Prepared by: Charles J. Bryant, C.I.H., Industrial Hygienist, Industrial Hygiene Section; Richard Gorman, C.I.H., Industrial Hygiene Engineer, Industrial Hygiene Section; John Stewart, B.S., Research Microbiologist, NIOSH, Morgantown, West Virginia; Wen-Zong Whong, Ph.D. Geneticist, NIOSH, Morgantown, West Virginia.

Field Assistance: James Boiano, C.I.H., Industrial Hygienist, Industrial Hygiene Section.

Originating Office: Hazard Evaluations and Technical Assistance Branch, Division of Surveillance, Hazard Evaluations, and Field Studies.

Report typed by: Kathy Conway, Clerk-typist, Industrial Hygiene Section.

XI. DISTRIBUTION AND AVAILABILITY OF REPORT

Copies of this report are currently available upon request from NIOSH, Division of Standards Development and Technology Transfer, Publications Dissemination Section, 4676 Columbia Parkway, Cincinnati, Ohio 45226. After 90 days, the report will be available through the National Technical Information Service (NTIS), 5285 Port Royal, Springfield, Virginia 22161. Information regarding its availability through NTIS can be obtained from NIOSH Publications Office at the Cincinnati address. Copies of this report have been sent to: 1. John E. Gatti, M.D., J. Brien Murphy, M.D. and R. Barrett Noone, M.D.; 2. NIOSH, Boston Region; 3. OSHA, Region III.

For the purpose of informing affected employees, copies of this report shall be posted by the employer in a prominent place accessible to the employees for a period of 30 calendar days.

Source: HETA 85-126-1932, September 1988, Bryn Mawr Hospital, Bryn Mawr, Pennsylvania. NIOSH Investigators: Charles Bryant, M.S., C.I.H.,
Richard Gorman, M.S., C.I.H., John Stewart, B.S., and Wen-Zong Whong, Ph.D.

(Mention of company names of products does not constitute endorsement by the National Institute for Occupational Safety and Health.)

PART 4

Laser Hazards in the Health Care Industry

Lindsey Kayman

Lasers have opened new vistas in many branches of health care. As with any technology, lasers are not without risks. Therefore, special precautions must be taken to prevent serious potential health hazards which can pose a threat to both medical staff and patients. This guide will summarize major laser safety issues, discuss regulation of lasers and provide resources for further information.

BACKGROUND INFORMATION ON LASERS

The word "laser" means light amplification by the stimulated emission of radiation. Lasers are now used in many industries, including building construction and telecommunications, and in consumer products such as compact disk players and office printers. In health care settings lasers are widely used for microscopic surgery and for measuring immunoglobulins and other elements in the blood.

Lasers create biological effects in tissue because they focus large amounts of light energy on a small surface area. The precise effect depends on the length of exposure time. When tissue is exposed to very short laser pulses, it becomes extremely hot and promotes a micro-explosion with a shock wave. The laser energy causes coagulation, cutting or vaporization of the tissue. At longer exposure times, the absorbed energy is spread out over a larger area, and the effect is smaller.

Commonly used lasers in health care settings are the carbon dioxide laser and argon laser which use gas, and the Nd:Yag laser which uses a crystal medium. Continuous wave lasers deliver energy as long as the laser is activated; pulsed lasers deliver short bursts of energy. Each laser system has its own unique set of optical properties, controls and output characteristics. The chart below lists some lasers used in health care and their application.

Lasers are being used in health care settings precisely for their ability to heat and cut biological tissue. Yet, this ability of lasers to damage biological tissue is also the reason why lasers are potentially dangerous to users and patients, unless lasers are properly operated and maintained.

Various professional organizations and government agencies have recognized the potential dangers of laser use. They have categorized lasers into four major hazard classes according to the wavelength and the power density of the laser beam. Most lasers used in surgery are considered Class 4 high-risk laser products.

1-56670-083-3/94/$0.00+$.50

Types of Lasers Used in Health Care	
Type	**Application**
Carbon dioxide	Incision and excision by vaporization in neurosurgery, otolaryngology, gynecology, podiatry, general surgery
Argon	Coagulates tissue, sealing blood vessels in retina. Also used in plastic surgery, gastroenterology and dermatology
KTP 532 (Green)	Cutting, coagulation, vaporization of tissue. Used in otolaryngology, gynecology, neurosurgery, urology, podiatry, general surgery, etc.
Nd:YAG (continuous wave)	Thermal effect and deep penetration used in gastrointestinal, urological, gynecological, general surgery
Nd:YAG (Q switched)	Breaking apart a target tissue. Used primarily in ophthalmology
Helium-neon	For aiming the invisible beams of carbon dioxide and Nd:YAG lasers
Ruby	Destroying tissues in dermatology and plastic surgery
Tunable dye	Treating malignant tissues

Source: Abstract from Emergency Care Research Institute (ECRI), Health Care Environmental Management Systems, "Special Issues in Healthcare Safety" Volume 3, April 1991.

Every employer that utilizes Class 3B or Class 4 lasers must have a trained Laser Safety Officer (LSO) who has the knowledge and authority to monitor and enforce the control of laser hazards. The LSO must:

- Confirm the classification of lasers used at the workplace
- Approve standard operating procedures
- Ensure that workers exposed to lasers have received proper training
- Limit the access of non-essential personnel to laser work areas
- Maintain laser equipment properly
- Ensure that laser equipment is properly installed
- Reduce or eliminate other risks in the work area that could make the use of lasers more dangerous to users
- Recommend appropriate protective equipment such as eye wear, protective clothing
- Specify appropriate warning signs

LASER HAZARDS

Several hazards are associated with lasers in the health care industry: damage to the eyes, skin burns, inhalation of toxic chemicals and pathogens, fires and electrical shock. Each hazard and how it can be controlled is discussed below.

EYE HAZARDS

The eye is especially vulnerable to injury from a laser beam; it is generally considered to be the organ at greatest risk. Exposure of unprotected eyes for a fraction of a second, as well as chronic lower-power exposures from scattered, diffused and reflected laser beams (such as from surgical instruments and tissue) can cause serious, irreversible damage. The damage will be located in the part of the eye where the laser energy is absorbed. The invisible beam of the laser which operates in the infrared region of the light spectrum is absorbed in the cornea and may cause corneal scarring and loss of vision. Lasers which transmit light in the visible or near infrared regions are focused by the lens of the eye to produce an intense concentration of light energy on the retina. The energy is converted to heat and causes a burn and loss of vision, especially if the burn is located in the macula, the center of visual acuity of the retina. Reflected laser beams can also injure the eye.

Eye Protection Devices

Engineering controls, which are built into laser equipment by the manufacturer, are an important means of controlling laser hazards. Examples of engineering controls are protective housing, fail-safe interlocks, master switch controls and beam stops or attenuators to reduce output emissions. However, these measures may not completely control the beam stray reflections in many situations. Therefore, OSHA requires eye wear in addition to engineering controls whenever accessible emission levels exceed maximum permissible exposure levels.

Unfortunately, eye wear creates many problems for users, which explains why many workers are reluctant to wear it: eye wear is often uncomfortable. It also reduces visibility—it can fog up; it can cause tunnel vision, making it difficult to see the laser beam and the patient. For these and other reasons the LSO must select eye wear that is effective and also "user-friendly"; i.e., eye wear that has good visibility and is comfortable to wear.

Several complex factors must be taken into consideration when selecting eye wear. For example, appropriate eye wear is dependent on the wavelength and power density of the laser in use: eye wear for the carbon dioxide beam may not protect against the Nd:YAG laser. Many other considerations affect the quality and appropriateness of eye wear. These are discussed below.

Optical Density

The filtering ability of laser eye wear is rated by a factor called optical density (OD). The laser manufacturer or LSO should determine the OD needed for protective eye wear. A high OD at a given wavelength indicates greater laser beam absorbency. The OD and the wavelength the eye wear is designed for must be imprinted on the lens or frame. Note: The OD does not take into account the mechanical strength of the eye wear.

Although it is more protective from a safety viewpoint, eye wear with a high OD causes reduced visibility, difficulty seeing a beam which normally may be visible and eye fatigue.

The visibility allowed by protective eye wear is rated by its visible luminous transmission (LT) expressed as the percentage of light seen through the glasses; the higher the LT, the better the visibility. Eye wear used during procedures should have the highest LT at the safest OD.

Goggles vs. Glasses

Goggles rather than glasses are generally recommended to protect against back reflection or side entrance of a stray beam. With side shields, properly fitted, lightweight laser protective spectacles provide a good alternative to goggles if discomfort might result in their not being worn. Side shields made of the same material as the lenses allow for adequate peripheral vision as well.

Reflective vs. Absorptive Laser Protective Eye Wear

The lenses of protective eye wear are either reflectors or absorbers. Both can be made out of plastic or glass. Reflective lens filters have a thin surface coating which is designed to reflect a laser beam away from the eye. An advantage of reflective coatings is good visibility. They can be designed to selectively reflect a given wavelength while transmitting as much of the remaining visible spectrum as possible.

A surface scratch is a serious problem in a reflective filter because it could allow penetration of the laser beam, causing possible injury to the eye. In addition the angle at which the laser beam hits the lens could affect the protection afforded by the eye wear. Another problem with reflective lenses is that the laser beam remains a safety hazard when it is reflected. Reflected beams are the most common type of safety hazard from lasers. Co-workers may be exposed to the reflected beam. There is a danger of fire if the beam hits a combustible or flammable item.

Rather than reflecting the beam, absorptive filters convert the incoming laser energy to a heat which is harmlessly diffused through the lens. Absorptive lenses have many benefits compared to reflective lenses:

> The protection afforded by absorptive filters is not affected by surface scratches; they are less likely to create hazardous beam reflections; and, the protection they afford is not affected by the angle at which the beam hits the lens. However, many absorbing glass filters cannot be easily annealed (thermally hardened). Consequently, they do not provide adequate impact resistance. In some goggle designs,

impact-resistant polycarbonate filters are used together with non-hardened glass filters to provide good impact resistance. Another problem with absorbing plastic filter materials is that the organic dyes which are used as absorbers are affected by heat and/or ultraviolet radiation which can cause the filter to darken or decrease its ability to absorb laser energy.

Plastic vs. Glass Lenses

There are pluses and minuses of both glass and plastic (polymeric) protective lenses. Unless thermally hardened, glass does not resist physical impact as well as plastic. Glass is more easily scratched than many polymers. Although glass can usually withstand higher laser exposure levels than plastic, plastic boasts a high heat-deflection temperature which enables them to withstand laser beams with high energy densities.

Plastic is more lightweight than glass, and it may be molded into comfortable shapes. Plastic materials generally also display a lower threshold for laser beam penetration. According to OSHA Guidelines, plastic eye wear is only appropriate when the wearer is more than an arm's length from the target area where the beam is focused (typically 0.5 meters, or about 18 inches); when the direct "raw beam" exposure cannot exceed 20 watt level.

Plastic eye wear should be adequate for support staff standing at a distance from the laser, but plastic filters are not considered protective enough for technicians servicing the laser. A 20 watt "raw beam" exposure would be more likely to occur to workers during servicing of the laser or to the operator of a laser while working at a close distance.

Other Considerations

- Some eye wear frames cannot withstand the same exposures that the lenses are designed to tolerate. Certain frames are available which are coated with a laser-absorbent material to correct for this.
- Dyes used in eye wear to absorb laser radiation can be bleached or darkened by long term exposure to light and heat or can simply deteriorate over time.
- Buildup of humidity within tight fitting eye wear is a problem. OSHA has cited a hospital for use of eye wear with side shields that had air circulation holes since open holes can increase the risk of beam exposure. Covered vents are more acceptable. Fogging may be reduced by the use of anti-fogging cleaning solutions.
- If protective eye wear which provides multi-wavelength eye protection is to be used, each wavelength and corresponding optical density should be confirmed in advance and understood by each wearer.
- Some brands of eye wear are designed to improve visibility by selectively altering and reflecting the hazardous wavelengths while transmitting a great deal of the remaining visible spectrum.

- Certain eye wear filters make it difficult to see certain colors found in the beam, blood and tissue, colored warning lights, laser emission indicators and other important instrument displays. Alignment eye wear may be useful in these situations. It is used for low power visible laser beams which align the high power or invisible beam. Wearing alignment eye wear with a low power visible beam allows beams, blood and instrument displays to be seen while providing some protection from diffuse radiation. Alignment eye wear should never be worn during the operation; it is also not meant for use with power or invisible beam lasers.

Controlling Eye Hazards

The employer's LSO should develop a written safety program including a protocol for eye hazards. The protocol should include the following:

- The determination of a Nominal Hazard Zone (NHZ). The NHZ is the space within which the level of direct, reflected or scattered radiation during normal operation exceeds the applicable maximum permissible exposure; it is the area around the laser where eye protection is required when the laser is activated. The NHZ may be determined by using the tables in the American National Standards Institute's (ANSI) laser standard, Z136.1 Section 8 and Appendix B. (See Resources Section for information about ANSI). The NHZ can also be derived from information supplied by the laser manufacturer.
- Periodic eye examinations. The frequency and content of eye exams are found in the eye examination chart.
- Selection of eye wear. A variety of high quality models that provide good visibility and comfort should be available. Joint labor/management safety committees or product evaluation committees can be helpful to the LSO in evaluating and selecting appropriate eye wear.
- Eye wear inspection and maintenance program. Eye wear should be properly stored when not in use. It should be periodically cleaned and inspected and should be replaced if discolored, pitted, or cracked.
- Using additional controls to reduce exposure. Eye wear is not the only protection against vision hazards. The LSO should also make sure that additional protective measures are also in place to reduce potential exposures. Here are a couple of examples:

  A thorough examination of all operating room materials and instruments must be conducted to, reflecting surfaces. Those that could be in the path of the laser must be replaced, modified or covered. Sometimes, instruments that appear to be non-reflective may, in fact, be reflective to infrared radiation. Only special non-reflective laser instruments should be used for laser surgery.

  Laser absorbent filters are available to shield OR observation windows and other areas. Use of filters allows observers to view laser procedures through a window without wearing protective eye wear.

EYE EXAMINATIONS

Eye examinations should be given before starting to work with lasers, periodically, following any incidental exposure to direct or scattered laser beam, and at the termination of a job involving lasers. Eye exams should include:

OCULAR HISTORY: The past history of the subject's eyes, including injuries, diseases which may affect the eyes, ocular problems which could be confused with laser beam induced injury, and vision correction.

VISUAL ACUITY: The acuteness or clearness of vision as tested by the standard eye chart.

MACULAR FUNCTION: The macula in the retina is responsible for central vision. Macular function is tested for distortions and scotomas.

CONTRAST SENSITIVITY: A test that determines the ability to distinguish objects by light and dark contrast.

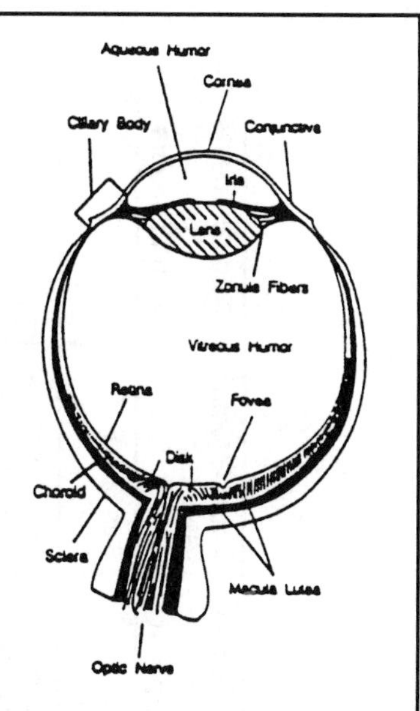

Source: ECRI, *Health Devices* 1989 Nov; 18(11):373.

OCULAR FUNDUS EXAMINATION: Given to individuals with abnormal acuity, macular function or contrast sensitivity, this ophthalmoscopic exam detects scarring and blood flow as well as other problems.

Source: Abstracted from ANSI Z136.3 Sections 6.2, 6.3, 6.4

SKIN DAMAGE

Damage to skin from laser beam exposure may range from localized reddening to charring and deep incision. The amount of damage is largely dependent upon wavelength and energy density of the beam and duration of exposure.

Skin effects for exposure to various wavelength radiation are detailed in Table 1.

Surgical gloves, gown, cap, mask and laser safety eye wear are considered adequate attire when working near lasers. The LSO should determine whether special fire-resistant clothing is necessary.

Table 1. Effects of laser exposure to eyes and skin.

Type of Laser	Wave Length (in micrometers)	Affected Part of Eye	Acute Effects of Eye Exposure	Chronic Effects of Eye Exposure	Acute Effects of Skin Exposure	Chronic Effects of Skin Exposure
CO_2	Far IR invisible beam 10.6	Cornea, Lens	Burns; ulcers; intense pain	Cataracts; blindness	Skin burns	--
Nd:YAG	Near IR invisible beam 1.064	Retina	Because retina has no pain sensing nerve, exposure to invisible beam may go unnoticed until considerable damage is done. Q-switched Nd:YAG eye exposure causes audible pop. Retinal coagulation, blind spots, ulcers	Cumulative retinal injuries: (at first, difficulty detecting blue-green); progressive damage; blindness	Skin burns; photosensitive reactors	--

Table 1. Continued.

Type of Laser	Wave Length (in micrometers)	Affected Part of Eye	Acute Effects of Eye Exposure	Chronic Effects of Eye Exposure	Acute Effects of Skin Exposure	Chronic Effects of Skin Exposure
Argon	Blue-green .488 and .514 visible	Retina	Bright flash of laser beam color followed by after-image of complementary color. Retinal coagulation; blind spots; blindness	Cumulative retinal injuries; progressive damage; blindness	Skin burns; photosensitive reactions	Retinal changes, particularly to color vision and small angle acuity
Helium-Neon	Red .632 visible	Retina	Same as above	Same as above	Same as above	--
Gold Vapor	Red .632 visible	Retina	Same as above	Same as above	Same as above	--
KTP/532	Green .532 visible	Retina	Same as above	Same as above	Same as above	--
Argon fluoride	.193 UVC	Cornea Conjunctiva	Photo keratitis: inflammation of outer layer of cornea; redness, tearing, discharge from conjunctiva; corneal surface cell layer splitting; stromal haze	--	Skin burns; skin tanning	Skin cancer

Table 1. Continued.

Type of Laser	Wave Length (in micrometers)	Affected Part of Eye	Acute Effects of Eye Exposure	Chronic Effects of Eye Exposure	Acute Effects of Skin Exposure	Chronic Effects of Skin Exposure
Krypton chloride	.222 UVC	Cornea Conjunctiva	Same as above	--	Same as above	Same as above
Krypton fluoride	.248 UVC	Cornea Conjunctiva	Same as above	--	Same as above	Same as above
Xenon chloride	.308 UVB	Cornea	Photokeratitis	--	Increased accelerated pigmentation; skin burns	Skin aging; skin cancer
Xenon fluoride	.351 UVA	Cornea, lens	Redness and tanning of eyelid skin	Cataracts; blindness	Tanning; skin burns	--

LASER PLUMES-JUST A BAD SMELL?

Plume Hazards

Medical lasers work by vaporizing, coagulating, etc. human tissues. The resulting vapors, smoke and particle debris, called "laser plume," is composed of both gaseous and particulate pollutants. The amount and type of plume is dependent on the type of surgery being done, the type of laser, the surgeons technique and other factors.

Laser surgery team members may suffer acute (short term) health effects from laser plume exposure including: eye, nose and throat irritation, tearing of the eyes, abdominal cramping, nausea, vomiting, nasal congestion, poor inspiratory effort, chest tightness, flu-like symptoms and fatigue. Symptoms may persist for 24 to 48 hours after exposure.

There are no studies at this time documenting chronic health effects caused by long term exposure to plume. However, carcinogens, mutagens, irritants and fine dusts have been found in laser plumes, as well as viable bacteria spores, cancer cells and viral DNA. A discussion of these follows here.

Plume Contents

1. Lung damaging dust

In one study, smoke was collected from nearby patients' abdomens during carbon dioxide laser laparoscopic treatment of endometriosis. Smoke particles were found to be spheres with the size range 0.1–.8 micrometer (μm). When particles of this size are inhaled they can penetrate to the alveoli, the deepest regions of the lung. There is concern that high exposure to these particles in the laser plume over time can cause lung problems similar to that found with other similar sized dust, including coal, cotton and grain dusts and cigarette smoke particles. Surgical masks are not able to filter out such small particles. Studies of the effects on laser smoke in the lungs of rats found identifiable lesions as a result of chronic exposure.

2. Toxic chemicals

Some of the chemicals which have been documented from laser beam contact with human and animal tissue are listed below. The number and type of chemicals depend on the type of surgery and other factors.

- Benzene
- Formaldehyde
- Acrolein
- Aldehydes

- Polycyclic hydrocarbons
- Methane
- Hydrogen cyanide and cyanide compounds
- Water

Benzene and formaldehyde are known carcinogens. Acrolein produces a bad odor and is an irritant at extremely low concentration. Hydrogen cyanide and cyanide compounds are irritants.

If the laser beam contacts material other than tissue, additional chemicals will be given off. Using the laser to remove methyl methacrylate bone cement generates formic acid formaldehyde, acrolein and methyl methacrylate monomer.

Teflon®-coated products should never be used in the vicinity of the laser. Lethal hydrofluoric acid can be given off if the laser beam touches the material or if the material was involved in a fire involving an oxygen enriched atmosphere.

3. Biological agents

The possibility of transmitting pathogens through the particles in the laser smoke is still being researched. However, in various studies, intact cells, identifiable cell parts, bacterial spores and intact viral DNA have been collected from the plume. Here are some examples:

- Human Papilloma Virus (HPV) is associated with warts, other lesions as well as benign and malignant skin tumors. Lasers are commonly used to vaporize warts and other lesions caused by HPV. Intact DNA from HPV has been found in the plume. However, a study found that most doctors who contracted warts got them on their hands, probably from direct contact, rather than from the laser plume.
- A CO_2 laser was used to vaporize HIV culture medium in a laboratory. HIV DNA was found in the smoke evacuator hose used to collect the laser plume. The HIV particles were found to be infective, although long-term replication appeared to be impaired.
- At low irradiance levels, viable biological spores were found in the plume from CO_2 lasing of bacteria-treated skin.

Further studies are certainly needed for a better understanding of the potential infectivity of laser plumes with varying characteristics.

Controlling Plume Exposures

If smoke is not adequately scavenged, contaminants will buildup in the air. Laser plume buildup not only contains materials which can be hazardous to health; the plume can also obstruct workers' field of vision during surgical procedures. If there is inadequate room ventilation, contaminants may remain in the air for significant time periods even after the surgery is completed, which could result in exposure to employees and to subsequent patients. Therefore, it is essential to reduce the plume as much as possible, if not to eliminate it altogether.

1. Ventilation

In-House vs. Dedicated Exhaust Systems

The most important control measure for gaseous and particulate emissions is local exhaust ventilation. Local exhaust ventilation eliminates the plume at its source. Many hospitals use the wall suction system for this purpose. In-line filters must be used, or the suction lines will become quickly clogged. However, the high efficiency filters which are necessary to capture sub-micron sized particles may offer too much resistance for the system and may result in too little suction capability. In addition, in-house vacuum lines may terminate in a machine room in some very old units. In such situations maintenance employees may become inadvertently exposed to plume contaminants.

A suction hose which is attached to a dedicated exhaust (a duct which exhausts air directly out of the building rather than recirculating it) is preferable to the in-house vacuum system. The blower should be located on top of the roof. This will keep the ductwork under negative pressure, preventing leakage of toxic or malodorous contaminants. Finally, various nozzles should be available for improving collection efficiency of plume contaminants for different procedures.

Smoke Evacuation Systems

If a dedicated exhaust system is unavailable, plume buildup and exposures can be adequately controlled by the proper and diligent use of a smoke evacuation system. Smoke evacuators are used to suction the smoke generated by laser procedures. Activated charcoal beds are used to reduce odors and remove certain organic vapors. High Efficiency Particulate Accumulator (HEPA) or Ultra Low Penetration Air (ULPA) filters are used to remove airborne particles. HEPA filters generally remove 99.97% of 0.3 micron particles. ULPA filters generally remove 99.99% of 0.12 micron particles.

Studies have shown that smoke evacuator performance can be affected by:

- Angle placement of the nozzle. The nozzle should be placed in the same direction as external air flow of plume production.
- How close the nozzle is to the treatment site. An orifice 1 cm from the treatment site is 98.6% efficient. If the laser collection device is moved only 2 cm away from the exposed tissue, the collection efficiency is reduced to 50% and drops further as the distance is increased.
- High power versus low power laser procedures. High power laser procedures emit smoke in many directions at a greater velocity compared to low power

procedures. Higher smoke evacuator air flow speeds are necessary for high power procedures.

2. Respiratory Protection

Respirators should never replace air cleaning. Respirators should be used as additional protection, where needed. Like other protective equipment, such as eye wear discussed above, respirators come with a host of problems. These problems are discussed here.

Standard surgical masks are designed to protect patients from the germs of their health care provider, but surgical masks do not provide adequate employee protection against plume contaminants including:

- Chemical vapors such as formaldehyde, acrolein, cyanide which may be found in the plume.
- Very tiny dust particles, including viruses, which once inhaled, can reach the lung's alveoli and possibly cause damage.
- Cellular debris and bacteria. Surgical masks may filter much of these larger particles; however, particles can enter the breathing zone at the nose piece and other loose-fitting areas.

More research is needed on which respirators should be worn to protect employees from inhaling laser plume. One possible alternative to surgical masks in use at some institutions is industrial-type disposable respirators, which are available from industrial respirator suppliers (e.g., 3M or Moldex). Disposable respirators contain materials which are more effective at filtering out small particles than the paper used for surgical masks. Disposable respirators also provide a better fit than surgical masks, and they are comfortable to wear. However, at this time it is not known if their use would offer significantly more protection than standard surgical scrub masks.

Rubber or silicone half mask respirators with disposable cartridges are another alternative. Such respirators are even more protective than disposable respirators, but they are considered impractical for the surgical setting. They are uncomfortable, difficult to wear with eye projection, and they cannot be worn by persons with moustaches, beards or sideburns. They may also be too distracting for the intense concentration needed during laser surgery. Half-mask respirators should be considered if there is the likelihood of exposure to biohazardous aerosols. The cartridges should contain a combination of HEPA filters and organic vapor adsorbent materials to prevent exposure to both particles and chemical vapors.

If disposable respirators, half-mask respirators or any other type of respirator is to be used by employees, OSHA requires that a comprehensive respirator program must be put in place. OSHA's respirator program includes fit-testing training and proper maintenance. (See OSHA Standard 29 CFR 1910.134 for more details.)

TIPS ON USING SMOKE EVACUATORS

- Alternate nozzles may be useful for different surgical procedures.
- The smoke evacuation system should be left on for a short time after lasing is completed in order to clear remaining pollutants from the air.
- The smoke evacuator must be regularly maintained. Charcoal and particulate filters must be changed when they have reached capacity.
- Whenever possible, the air leaving the smoke evacuator should be exhausted directly outdoors to prevent contaminants which have not been adsorbed or filtered from reentry into the building.
- Smoke evacuator filters must be changed once they have reached their capacity. This is best determined by a pressure drop across the filter. Charcoal filters are used to remove gaseous hydrocarbons. Higher grade filters are able to adsorb more carbon tetrachloride. A CTC-60 grade charcoal is capable of adsorbing 60% carbon tetrachloride by weight, whereas a CTC-80 grade charcoal can adsorb 80% carbon tetrachloride by weight.
- All interconnecting parts of the smoke evacuator system must be kept free of kinks and clogs, and they must be maintained at optimal efficiency at all times.
- Collection efficiencies of filters may vary with the manufacturer. Leakage around the filter can occur if the filter is not properly fitted in its housing, or if the filter has been damaged. Ability to suck will be reduced when the filter has reached its capacity.

3. Skin Protection

Surgical gloves, gown and cap are considered adequate attire when working near lasers. These garments will protect skin from exposure to laser plume and tissue debris.

Most Frequently Reported Laser Accidents

- Unanticipated eye exposure during alignment
- Misaligned optics
- Available eye protection not used
- Equipment malfunction
- Improper methods of handling high voltage
- Intentional exposure of unprotected personnel
- Use of unfamiliar equipment
- Lack of protection from ancillary hazards
- Improper restoration of equipment following service

Source: Abstracted from D. Terrible, "Preview of the 21st Century." *Ohio Monitor,* February, 1986, p. 6.

FIRES

Causes of Laser Fires

Fires caused by lasers are uncommon, but they do happen. A fire can be started by inadvertently misdirecting the beam or the reflection of the beam onto a combustible material. Reflective surfaces that are concave in shape can focus the beam, making it potentially more harmful.

If oxygen or nitrous oxide is being used, there will be extra oxygen in the air. The oxygen enriched atmosphere will allow many materials to ignite that are not normally flammable—even those marketed as fire resistant. In an oxygen enriched environment the flames will be especially intense.

Conventional fires can also be a problem. When the laser is in the standby mode, the hot tip can start a fire if it touches surgical drapes or other flammable materials. The toxic fumes produced when certain plastics and fabrics burn can be life-threatening for patients and staff.

Items which may be ignited by a laser beam include:

- Intestinal gas (flatus)
- Hair (facial hair is particularly vulnerable due to its proximity to anesthetic gases)
- Skin that has been treated with acetone or alcohol based skin preparation solutions
- Paper products
- Surgical drapes (even if they are marketed as fire resistant)
- Rubber and plastic instruments including nonmetallic endotracheal tubes: PVC, silicone red rubber
- Human tissue, miscellaneous materials including plastic adhesive tapes, oil-based lubricants, ointments, gloves, dry gauze, cotton

Preventing Laser Fires

- All personnel should be trained in the causes and prevention of fires as well as appropriate responses to a fire involving a laser.
- Precise control of the laser beam must be maintained at all times.
- Rooms, instruments and equipment should be inspected to detect surfaces which can cause unwanted reflection of the beam.
- During surgery, the laser should be left in the stand-by mode at all times except when the handpiece is in the hands of the surgeon.
- Skin preparation solution vapors should fully vaporize before covering the area with surgical drapes.
- When in the stand-by mode, the hot tip of the laser should not be allowed to touch combustible items.
- During surgery in or near the bowel, proper bowel preparation is necessary to avoid ignition of rectal gas.
- During laryngotracheal surgery an endotracheal tube is sometimes used. Special precautions must be taken to reduce the risk of combustion while keeping the patient well-oxygenated. Silicone, PVC, red rubber and even

specially designed laser-resistant endotracheal tubes may be ignited in oxygen rich atmospheres. Stainless steel endotracheal tubes or non-endotracheal methods to ventilate the patient should be used.

Controlling Laser Fires

- A basin of sterile saline and a syringe should be kept on hand to douse a small fire and to keep protective dressings wet.
- Portable fire extinguishers should be conveniently located. Personnel should be trained in how to use them.
- National Fire Protection Association Standard 99 (1990) Appendix C-12.4 "Suggested Procedures in the Event of a Fire or Explosion-Anesthetizing Locations" should be in effect.
- A Halon or CO_2 fire extinguisher should be used rather than water if power cannot be disconnected during a fire.

ELECTRICAL HAZARDS

Electrical hazards are the largest cause of fatalities in accidents involving lasers. Many lasers use high voltage and high amperage currents. Electrical hazards are usually minimized by enclosures around high voltage devices within the laser cabinet. Electrocutions have primarily involved technicians who opened the protective covering of the laser. The high voltage DC capacitors of some lasers can remain energized for an extended period of time after the laser has been unplugged from the wall outlet. Danger arises mainly when an untrained or unauthorized person attempts internal laser maintenance.

The American National Standard Institute (ANSI) recommends special precautions for equipment servicing within 24 hours after the presence of high voltage within the unit. A special grounding rod is used to ensure discharge of high voltage capacitors. Additional details for servicing equipment are included in ANSI Z 136.3 section 7.1.

ANSI Z 136.3 section 7.1 contains additional electrical safely requirements:

- Grounding of metallic parts of laser equipment
- Short circuit tests for combustible components in electrical circuits
- Prevention of shock hazard
- Prevention of electrical hazards from gas-laser tubes and flash lamps
- Labeling lasers with electrical rating, frequency and watts
- Preventing electromagnetic interference
- Preventing explosions in high pressure arc lamps and filament lamps

LASER SAFETY POLICY

Every employer utilizing a laser should have a written policy. The elements of a policy are outlined here.

LASER SAFETY PROGRAM

The employer has a written laser safety program which includes:

- Formation of a laser safety committee (LSC)
- Credentialing of medical personnel
- Education and training
- Safety precautions including general precautions, eye protection, eye examinations, pre-operative setup, operating precautions, anesthesia guidelines
- Maintenance and service
- Scavenging of plume contaminants
- New laser acquisitions
- Appointing a laser safety officer.

LASER SAFETY PERSONNEL

1. A laser safety committee (LSC) has been established which:

    - Reviews and approves new laser technology and the design of laser facilities
    - Reviews and approves protocols and safety measures for lasers used in medical treatment
    - Appoints a laser safety officer
    - Investigates unusual occurrences related to laser uses and reports the results to the Quality Assurance Program

2. The Laser Safety Officer (LSO) is trained in laser operation, clinical applications, safety measures. The LSO has the authority to monitor and enforce the control of laser hazards and effect the knowledgeable evaluation and control of laser hazards. The LSO acts as the laser education coordinator.

3. Laser Technicians report to the LSO on all technical and clinical issues. Laser technicians are responsible for the day to day operation of surgical lasers.

4. The laser user is responsible for performing a daily visual inspection of the laser system to check for apparent abnormalities.

LASER MAINTENANCE

1. All persons who install, maintain and/or service lasers have been properly trained and approved by the laser safety committee.

2. Enclosed lasers have built-in access panel interlocks and automatic shuttering to protect maintenance persons.

3. Maintenance on a laser is only done when another person is present to render emergency medical aid or to call for assistance in the event of an injury.

4. Initial installation and subsequent maintenance or servicing of lasers that may affect their performance is followed by performance and safety testing prior to the laser's use in the operating room.

TRAINING AND APPROVAL OF LASER OPERATORS

1. Only personnel with specific credentials are authorized by the laser safety committee to operate lasers.

2. All personnel who are present during laser operation or maintenance have received appropriate training on the principles of operating lasers, their applications, attendant risks to patient and staff, safety control measures and equipment care.

   Authorization is given to specified physicians (and specified other persons to use non-surgical lasers) for specific types of laser delivery systems and procedures. Authorization is dependent on at least one of the following criteria:

   - Documented attendance and completion of a formal laser training course offered by a recognized authority
   - Documented completion of a formal training program offered by the manufacturer of the laser
   - Residency or on the job training with proven proficiency while assisting a certified laser physician

GENERAL OPERATIONAL GUIDELINES

1. A nominal hazard zone has been established by the laser safety officer. Outside the nominal hazard zone, the level of direct, reflected or scattered radiation is not expected to exceed the applicable Maximum Permissible Exposure Limit.

2. The laser is locked when not in use or unattended to prevent unintentional or unauthorized activation. The key is stored in a secure location.

3. The laser technician ensures that all necessary equipment is ready for operation, and that all required safety precautions have been implemented, including visual checks of the surgical laser and its control settings.

4. All reflective surfaces which are likely to be contacted by the laser beam are removed from the path of a fixed laser or from the operating room prior to laser surgery.

5. All windows shall be covered as necessary.

6. Where necessary, safety latches or interlocks are used to prevent unexpected entry into laser controlled areas.

7. All access doors to laser operating rooms are posted with a laser warning system which includes a warning light and sign to prevent unexpected interruption of the laser operator.

PERSONNEL PROTECTION

1. Laser users and support staff have pre-placement and periodic eye examinations for adverse effects.

2. Appropriate laser eye protection is worn by all persons present prior to powering the laser system and throughout the entire procedure. This includes all staff, bystanders and the patient. The eye protection is labeled with its optical density and the wavelength it protects against.

3. Eye wear is appropriately cleaned after each use.

4. Respirators which are effective at filtering out particles as small as 0.3 μm with 99 percent efficiency are used in conjunction with scavenging of the laser smoke when there is a possibility of generating biohazardous aerosols.

OPERATING PRECAUTIONS

1. The laser is not switched to the operating mode until the procedure is ready to begin and the laser has been aligned and positioned.

2. The laser beam is never aimed at a person, except for the therapeutic purposes.

3. The laser is positioned to avoid placing the beam at eye level whenever possible.

ENGINEERING CONTROL MEASURES

1. Local exhaust ventilation is used to control airborne contaminants. Whenever possible, the exhaust is vented directly out of the building. Otherwise, smoke evacuators employing HEPA and charcoal filtration are used. Filters are changed regularly in accordance with the manufacturers directions.

2. Microscopes have a fail-safe method to project the users eyes against laser beam reflections (i.e., built-in filters, separate optical paths for intermittent viewing and firing, shutters that automatically close prior to firing).

FIRE SAFETY

1. Special precautions are taken if laser energy is to be used near an endotracheal tube.

2. Patients are ventilated with non-oxygen enriched room air when possible. If oxygen therapy is necessary it is administered at the minimum concentration necessary to properly support the patient.

3. Staff is knowledgeable about procedures to follow in the event of various types of fires.

4. Flammable prepping solutions are not used.

5. Liquids are never placed on top of the laser to avoid short circuits.

6. A basin of saline or water is on hand for patient-related fires.

7. A UL-approved fire extinguisher is readily available in the event of equipment or material fire.

8. Combustible material such as OR gowns, drapes and towels are kept out of the laser path to avoid combustion.

LASER SAFETY TRAINING

The Laser Safety Officer must ensure that an appropriate training program is in place for each group of employees who operate lasers or who are exposed to lasers. Here is an outline of topics to be included in a training program for health care support staff as recommended by OSHA and ANSI. (For information about OSHA and ANSI, see Laser Resources Section.)

I. The Laser Beam

a. What it is and what it can do
b. The hazards of lasers
c. Eye and skin hazards
d. Other laser hazards

Laser Safety Checklist

Date: ____________ **Location:** ______________
Laser status: A) not in use B) after use C) in use

Requirement:	YES	NO	N/A
Room:			
Appropriate signs posted outside procedure room when laser is in use (sign includes wavelength, class, "Warning," CW or pulsed)			
All windows and doors covered with a non-transparent, non-reflecting material			
Appropriate fire extinguisher present			
Gas tanks stored properly			
Laser keys locked in designated place when not in laser			
Auxiliary accessories stored properly			
Laser Users:			
All persons including patient, wearing appropriate eye wear when laser in use			
Laser:			
Electrical cords intact			
Outlets intact			
Water pressure adequate for cooling			
No leakage from laser hoses			
Operating procedures on laser			
Maintenance up-to-date			
Properly stored			
Laser log complete			
Laser log accurate			
Smoke evacuator:			
Used with laser			
Time recorded on filter			
Filter changed as necessary			
Valid safety sticker			
Properly stored			

COMMENTS ____

____

Name of Inspector ____

II. Safety Measures In Laser Surgery

a. Eye protection
b. Reflected beam hazards
c. Explosions
d. Smoke evacuation
e. Fire hazards
f. Details of standard operating procedures (SOPs) for the operating room.

III. Methods and Procedures to Assure Safety

a. Boundary of Nominal Hazard Zone
b. Laser area warning signs
c. Entry way controls
d. Availability of personal protective equipment: eye wear, non-flammable gowns, etc.
e. Control of unauthorized personnel to prevent access to the laser
f. Techniques for safety
g. Use of surgical drapes in laser surgery procedures
h. Proper laser system controls (e.g., foot switch)

GOVERNMENT REGULATION OF LASERS

1. The Occupational Safety and Health Administration (OSHA). (Look in the blue pages of your telephone book under U.S. Government, Department of Labor for OSHA office nearest you.)

OSHA is a regulatory agency within the U.S. Department of Labor that is responsible for overseeing the health and safety of America's private sector workers. (Public employees in many states are also covered by state OSHA plans that are administered by their state Department of Labor.) OSHA promulgates worker safety regulations and has the authority to enforce them. Employers that do not comply with OSHA regulations must correct the problem and pay fines.

Although OSHA has not yet set a formal standard for medical lasers, OSHA does have very detailed guidelines on lasers, which its inspectors use. These guidelines, which were issued on August 5, 1991, are called Guidelines for Laser Safety and Hazard Assessment (OSHA Instruction PUB 8-1.7, Directorate of Technical Support). The guidelines incorporate many of the recommendations for laser safety written by a professional organization called ANSI (see Resources below). The guidelines include detailed information in the following areas:

- The principles of laser operation and use
- Effects of laser light on the eye and skin
- Standards for laser safety

- Hazard evaluation
- Control measures
- Personal protective equipment
- Laser training requirements for various classes of lasers.

In addition to the guidelines, OSHA has cited employers for unsafe conditions under a variety of its existing standards. For example:

- Eye & Face Protection Standard (29 CFR 1910.133). OSHA can issue citations to health care employers
  - If the wrong type of eye wear is provided for the surgery being done
  - If employees fail to use eye wear
  - It the eye wear is not protective enough. For example, OSHA has issued a citation against a hospital for providing eye wear without side shields and for providing eye wear with vented side shields (because vents can allow the laser beam to penetrate through the opening).
- Respiratory Protection Standard (29 CFR 1910.134). OSHA can issue citations to health care employers
  - If the employer is using respirators without determining that wearers are medically able to wear them
  - If workers have not been trained how to use respirators
  - If respirators have not been properly fit on the wearer
  - If respirators are not properly stored
  - If reusable respirators are not cleaned and disinfected
- Fire Protection (29 CFR 1910.155-165). There are several standards for fire protection, including alarms, detection systems, extinguishing systems.
- General-Duty Clause, Section 5(a)(1) of the OSHA Act of 1970. The General Duty Clause states that it is the employers duty to provide a safe workplace. The clause is a catch-all for the enforcement of health and safety measures that are not specifically included in its regulations. OSHA has used the General Duty Clause to issue citations against hospitals that do not take proper precautions to prevent the laser from being inadvertently activated.
- Air Contaminants (29 CFR 1910.1000 Subpart Z). Despite the fact that the plume may contain ingredients that pose a real hazard to health it is unlikely that OSHA will find violations of any of its chemical standards. In most cases the specific chemical contaminants in the plume will be present in concentrations which are allowed by OSHA.

Furthermore, OSHA does not currently regulate exposure to biological agents in laser plumes. However, the ANSI standard which is incorporated into the OSHA Guidelines, includes a recommendation that effective smoke evacuators be used during laser surgery.

A more appropriate agency to investigate both chemical and biological hazards of laser plumes is the National Institute of Occupational Safety and Health (see below.)

2. National Institute for Occupational Safety and Health (NIOSH). U.S. Department of Health and Human Services, U.S. Public Health Service, Centers for Disease Control, 4676 Columbia Parkway, Cincinnati OH 45226 (513)841-4382

NIOSH was created by the OSHAct to be the research arm of OSHA. NIOSH does research on occupational hazards, evaluates control measures, and makes recommendations to OSHA for standards. Although its recommendations do not have the force of law, health care workers can call in NIOSH to perform a health hazard evaluation of the workplace, including the hazards of laser plume.

3. Center for Devices and Radiological Health (CDRH). Office of Compliance (HFZ-300), 8757 Georgia Ave., Silver Spring, MD 20910 (301)443-4190

The CDRH is a regulatory bureau within the Federal Food and Drug Administration of the U.S. Department of Health and Human Services. All laser products manufactured since 1976 must comply with CDRH specifications. Manufacturers obtain pre-market approval or clearance of their laser surgical devices through CDRH. The CDRH issues "The Compliance Guide for Laser Products", which summarizes the requirements of the U.S. Federal Laser Product Performance Standard (21 CFR Part 1000, 1040.10, 1040.11). Manufacturers should use this performance standard in order to comply with CDRH requirements for labeling and classifying lasers.

4. State Regulations

Laser regulations vary considerably from state to state. Such regulations are generally concerned with the registration of lasers and the licensing of operators and institutions.

New York laser regulations, Code Rule 50, are enforced by the Department of Labor. Massachusetts regulations, 105 CMR 21, are administered by the Department of Health. New Jersey and Connecticut do not have state regulations on lasers at this time. "Suggested State Regulations for Lasers" has recently been promulgated by the Conference of Radiation Control Program Directors. This may lead to changes in state regulations.

LASER RESOURCES

1. Government Agencies OSHA, NIOSH, FDA. See previous section on Government Regulation of Lasers.

2. American National Standards Institute (ANSI), 11 West 42nd St., 13th floor, New York, NY 10036 (212)642-4900

ANSI is a professional organization of engineers from many fields which has issued the leading consensus standard on the safe use of lasers (ANSI Z 136.1) and on safe use of lasers in medicine (ANSI Z 136.3) OSHA relies in large part on ANSI's research for its policies regarding laser safety.

3. American Conference of Governmental Industrial Hygienists (ACGIH), 6500 Glenway Avenue, Building D-7 Cincinnati, OH 45211 (513)661-7881

ACGIH is a professional organization of industrial hygienists and safety experts. The ACGIH has established maximum exposure limits (MPE) known as threshold limit values for employee eye and skin exposure to laser radiation. These MPE are used to select protective eye wear, determine nominal hazard zones and other safety precautions for laser use.

4. Joint Commission of Accreditation of Health Care Organizations (JCAHO), One Renaissance Blvd., Oakbrook Terrace, IL 60181 (708)474-7028

The JCAHO is a private, professional organization that accredits health care facilities. JCAHO is in the process of adopting the ANSI standard for medical lasers - ANSI Z-136.3. An institution's accreditation can be held up if it is not in compliance with ANSI.

5. The American Society for Laser Medicine and Surgery, Inc. 2404 Stewart Square, Wausau, WI 54401 (715)845-9283

This organization is composed of scientists, physicians, nurses and paramedical personnel. It holds annual meetings and publishes a bimonthly journal, *Lasers in Surgery and Medicine.*

6. Laser Institute of America, 12424 Research Parkway, Suite 130, Orlando, FL 32826 (407)380-1553

The Laser Institute of America is a non-profit educational society which conducts continuing education courses, holds technical symposia, offers educational materials and publishes the peer-reviewed journal, *Laser Topics.* It promotes the advancement of laser technology and applications.

7. Association of Operating Room Nurses (AORN), 2170 S. Parker Road, Denver, CO 80231 (303)755-6300

This professional organization publishes general information of interest to operating room personnel. See "Proposed recommended practices: Laser safety in the operating room", *AORN Journal,* 1989 49(1):284-91.

8. The Emergency Care Research Institute (ECRI), 5200 Butler Pike, Plymouth Meeting, PA 19462 (215)825-6000

ECRI is an independent, non-profit research and consulting organization which, for a fee, provides training seminars and publishes information on health technology, including: surgical lasers and accessories, laser fires, credentialing and training recommendations and safety programs. It also performs comparative evaluations of surgical lasers, laser resistant endotracheal tubes, smoke evacuators and laser protective eye wear.

Source: This document is reprinted courtesy of Communications Workers of America District 1

PART 5

Proposed Recommended Practices: Electrosurgery

Association of Operating Room Nurses

The following draft is being published for review and comment by Association of Operating Room Nurses (AORN) members. The AORN Recommended Practices Committee (RPC) is interested in receiving comments on this proposal from members and others.

These recommended practices are intended as achievable recommendations representing what is believed to be an optimal level of practice. Policies and procedures will reflect variations in practice settings and/or clinical situations that determine the degree to which the recommended practices can be fulfilled.

AORN recognizes the numerous different settings in which perioperative nurses practice. The recommended practices are intended as guidelines adaptable to various practice settings. These practice settings include traditional operating rooms, ambulatory surgery units, physicians' offices, cardiac catheterization laboratories, endoscopy rooms, radiology departments, emergency departments and all other areas where surgery may be performed.

Although nonmembers may submit comments, the intent of the committee is to reach a consensus among AORN members. All comments will be acknowledged and considered by RPC before final approval of these recommendations by the committee and the AORN Board of Directors. Comments should be sent to: Recommended Practices Committee, AORN, Inc., 2170 S. Parker Road, Suite 300, Denver, CO 80231-5711, Attention: Mary O'Neale, RN, BS, CNOR

PURPOSE

These recommended practices provide guidelines to assist perioperative personnel in the use of electrosurgical equipment in their practice settings. Proper care and handling of electrosurgical equipment is essential to patient and personnel safety. Electrosurgery is used routinely to cut and coagulate body tissue with high radiofrequency electrical current. These recommended practices do not endorse any specific product. Biomedical services in practice settings should develop detailed, routine safety and preventive maintenance inspections and maintain records.

RECOMMENDED PRACTICE I

The electrosurgical unit (ESU), dispersive electrode and active electrode selected for use should meet performance and safety criteria established by the practice setting.

1-56670-083-3/94/$0.00+$.50

Interpretive Statement 1

Information regarding adequate safety margins, in-factory testing methods, warranties and a manual for maintenance and inspections should be obtained from the manufacturer.

Rationale

Equipment manuals assist in developing operational, safety and maintenance guidelines.[1] The ESU should be used according to the manufacturer's written instructions.[2]

Interpretive Statement 2

The ESU should be designed to minimize unintentional activation.

Rationale

Unintentional activation may result in patient and personnel injury.[3]

Interpretive Statement 3

The ESU cord should be of adequate length and flexibility to reach the outlet without stress or use of an extension cord. Kinks, knots, and curls should be removed from the ESU cord before it is plugged into the wall outlet.

Rationale

Tension increases the risk that the cord will become disconnected or frayed, which may result in injury to patients and personnel.[4] Use of extension cords may result in macroshock or microshock.[5] Cords that do not lie flat on the floor produce a potential for tripping and/or accidental unplugging.[6]

Interpretive Statement 4

The ESU plug, not the cord, should be held when it is inserted into or removed from an electrical outlet.

Rationale

Pulling on the ESU cord may cause it to break at the point where the wire is attached to the plug.[7] Cord breakage is dangerous to patients and personnel and is inconvenient, and replacements are costly.[8]

Interpretive Statement 5

The ESU should be inspected before each use. An ESU that is not working properly or is damaged should be reported, labeled, and removed immediately to be checked by the biomedical department.

Rationale

Equipment is checked to ensure it is in good working order.[9] The manufacturer's written safety precautions are followed for the well being of the patient and personnel involved with the procedure.[10]

Interpretive Statement 6

The ESU should be grounded properly.

Rationale

Proper grounding reduces the risk of electrical shock to the patient and perioperative personnel.[11]

Interpretive Statement 7

The ESU should be mounted on a movable stand that will not tip.

Rationale

Safety measures for perioperative personnel and patients to prevent injury and damage to the ESU require the stand to be tip resistant and moved carefully.[12]

Interpretive Statement 8

The ESU and all reusable parts are cleaned with care following use according to the manufacturer's written instructions.

Rationale

The ESU surface should not be saturated or have fluid poured over it because this could permit chemical germicide into the generator and cause malfunction.[13]

Interpretive Statement 9

When the ESU foot switch is used, perioperative personnel should cover it with a clear, impervious cover if recommended by the manufacturer.

Rationale

Placement in a clear, impervious cover protects the foot switch from fluid spillage.[14]

Interpretive Statement 10

During the procedure, perioperative personnel should check the entire circuit if higher than normal power settings are requested by the operator.

Rationale

The dispersive electrode, generator or connecting cords may be at fault and should be checked for any possible malfunction or hazard. Shock to those touching the patient may result. The patient and/or perioperative personnel may be burned.[15]

Interpretive Statement 11

Each ESU should be assigned an identification number/serial number.

Rationale

An identification number/serial number allows for documentation of inspections, routine preventive maintenance and tracking of equipment function and problems.[16]

RECOMMENDED PRACTICE II

Perioperative personnel should demonstrate competency in the use of the ESU in the practice setting.

Interpretive Statement 1

Perioperative personnel should be instructed in the proper operation, care and handling of the ESU before use.

Rationale

Instruction and return demonstration in proper usage helps prevent injury and extends the life of the ESU.[17]

Interpretive Statement 2

A detailed manual of operating instructions should be obtained from the manufacturer and be available in the practice setting. A brief set of operational directions should be on or attached to the ESU.

Rationale

Each type of ESU has specific manufacturer's written operating instructions that should be followed for the safe operation of the unit.[18]

RECOMMENDED PRACTICE III

The ESU, active electrode and dispersive electrode should be used in a manner that reduces the potential for injury.

Interpretive Statement 1

The ESU should

- Not be used in the presence of flammable agents (e.g., alcohol, tincture-based agents)
- Have safety features (e.g., lights, activation sound) and be tested before each use
- Have the cord, plug and foot switch cord checked for exposed wires or frays in the insulation
- Have power settings confirmed orally with the operator before activation and determined in conjunction with the manufacturer's recommendations
- Be protected from spills
- Be operated at the lowest effective power settings for coagulation and/or cutting

Rationale

Inspections of the ESU and all safety features should be performed before each use because of potential hazards.[19] The volume of the activation indicator should be adjusted to an audible level to alert perioperative personnel immediately when an ESU is activated inadvertently.[20] Ignition of flammable agents by the active electrode has resulted in patient and perioperative personnel injury.[21] Fluids should not be placed on top of the ESU, because unintentional activation or device failure may occur if liquids enter the ESU generator.[22]

Interpretive Statement 2

The active electrode should

- Fasten directly into the ESU in a labeled, stress-resistant receptacle (if an adapter is used, it should be one that is approved by the manufacturer and does not compromise the generator's safety features)
- Be inspected at the field for damage before use
- Be placed in a clean, dry, well-insulated safety holster (i.e., recommended by the manufacturer for use with the ESU) in a highly visible area when not in use during a procedure
- Be impervious to fluids
- Be disconnected from the ESU if allowed to drop below the sterile field
- Have a tip that is secure and easy to clean of charred tissue

Rationale

Incomplete circuitry, unintentional activation and incompatibility of the active electrode with the generator may result in patient injury.[23]

Interpretive Statement 3

The dispersive electrode should

- Be inspected before each use for wire breakage or fraying
- Be the appropriate size for the patient (i.e., neonate/infant, pediatric, adult) and never be cut to reduce size
- Be placed on the positioned patient on a clean, dry skin surface, over a large muscle mass and as close to the operative site as possible (i.e., bony prominences, scar tissue, skin over an implanted metal prosthesis, hairy surfaces, pressure points should be avoided)
- Fasten directly into the ESU in a labeled stress-resistant receptacle if an adapter, which is approved by the manufacturer and does not compromise the generator's safety features, is used

- Have connections that are intact, clean and make effective contact
- Maintain uniform body contact (potential problems include tenting, gaping and liquids that interfere with adhesion)

Perioperative personnel should check the status of the dispersive electrode and connection of the cable if any tension is applied to the cord or if the surgical team repositions the patient.

If reusable, the electrode should have periodic inspections by the biomedical service for electrical integrity and as recommended by the manufacturer.

Rationale

Wire breakage and frays can deviate current flow.[24] Incomplete circuitry may lead to patient injury.[25] Adequate tissue perfusion promotes electrical conductivity in the area and dissipates heat at the electrode contact surface.[26] Hair should be removed before applying the dispersive electrode according to the manufacturer's written instructions. Hairy surfaces have poor adhesion and tend to insulate.[27]

There is potential for superheating if a dispersive electrode is placed on the skin over an implanted metal prosthesis. The important factor in the dispersive electrode is the actual surface area in contact with the patient. The amount of surface area affects heat buildup at the dispersive site.[28]

Discussion

During some surgical procedures, it may be desirable to use two ESUs simultaneously on the same patient. Perioperative personnel should place each dispersive electrode as close as possible to the respective surgical sites and ensure that there is no possibility of the two dispersive electrodes touching. The two ESUs must be of the same technology (e.g., both grounded ESUs, both isolated ESUs). The biomedical service should test ESUs to ensure that simultaneous operation will not create any microshock hazards.

Interpretive Statement 4

The bipolar ESU should be used with its foot switch or a hand switching forceps according to the manufacturer's written instructions.

Discussion

In bipolar electrosurgery, a forceps is used for the coagulation of body tissue. One side of the forceps is the active electrode and the other side is the inactive electrode or ground. A dispersive electrode is not needed because current flows between the tips of the forceps rather than through the patient. The operator uses a foot switch to control the bipolar unit to provide precise hemostasis without stimulation or current spread to nearby structures.[29]

Interpretive Statement 5

Patients with pacemakers should have continuous electrocardiogram (ECG) monitoring when an ESU is being used.

Rationale

Use of the ESU may interfere with pacemaker circuitry. The bipolar unit may be used when operating on a patient with a pacemaker.[30]

Discussion

Modern pacemakers are subject to interference; most are designed to be shielded from radio frequency current during ESU use. Perioperative personnel should implement additional actions for the pacemaker patient that include, but are not limited to

- Making the distance between the active and dispersive electrodes as close as possible and placing both as far from the pacemaker as possible
- Ensuring that the current path from the surgical site to the dispersive electrode does not pass through the vicinity of the heart
- Keeping all ESU cords and cables away from the pacemaker and the leads
- Having a defibrillator available in the room
- Checking with the pacemaker's manufacturer regarding its function during use of ESUs
- Evaluating the pacemaker postoperatively for proper function[31]

Interpretive Statement 6

A patient with an automatic implantable cardioverter defibrillator (AICD) should have the device deactivated before the procedure and have his or her ECG monitored continuously if an ESU will be used.

Rationale

Electrosurgery must not be used on a patient with an activated AICD because it may trigger the device to shock.[32]

Interpretive Statement 7

The patient's skin integrity should be evaluated and documented before and after ESU use. Particular areas to observe are under the dispersive electrode, under ECG leads and at temperature probe entry sites.

Rationale

Assessment will allow evaluation of skin condition for possible injury. Alternate pathway burns have been reported at ECG electrode sites and temperature probe entry sites.[33]

Interpretive Statement 8

If an adverse skin reaction or injury occurs the ESU and active and dispersive electrodes should be sent with their packages to the biomedical service for a full investigation. Device identification, maintenance/service information and event information should be included in the report from the practice setting.

Rationale

Retaining the ESU and electrodes allows for a complete systems check to determine system integrity.[34]

RECOMMENDED PRACTICE IV

Patients and perioperative personnel should be protected from inhaling the smoke generated during electrosurgery.

Interpretive Statement 1

An evacuation system should be used to remove surgical smoke.

Discussion

There may be a potential hazard from exposure to smoke generated during electrosurgery.[35] Further research must be performed to determine the actual magnitude of smoke exposure under practical electrosurgical conditions.[36]

Interpretive statement

Smoke evacuation systems should be used according to manufacturer's written instructions.

Discussion

Health care facilities may use the AORN "Recommended Practices for Product Evaluation and Selection for Patient Care in the Practice Setting" and AORN "Recommended Practices for Laser Safety in the Practice Setting" to assist in selecting a smoke evacuation system.

When the evacuation system is used for the filtration of electrosurgical smoke, placement of the evacuator suction tubing should be as close to the source of the smoke as possible. This will maximize smoke capture and enhance visibility at the surgical site.

Research findings suggest that there is little difference between the smoke generated from electrosurgery and from lasers. There is an undefined potential for bacterial and viral contamination of smoke. Toxicity and mutagenicity of the gaseous byproducts exist.[37] High filtration surgical masks may be worn by perioperative personnel during procedures that generate surgical smoke.

RECOMMENDED PRACTICE V

Policies and procedures for electrosurgery should be developed, reviewed annually and available within the practice setting.

Discussion

These policies and procedures should include, but are not limited to,

- Equipment maintenance programs
- Reporting of injuries
- Sanitation of ESU
- Documentation of the ESU brand name, ESU identification number/serial number, settings used, dispersive electrode and ECG pad placement, patient skin condition before and after electrosurgery and other electrical devices used

These recommended practices should be used as guidelines for the development of policies and procedures in the practice setting. Policies and procedures establish authority, responsibility and accountability. They also serve as operational guidelines.

An introduction and review of policies and procedures should be included in orientation and ongoing education of personnel to assist in the development of knowledge, skills and attitudes that affect patient outcomes. Policies and procedures also assist in the development of quality assessment and improvement activities.

GLOSSARY

Active electrode: The accessory that directs current flow to the operative site. Examples include pencils with various tips, resectoscopes and fulguration tips.

Current: A movement of electricity analogous to the flow of a stream of water.

Dispersive electrode: The accessory that directs current flow from the patient back to the generator (often called the patient plate, return electrode, inactive electrode or grounding plate/pad).

Electrosurgery: The cutting and coagulation of body tissue with a high radiofrequency current.

Electrosurgical unit (ESU): For the purposes of this document, the ESU is defined as the generator, the foot switch and cord (if applicable), and the electrical plug, cord and connections.

Generator: The machine that produces radiofrequency waves (often called a cautery unit, power unit or Bovie).

Grounded electrosurgery: The dispersive electrode is grounded to the metal chassis of the generator. Current will flow from the active electrode when it touches any grounded object in the room.

Isolated electrosurgery: No reference to ground. For current to flow, there must be a complete circuit path from the active terminal to the patient terminal.

Macroshock: Occurs when current flows through a large skin surface, as during inadvertent contact with moderately high voltage sources, such as electrical wiring failures, that allow skin contact with a live wire or surface at full voltage.

Microshock: Occurs when current is applied to a small area of skin, as when current from an exterior source flows through the cardiac catheter or conductor.

REFERENCES

1. Shaffer MJ, Gordon MR: "Clinical engineering standards, obligations, and accountability," *Medical Instrumentation* 13 (July/August 1979) 209–215.

2. Atkinson LJ: *Berry and Kohn's Operating Room Technique,* seventh ed (St Louis: Mosby-YearBook, Inc, 1992) 253.
3. Emergency Care Research Institute. "Electrosurgical units," *Technology for Surgery* 8 (November 1987) 3; E. Moak, "Electrosurgical unit safety: The role of the perioperative nurse," *AORN Journal* 53 (March 1991) 745.
4. Groah LK: *Operating Room Nursing: Perioperative Practice,* second ed (Norwalk, Conn: Appleton & Lange, 1990) 299.
5. Moser ME: "Electrical shock: An orientation study guide," *Point of View* 23 (May 1, 1986) 4–5.
6. Groah LK: *Operating Room Nursing, Perioperative Practice,* 299.
7. Meeker MH, Rothrock JC: *Alexander's Care of the Patient in Surgery,* ninth ed (St. Louis: Mosby-YearBook, Inc, 1991) 43: Atkinson, *Berry and Kohn's Operating Room Technique,* 253.
8. Meeker MH, Rothrock JC: *Alexanders's Care of the Patient in Surgery,* 43.
9. Kneedler JA, Dodge GH: *Perioperative Patient Care: The Nursing Perspective,* second ed (Boston: Jones and Bartlett Publishers, 1991) 382.
10. *Ibid.*
11. Schellhammer PF: "Electrosurgery: Principles, hazards, and precautions," *Urology* 3 March 1974) 261–267.
12. Atkinson LJ: *Berry and Kohn's Operating Room Technique,* 170.
13. *Ibid,* 232; Groah LK: *Operating Room Nursing: Perioperative Practice,* 300.
14. Moak E: "Electrosurgical unit safety: The role of the perioperative nurse," 746.
15. Atkinson LJ: *Berry and Kohn's Operating Room Technique,* 172; Moak, "Electrosurgical unit safety: The role of the perioperative nurse," 746; Groah, *Operating Room Nursing: Perioperative Practice,* 300.
16. Shaffer MJ, Gordon MR: "Clinical engineering standards, obligations, and accountability," 214; Atkinson, *Berry and Kohn's Operating Room Technique,* 170–171.
17. Skreenock JJ: "Electrosurgical quality assurance: The view room the OR table," *Medical Instrumentation* 14 (September/October 1980) 261–263; Meeker, Rothrock, *Alexander's Care of the Patient in Surgery,* 43.
18. Skreenock JJ: "Electrosurgical quality assurance: The view from the OR table," 261–263; Atkinson, *Berry's and Kohn's Operating Room Technique,* 170.
19. Skreenock JJ: "Electrosurgical quality assurance: The view from the OR table," 261–263.
20. Emergency Care Research Institute, "Update: Controlling the risks of electrosurgery," *Health Devices* 18 (December 1989) 431.
21. Bowdle TA, et. al.: "Fire following use of electrocautery during emergency percutaneous transtracheal ventilation," *Anesthesiology* 66 (May 1987) 697-698; P.R. Freund, H.M. Radke, "Intraoperative explosion: Methane gas and diet," *Anesthesiology* 55 (December 1981) 700–701.
22. Skreenock JJ: "Electrosurgical quality assurance: The view from the OR table," 262; Moak, "Electrosurgical unit safety: The role of the perioperative nurse," 746; Groah, *Operating Room Nursing: Perioperative Practice,* 300.
23. Reeter AK: "Bipolar forceps misconnection hazardous," *OR Manager* 6 (February 199U) 13; Atkinson, *Berry and Kohn's Operating Room Technique,* 172.
24. Emergency Care Research Institute, "Electrosurgical units," 3.
25. *Ibid.*
26. Neufeld GR, Foster KR: "Electrical impedance properties of the body and the problem of alternate site burns during electrosurgery." *Medical Instrumentation* 19 (March/April 1985) 83–87.

27. Groah LK: *Operating Room Nursing: Perioperative Practice*, 300; Atkinson, *Berry and Kohn's Operating Room Technique*, 171.
28. Emergency Care Research Institute, "Update: Controlling the risks of electrosurgery," 430–431.
29. Groah LK: *Operating Room Nursing: Perioperative Practice*, 301.
30. Moak E: "Electrosurgical unit safety: The role of the perioperative nurse," 748.
31. Groah, *Operating Room Nursing: Perioperative Practice*, 301.
32. Lee BL, Mirabal G: "Automatic implantable cardioverter defibrillator: Interpreting, treating ventricular fibrillation," *AORN Journal* 50 (December 1989) 1226, S.A. Moser, D. Crawford, A. Thomas, "Updated care guidelines for patients with automatic implantable cardioverter defibrillators," *Critical Care Nurse* 13 (April 1993) 70.
33. Schneider AJL, Apple HP, Braun RT: "Electrical burns at skin temperature probes," *Anesthesiology* 47 (July 1977) 72–74; B. Finley et. al., "Electrosurgical burns resulting from use of miniature ECG electrodes," *Anesthesiology* 41 (September 1974) 263-269; Moak, "Electrosurgical unit safety: The role of the perioperative nurse," 748–749; Groah, *Operating Room Nursing Perioperative Practice*, 299.
34. Gendron F: "Burns' occurring during lengthy surgical procedures," *Journal of Clinical Engineering* 5 (January-March 1980) 19–26; Moak, "Electrosurgical unit safety: The role of the perioperative nurse," 749, 752.
35. Emergency Care Research Institute, "ESU smoke—should it be evacuated?" *Health Devices* 19 (January 1990) 12.
36. *Ibid.*
37. Baggish MS, et al.: "Presence of human immunodeficiency virus DNA in laser smoke," *Laser in Surgery and Medicine* 11 (1991) 202–203; Y. Tomita et al., "Mutagenicity of smoke condensates induced by CO_2 laser irradiation and electrocauterization," *Mutation Research* 89 (1981) 145; W.S. Sawchuk et al., "Infectious papillomavirus in the vapor of warts treated with carbon dioxide laser or electrocoagulation: Detection and protection," *Journal of the American Academy of Dermatology* 21 (July 1989) 41.

Suggested reading

Becker, CM; Malhotra, IV; Hedley-Whyte, J. "The distribution of radiofrequency current and burns." *Anesthesiology* 38 (February 1973) 106–122.

Buczko, GB; McKay, WPS. "Electrical safety in the operating room." *Canadian Journal of Anesthesia* 34 no 3 (1987) 315–322.

Emergency Care Research Institute. "ESU monitoring systems." *Technology for Surgery* 8 (December 1987) 1–3.

Gatti, JE, et al. "The mutagenicity of electrocautery smoke." *Plastic and Reconstructive Surgery* 89 (May 19920 781–784.

National Institute for Occupational Safety and Health. *Health Hazard Evaluation Report*. publ no HETA 85-126-1932. Washington, DC: U.S. Department of Health and Human Services, 1988.

Pearce, J. "Current electrosurgical practice hazards." *Journal of Medical Engineering and Technology* 9 (May/June 1985) 107–111.

Soderstrom, RM. "Electrosurgery's advantages and disadvantages." *Contemporary OB/GYN* 35 (Oct 15, 1990) 35–47.

Tucker, RD; Ferguson S. "Do surgical gloves protect staff during electrosurgical procedures?" *Surgery* 110 (November 1991) 892–895.
Voyles, CR; Tucker, RD. "Education and engineering solutions for potential problems with laparoscopic monopolar electrosurgery." *American Journal of Surgery* 164 (July 1992) 57–62.

Reprinted from the *AORN Journal,* Official Journal of the Association of Operating Room Nurses, Inc., 2170 South Parker Road, Suite 300, Denver, Colorado 80231. July 1993. Vol. 58, No. 1, pp. 131–138.

PART 6

Laser Plume Quantification

William Charney

Quantification data of laser plume to test the efficiency of a laser scavenging system are presented. These data were compiled during a simulated case and during a real case of removal of a condyloma. The scavenging device tested was a Baxter Class 1 Smoke Evacuator equipped with both HEPA and ULPA filters with an efficiency to 0.01 μ.

SAMPLING INSTRUMENTATION

A Miniram PDM 3, a light screening aerosol monitor of the nephelometric type that continually senses the combined scattering from the population of particles present, was used. The Miniram uses a GaA1As light-emitting source, which generates a narrow-band emission. The radiation scattered by airborne particles is sensed over an angular range of approximately 45 to 90° by a silicon-photovoltaic hybrid detector.

- Measurement ranges: 0.01 to 10 mg/m^3 and 0.01 to 100 mg/m^3
- Particle size range: 0.01 to 10 μ

METHOD

During simulation and real case the hose of the scavenging device was held approximately 2 cm from the burn site. The wattage of the laser power was set at 10 and the burn times were normally 10 s with a 20-s burn. The scavenger was turned on simultaneously with the burn by the use of a foot activator switch.

DISCUSSION

During both the simulation and the real case all the smoke was evacuated and zero level exposure was quantified. During simulation when the scavenging device was turned off at a median level of 12 mg/m^3 particulate, staff without proper respirators complained of symptoms.

RECOMMENDATIONS

Use of the scavenging system reduces quantified levels of particulate to zero. The scavenger hose when held within the range of 2 cm from the burn site has

1-56670-083-3/94/$0.00+$.50

good capture and it is assumed that capture capacity decreases as the distance from the burn site increases. Therefore, the scavenging system should be used for all laser cases that produce a plume to protect the staff and the patient from the content of the plume. (See Tables 1 and 2.)

Table 1. Laser plume quantifications: simulation. April 29, 1992.

Type	Position	Laser (W)	No. of Seconds	mg/m^3
1 Background (9:29)	5 ft	Off	Off	0.00
2 Background (9:30)	5 ft	Off	Off	0.00
3 Background (9:31)	5 ft	Off	Off	0.00
4 Background (9:34)	5 ft	Off	Off	0.00
5 With scavenging	5 ft (SBZ)	10 W	10	0.00
6 With scavenging	5 ft (SBZ)	10 W	10	0.00
7 With scavenging	5 ft (SBZ)	10 W	10	0.00
8 With scavenging	5 ft (SBZ)	20 W	10	0.00
9 With scavenging	5 ft (SBZ)	20 W	10	0.00
10 With scavenging	5 ft (SBZ)	20 W	20	0.00
11 With scavenging	5 ft (SBZ)	20 W	20	0.00
12 With scavenging	5 ft (SBZ)	10 W	10	0.00
13 With scavenging	5 ft (SBZ)	10 W	10	0.00
14 With scavenging	5 ft (SBZ)	10 W	10	0.00
15 With scavenging	5 ft (SBZ)	10 W	10	0.00
16 With scavenging	5 ft (SBZ)	10 W	10	0.00
17 With scavenging	5 ft (SBZ)	10 W	10	0.00
18 With scavenging	5 ft (SBZ)	10 W	10	0.00
19 With scavenging	5 ft (SBZ)	10 W	10	0.00
20 With scavenging	5 ft (SBZ)	10 W	10	0.00
21 No scavenging	5 ft (SBZ)	10 W	10	12.7
22 No scavenging	5 ft (SBZ)	10 W	10	13.8
23 No scavenging	5 ft (SBZ)	10 W	10	14.2
24 No scavenging	5 ft (SBZ)	10 W	10	13.7
25 No scavenging	5 ft (SBZ)	10 W	10	12.7
26 No scavenging	5 ft (SBZ)	10 W	10	15.8
27 No scavenging	5 ft (SBZ)	10 W	10	12.7
28 No scavenging	5 ft (SBZ)	10 W	10	12.7
29 No scavenging	5 ft (SBZ)	10 W	10	12.7
30 No scavenging	5 ft (SBZ)	10 W	10	12.7
31 With scavenging	5 ft (NBZ)	10 W	10	0.00
32 With scavenging	5 ft (NBZ)	10 W	10	0.00
33 With scavenging	5 ft (NBZ)	10 W	10	0.00
34 With scavenging	5 ft (NBZ)	10 W	10	0.00
35 With scavenging	5 ft (NBZ)	10 W	10	0.00
36 With scavenging	5 ft (NBZ)	10 W	10	0.00
37 With scavenging	5 ft (NBZ)	10 W	10	0.00
38 With scavenging	5 ft (NBZ)	10 W	10	0.00
39 With scavenging	5 ft (NBZ)	10 W	10	0.00
40 With scavenging	5 ft (NBZ)	10 W	10	0.00

Note: SBZ = surgeon breathing zone; NBZ = nurse breathing zone

Table 2. Laser quantifications: real case

Type	Position	Laser (W)	No. of Seconds	mg/m³
1 Background	(SBZ)	Off	Off	0.00
2 Background	(SBZ)	Off	Off	0.00
3 Background	(SBZ)	Off	Off	0.00
4 Background	(SBZ)	Off	Off	0.00
5 Laser on	(SBZ)	10	10	0.00
6 Laser on	(SBZ)	10	10	0.00
7 Laser on	(SBZ)	10	10	0.00
8 Laser on	(SBZ)	10	10	0.00
9 Laser on	(SBZ)	10	10	0.00
10 Laser on	(SBZ)	10	10	0.00
11 Laser on	(SBZ)	10	10	0.00
12 Laser on	(SBZ)	10	10	0.00
13 Laser on	(SBZ)	10	10	0.00
14 Laser on	(SBZ)	10	10	0.00
15 Laser on	(SBZ)	10	10	0.00
16 Laser on	(SBZ)	10	10	0.00
17 Laser on	(SBZ)	10	10	0.00
18 Laser on	(SBZ)	10	10	0.00
19 Laser on	(SBZ)	10	10	0.00
20 Laser on	(SBZ)	10	10	0.00
21 Laser on	(SBZ)	10	10	0.00
22 Laser on	(SBZ)	10	10	0.00
23 Laser on	(SBZ)	10	10	0.00
24 Laser on	(SBZ)	10	10	0.00
25 Laser on	(SBZ)	10	10	0.00

Note: SBZ = surgeon's breathing zone

CHAPTER 7

Back Injury Prevention in Health Care

PART 1

An Ergonomic Approach to Reducing Back Stress in Nursing Personnel

Bernice D. Owen and Arun Garg

Back pain in health care personnel has been around for a long time! For at least the last 25 years, epidemiologic studies have documented a high prevalence of back pain in this work group (Owen, 1984). Across occupations, the highest incidence and earliest appearance of work-related back problems have been found in heavy industry workers and nurses (Magora, 1970; Cust, Pearson and Mair, 1972). In fact, nursing personnel rank 5th nationally for filing workers' compensation claims; only heavy laborers such as miscellaneous workers, sanitation workers, warehouse workers and mechanics surpass nursing (Klein, Jensen, and Sanderson, 1984).

This problem may be even greater than published statistics indicate because through questionnaires, Owen (1989) found that 38% of 503 nurses stated they had episodes of occupationally related back problems but only one third of those with back pain actually filed an incident report with their employer; this group also averaged 6.5 days of their own sick days for unreported back pain perceived to be occupationally related. The units where significantly more low back pain (LBP) episodes occurred than were expected were intensive care, orthopedics and rehabilitation. Twenty percent of the nurses who said they had back pain stated they had made at least one transfer in order to decrease the amount of lifting/transferring of patients, e.g., they transferred to a different unit such as from surgical to obstetrics; changed employment settings such as from hospital to clinics or changed positions from staff nurses to educators. Another 12% indicated they were considering making a transfer and 12% stated they were thinking about leaving the profession of nursing due to occupationally related back pain. In England, Stubbs, Buckle and Hudson (1986) found 12% of all nurses intending to leave nursing permanently cited back pain as either a main or contributory factor.

1-56670-083-3/94/$0.00+$.50

CAUSES OF BACK PAIN

Many authors believe the back pain problem is resultant from a combination of biomechanical and postural stressors. Variables such as the heavy load, the distance of the load (patient) from the lifter's center of gravity, the duration of the lift, awkward lifting positions, confined work space, unpredictable patient behaviors and the amount of stooping and bending endured in the job, have an impact leading to excessive forces in the spinal area. Study into cause/effect relationships is important for this problem.

Lifting and transferring of patients have been perceived by nursing personnel to be the most frequent precipitating factors or causes of back problems (Harper, Billet and Gutowski, 1985; Jensen, 1985; Owen, 1989; Stobbe, Plummer and Jensen, 1988; Stubbs, Rivers and Hudson, 1981; Valles-Pankratz, 1989; Venning, Walter and Stitt, 1987). For example, Owen (1989) found that 89% of the back injury reports filed by hospital nursing personnel implicated a patient handling task (PHT) as a precipitating factor. (A PHT was any task that involved lifting, moving, or handling the patient in some manner.) Stubbs et al. (1981), through questionnaires, reported 84% of the nurses perceived PHTs as important factors in their back pain. Jensen (1985) (through workers' compensation records) found more than 73% of the back strain/sprain cases were reportedly triggered by these tasks.

Epidemiologic studies have reported that individuals who have more frequent back pain are more likely to report exposure to forward flexion, rotation and lateral bending than those who have less back pain because it has been determined that these body movements produce large loads on the lumbar spine and can be harmful to the disk (Andersson, 1981; Frymoyer, Pope and Costanza, 1980).

While observing the body movements of nursing assistants (NAs), Nordin, Ortengren and Andersson (1984) found the frequency and degree of trunk flexion to be high. A few studies have found high levels of biomechanical stress induced by patient lifting and transferring tasks (Gagnon, Sicard and Sirois, 1986; Garg, Owen, Beller and Banaag, 1991; Stubbs, Buckle, and Hudson, 1983; and Torma-Krajewski, 1986). In addition, high levels of postural stress (standing and stooping) are also cause for concern (Baty and Stubbs, 1987).

APPROACHES TO PREVENTION

Nursing personnel offered their ideas for prevention when asked through questionnaires and on incident reports (Owen, 1987). Responses of 244 nurses indicated 89% stated good body mechanics during lifting and transferring of patients as the most important preventive measure. The second most frequent suggestion (n = 142, 58%) was that nurses be more willing to ask for help when they had judged they should not attempt the PHT alone. Only 29 nurses spoke of preventive approaches external to themselves, e.g., adequate staff, reduced work load and increased patient participation with the task. Two nurses stated the side rails of the bed should have been lowered as well as the bed. The use of mechanical devices or "lifting aids" was mentioned but the only device specified was that of a draw sheet used for lifting or sliding the patient up in bed. Five nurses indicated efforts to prevent the LBP incident were futile because LBP is an "inevitable" part of nursing. Therefore, the vast majority of these nurses gave responses that focused on themselves such as body mechanics and their ability to ask for help.

Education and Training

Is training on body mechanics the answer? Most basic nursing textbooks have chapters pertaining to body mechanics. Unfortunately, some of the recommendations are confusing and even contradictory. For example, some authors suggest the nurse place at least one foot in the direction of the move when lifting or sliding a patient up in bed (Earnest, 1993; and Kozier, Erb, Blais et al., 1993); others say the nurse faces the side of the bed (Cole, 1991; and Christensen and Kockrow, 1991), another gives no information about placement of feet (Smith and Duell, 1992).

Snook, Campanelli and Hart (1978) studied the effectiveness of training as an approach to preventing occupationally related back injuries. They found training programs had no effect on reducing back injury rates; instead, those industries that had training programs had more back injuries than would be expected by chance. Some programs have experienced a decrease in back injuries after a training program, but with time the injury rates have returned to pre-training levels (Pheasant, 1991).

Knowledge and application of body mechanics are important but not the full answer to prevention of back injuries in health care workers. Some research is even suggesting that the more body mechanics are taught, the higher the injury rate (H. Knibble, personal communication). Apparently when we focus so much on teaching nurses *how* to manually lift and transfer patients, a message is relayed that nurses can do anything with their bodies as long as they do it correctly. However, this message negates the impact of characteristics of the patient such as weight and combativeness, or elements of the environment such as confined work space and unevenness in lifting surfaces.

Pheasant (1991, p. 295) very aptly states "many people (both within the nursing profession and elsewhere) take the view that nurses have back problems because they are undertrained. The reality is that they are physically overloaded by their work activities. *In situations of this kind, training alone is necessary but not sufficient.* To make further progress, we need to identify the features of the working system which are responsible for the physical overload."

The Ergonomic Approach

Therefore, an approach to prevention must also incorporate job design, workplace design and the impact that these factors have on patient care and the health and safety of health care workers. This ergonomic approach involves adjusting/changing the job to fit the capabilities and limitations of the worker rather than trying to change the worker in order to fit the job. The goals of this ergonomic approach are to identify those aspects of the job which are particularly hazardous and to redesign them so they are safer (Pheasant, 1991). This may be done through such avenues as redesign of the task, the product, the work station, the environment or the overall work organization.

Stubbs (1986) suggests the ergonomic elements relate to the interaction between equipment, environment, task and personnel. *Equipment* must be compatible with the strength and ability of nursing personnel. It must be compatible with other equipment such as the bathtub or bed and also with the environment such as with the floor/rug surfaces. Equipment should meet the needs of the user and a maintenance plan developed so it is available when needed. *Tasks* should not involve prolonged postural stress, there should be rest periods and high postural demands should be decreased through equipment that is comfortable for the nurse as well as the patient. Adequate staffing is also important to the task. The *environment* needs to be compatible with the equipment, the tasks and the capabilities/needs of the nursing personnel. All of these elements should fit the capabilities of *personnel;* the nursing staff should not be expected to fit into a poorly designed system but should have available to them the support, equipment, etc. needed to do the job.

The tasks that nurses carry out and the work station where these tasks occur must be studied. Minimally, the manual handling actions that should be designed out of the work setting include: lifting heavy loads, lifting away from the body, asymmetric lifting and rotation of the torso. Lifting loads from below the knees or over the shoulder should also be avoided. In addition, there are factors about the load that contribute to the hazard of manual lifting; these include unpredictability and instability of the load, bulkiness of load and inability to grasp the load securely because of lack of handles.

Nursing personnel who have been involved in the lifting and transferring of patients certainly must realize that many of those tasks should be redesigned based on the above discussion.

Therefore, in order to decrease the back stress problem in nursing, nursing personnel must begin to look at their own capabilities, the tasks which they feel are stressful to the upper and lower back, the environment in which the tasks are carried out and equipment that may be helpful for themselves and the patients. This group of health professionals should play a vital role in delineating approaches to decreasing that stress. They must be encouraged to problem solve and work with management in striving for changes that could impact on this problem which is costly in relation to human suffering, staffing and financial cost.

Example of Ergonomic Approach: Nursing Home

Through application of the ergonomic process the authors determined that biomechanical and perceived physical stresses could be reduced while lifting and transferring patients (Garg and Owen, 1992). This process involved determining the most physically stressful patient care tasks, evaluating these tasks to determine the problem areas, testing out approaches to decrease the stressfulness and evaluating the interventions to determine if stressfulness was decreased. The following is a summary of the process these authors used which further resulted in a reduction of back injuries and lost work days due to back injuries.

This study is important to hospital personnel because: the lifting and transfer needs of hospital and nursing home residents are similar, and the ergonomic approaches used in the study can be used in the hospital setting.

Setting: Two floors of a county nursing home were selected for this study. These were 70 patients on each floor and most required help with many tasks.

Subjects: Thirty-eight of the 57 NAs employed at least part-time volunteered as subjects; 36 were females and two were males. They ranged in age from 19 to 61 years, with a mean of 32.8 years. Their average length of employment was 7.8 years (SD = 4.7 years). Seventy-five percent stated that within the last three years they suffered from back problems perceived to be related to work. However 60% stated they lost no work time within that period due to back problems, 15% missed one to seven days and 25% lost eight days or more.

Patients: The 140 patients ranged in age from 56 to 98 years with a mean of 84.7 years and about 64% were female. They had an average weight of 133 pounds and height of 64 inches. Most patients could not stand up or walk independently and their mental status indicated unpredictiveness for willingness to help with the task. In addition, the average ability for balance, body flexibility and general physical ability was impaired. Patient characteristic data were collected at the beginning of the study and six months later to determine if the patient population had statistically changed; *t*-test findings indicated no significant differences between the first and second sets of data ($p > 0.05$).

Study Design: The major steps (goals) in the study design were the following:

Goal 1
Determination of Most Stressful Patient Handling Tasks

↓

Goal 2
Ergonomic Evaluation of These Tasks

↓

Goal 3
Testing Approaches to Decrease Stressfulness
(Locating Assistive Devices)

↓

Goal 4
Laboratory Study

Goal 5
Application of Findings to Clinical Area

GOAL 1: DETERMINING STRESSFUL TASKS

The NAs listed the following 16 PHTs as most stressful in their patient care duties (Owen and Garg, 1989):

- Transferring patient from toilet to wheelchair (WC)
- Transferring patient from WC to toilet
- Transferring patient from WC to bed
- Transferring patient from bed to WC
- Transferring patient from bathtub to WC
- Transferring patient from chairlift on bathtub to WC
- Weighing patient (transferring patient from WC to scale chair)
- Lifting patient up in bed
- Repositioning patient within bed (e.g., side to side)
- Repositioning patient in chair
- Changing the absorbent pad worn by patient
- Making bed with patient in it
- Undressing patient
- Tying "supports" to secure patient in WC
- Feeding bed-ridden patient
- Making bed when patient is not in it

The NAs then ranked these tasks according to stressfulness felt while performing the task. They also rated the amount of perceived physical exertion felt while carrying out each task; the Borg (1962) scale was used for rating exertion to low back, upper back, shoulder and whole body (this Likert-type scale progresses from six [very, very light] to 19 [very, very hard]).

The tasks ranked and rated as the most stressful were: transfer on and off the toilet, in and out of bed and transferring for the bathing and weighing processes (see Table 1). *Therefore, these tasks were selected for further study.*

Summary

It was not surprising that these transfer tasks were selected as most stressful because the lifting postures for these tasks are biomechanically very stressful.

Table 1. Ranking and rating of patient handling tasks for stressfulness.

Patient Handling Task	Rank Order[1]	Perceived Stress to Lower Back[2]	
		$\bar{x}$	(SD)
Transferring patient from toilet to WC	1	14.3	2.7
Transferring patient from WC to toilet	2	14.1	2.8
Transferring patient from WC to bed	3	14.2	3.0
Transferring patient from bed to WC	4	14.1	2.9
Transferring patient from bathtub to WC	5	13.3	2.9
Transferring patient from chairlift to WC	6	13.4	3.2
Weighing patient	7	13.8	3.9

1 = Most stressful
2 = Borg Scale; 6 = very, very light; 14 = very, very hard

Also, these findings are in agreement with researchers who have found (through interview, questionnaire and analysis of incident reports) that nursing personnel believe that much of their back stress results from the lifting and transferring of patients (Harper et al., 1985; Owen, 1987; Stobbe et al., 1988 and Venning et al., 1987).

GOAL 2: ERGONOMIC EVALUATION OF STRESSFUL TASKS

Over a six-month period the stressful tasks and the environment in which they were carried out were observed and videotaped (Garg, Owen, and Carlson, 1992). Analysis of the data (including analysis through use of a static biomechanical model [Garg and Chaffin, 1975]) revealed the following findings:

Patient transfer method. Patients who needed help with transfer were *manually lifted* 98% of the time. In executing this manual transfer, two nursing personnel stood facing the patient, each grasped the patient under the axilla with the upper arm, then lifted the patient up and carried him/her to a new location (Figure 1).

Assistive devices. There were gait belts and a mobile hydraulic hoist (Hoyer Lift) on each floor. The hoist was only used for two patients during these six

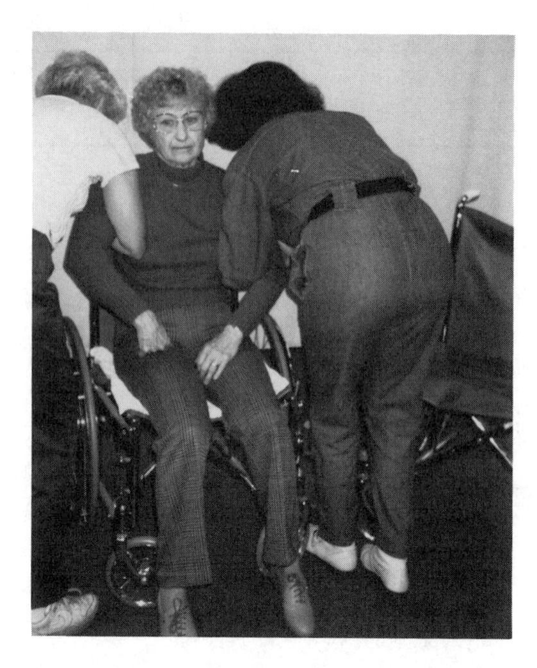

Figure 1. Manual transfer of patient using under-axilla technique.

months. A gait belt was observed being used with transfers of one patient. The reasons NAs gave for non use of assistive devices were: devices were not available, they took too much time, they were unstable or tipped over, the patients did not like them, they were not safe, they bruised the frail skin of some patients and some felt they did not have enough knowledge or skill to use them.

Six bathtubs had chairlifts (water pressure lifts shaped like a chair) which were used for the bathing process. The NAs transferred the patient from WC to chairlift and with water pressure guided the chairlift up over the edge of the bathtub and down into the water. To use the bathtubs without a chairlift, the NAs manually lifted the patient in and out of the water.

The patients were not showered in this nursing home because each shower had a drain guard that was six inches high which prevented a shower chair from being wheeled into the shower.

Additional Environmental Factors were found to contribute to the difficulty of carrying out patient lifting tasks. The floor of the bathrooms was 1/4 inch higher than the rest of the floor; this resulted in the NA having to push the WC or mechanical hoist several times against the elevation before rising over it. The beds were adjustable by manual crank only, so the NAs did not adjust the height of the bed to make it flush with the seat of the WC; the bed was nine inches higher than the seat of the WC necessitating the NA to lift the patient up the 9 inches to get the patient into the bed. The toilet seat was two inches lower than the WC seat. The arm rests of the WCs were not adjustable; the NAs lifted the patients out of the WC rather than using a pull technique to transfer the patient. The toilets had side rails on them for safety; to transfer the patient, the NA lifted the patient onto the front of the toilet and then eased the patient to the back and down onto the toilet seat.

Frequency of Patient Lifting/Transfer Tasks. Each NA carried out an average of 24 most stressful lifting/transferring tasks per eight-hour shift. Approximately half of these lifts were transferring patients on and off the toilet.

Postural Stresses. All the stressful tasks required multiple trunk flexions along with axial rotation of the spine and lateral bending. The mean trunk flexion for most tasks exceeded 30° and the average number of trunk flexions per occurrence of the task ranged from 2 to 6 (see Table 2).

Biomechanical Stresses. The greatest amount of compressive force L_5S_1 and flexion moments occurred during the transfers of toilet to WC and WC to bed (see Table 3). However, the estimated compressive forces for all tasks exceeded the action limit of 3430 Newtons recommended by U.S. Department of Health and Human Services (1981).

Summary

This study showed that the NAs were subjected to high levels of biomechanical and postural stresses. Most of the transfers were done manually, few assistive devices were used and environmental factors influenced stressfulness. The high levels of biomechanical stresses found are in agreement with those reported by Stubbs, Baty and Buckle (1986) based on intra-abdominal pressure, Gagnon et al. (1986) based on compressive force and Torma-Krajewski (1986) based on computations of action limits and maximum permissible limits recommended by the U.S. Department of Health and Human Services (1981).

GOAL 3: TESTING APPROACHES TO DECREASE STRESSFULNESS (LOCATING ASSISTIVE DEVICES)

One ergonomic approach recommended for decreasing back stress is redesign of the task. This could be done through the use of assistive devices. Hence, the criteria for selection of assistive devices were established and devices located (Owen and Garg, 1990).

Criteria for Selection of Devices

1. The device must be appropriate for the task to be accomplished. A device that could only be used for transfer of patient in a prone position could not be used for the stressful tasks of this study.

Table 2. Summary of trunk angles (in degrees) per task obtained through video tapes.

	Flexion (°)		Rotation (°)		Lateral bending (°)	
Task	Mean	SD	Mean	SD	Mean	SD
Toilet to WC	51	14	33	8	25	8
WC to toilet	53	15	7	8	11	7
WC to bed	57	14	21	9	15	8
Bed to WC	44	13	18	8	17	7
WC to chairlift	48	16	15	9	13	8
Chairlift to WC	46	11	9	8	12	8
Weigh patients	10	3	6	3	6	2

Table 3. Summary of biomechanical analysis of six selected patient handling tasks for 50th percentile patient weight.

	Patient Handling Task					
Variable	Toilet to WC	WC to Toilet	WC to Bed	Bed to WC	Chairlift to WC	WC to Chairlift
Applied hand force (N)	294	294	294	294	294	294
Compressive force (N)	4810	3680	4877	3991	4552	3680
Shear force (N)	788	886	805	801	792	699
Flexion moment (Nm)	202	153	217	161	196	139
Rotation moment (Nm)	85	61	59	82	64	92
Lateral bending moment (Nm)	49	65	46	54	49	45

2. The device must be safe for both patient and nurse. It must be stable, strong enough to secure and hold the patient and permit the nurse to use safe biomechanics.
3. The device must be comfortable for the patient; this may also help to allay fears. It should not produce or intensify pain, bruising or tear the skin.
4. The device should be understood and used with relative ease. Bell (1984) and Owen (1988) found nursing personnel were reluctant to use assistive

devices because they could not understand how to use them or lacked experience in their use.

5. The device must be efficient in the use of time. According to Bell (1984) and Owen (1988) the most frequent reason given for not using a device was the time needed for use.
6. Need for maintenance should be minimal. The above two authors found lack of proper functioning a major reason for non-use.
7. The device must be maneuverable in a confined work space. Owen (1988) and Valles-Pankratz (1989) found space to be a problem.
8. The device should be versatile. It could be inferred from Bell's findings (1984) that only a few assistive devices should be introduced at a time because the error rate and the need for time to execute the transfer increased when more than two devices were included in a teaching program.

Assistive Devices. The following assistive devices were selected for further study: *gait belt* (Figure 2), *MEDesign patient handling sling* (Figure 3), *Posey walking belt with handles* (Figure 4), *Hoyer lift* (Figure 5), *Trans-Aid lift* (Figure 6) and *Ambulift C_3 hoist* (Figure 7).

During the ergonomic study, it was found that several transfers could be eliminated for bathing if the patient was transferred from WC to a shower chair that could accommodate the patient for toileting and showering. Two shower/toileting chairs were selected for study. The heavy chair can be seen in Figure 8; the light weight chair with a plastic non-movable seat and no foot rests is not pictured.

Summary

Criteria for selection of assistive devices were established. A number of devices such as slings, belts and hoists were tested and the following were recommended for expanded study in the laboratory: gait belt, walking belt, MEDesign patient handling sling, shower chair (Figure 8), Hoyer lift, Trans-Aid lift, and C_3 lift.

GOAL 4: LABORATORY STUDY

The purpose of this laboratory study was to evaluate eight different methods for carrying out the most stressful patient handling tasks (Garg et al. 1991; Owen and Garg, 1991). Methods found to reduce back stress would then be taken into the clinical area so nursing personnel could apply these to patient care (Goal 5).

Subjects. Six female senior nursing students were subjects and all participated both as nurse and "patient." They ranged in age from 21 to 23 years with an

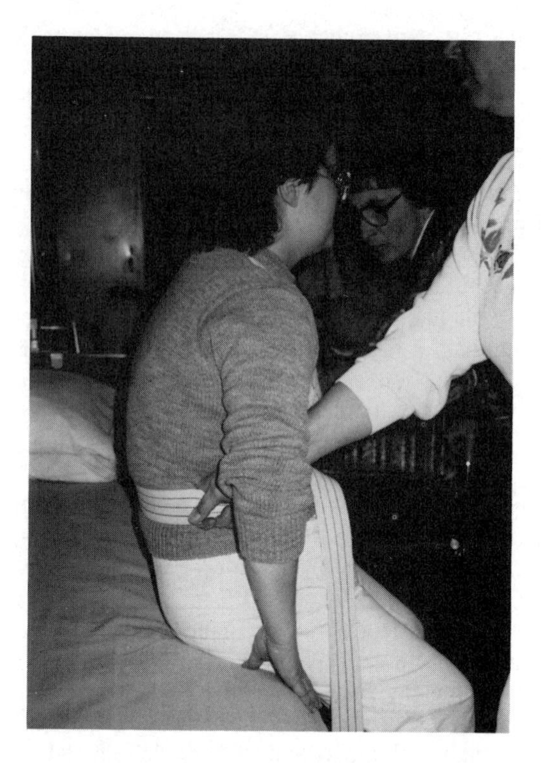

Figure 2. Gait Belt is about 2 inches wide, of varying lengths with adjustable belt-like loop or buckle closure, has no handles and is made of cotton-canvas or nylon material. It should fit securely around patient's waist and is grasped with hand.

average weight of 139 pounds and height of 65 inches. All stated they had no back problems. They were instructed not to support their own body weight while in a "patient" position.

Eight Different Methods. Five of the eight methods were manual transfers: the transfer method presently used in the nursing home (two NAs grasping the patient under the axilla area and lifting the patient to a new location); two NAs using a gait belt and with a gentle rocking movement pulling the patient to a new location; two NAs using the same rocking movement but pulling with the walking belt with handles; one NA using the walking belt with handles; one NA transferring via the MEDesign patient handling sling.

Three of the methods involved transferring the "patient" via mechanical lifts; Hoyer (H), Trans-Aid (T) and Ambulift - C_3 (A).

Procedure. Each subject was studied using each of the eight methods while transferring a "patient" from WC to toilet, toilet to WC, bed to WC, WC to bed, shower chair (SC) to WC and WC to SC.

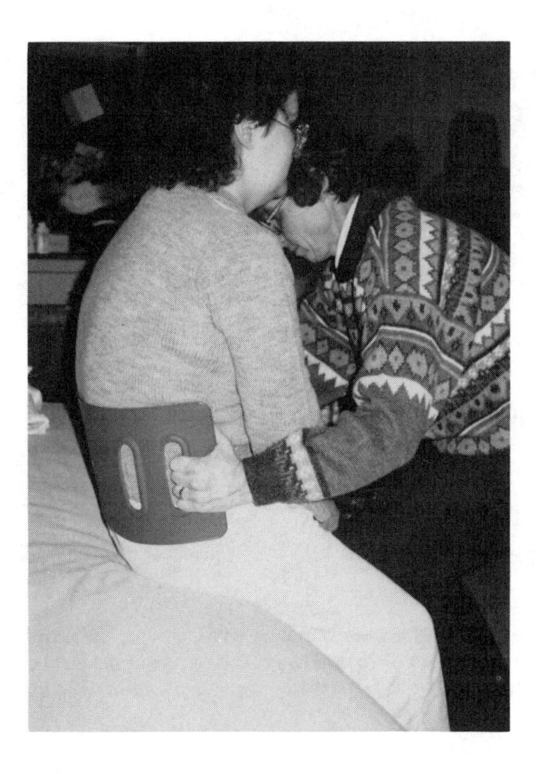

Figure 3. MEDesign Patient Handling Sling is 8 inches wide, 20 inches long, has a cut-out at each end allowing a hand grip, is made of flexible polymer material and is tucked securely around patient with bottom at buttock area.

The subjects were given time to learn the equipment and practice the methods. Immediately after each transfer, the subjects rated the physical stress felt for the shoulder, upper back, lower back and whole body using a ten point scale (0 = no stress, 9 = extreme stress). The "patients" rated their feelings of comfort and security using Likert scales of 0 = extremely comfortable and 7 = extremely uncomfortable; 0 = extremely secure and 7 = extremely insecure. After each transfer the subjects assumed their initial posture at the beginning of the transfer so body angles could be measured for biomechanical stresses and analyzed using the biomechanical model of Garg and Chaffin (1975). Tasks were also videotaped.

The following summarizes the combined data for all eight methods of transfer (Owen and Garg, 1991).

Perceived Physical Stress. The traditional method of the manual axilla lift was the most stressful to all four body parts (Table 4). Following this, the gait belt was most stressful. Lift H was more stressful to all body parts than the manual techniques using the Posey Walking belt. The least stressful were the Posey walking belt and Lift A.

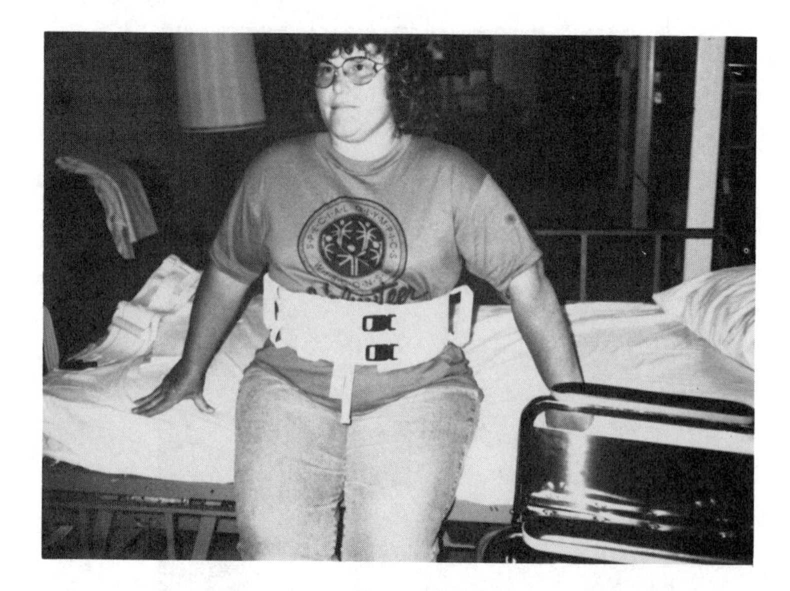

Figure 4. Posey Walking Belt is 5 inches wide, of varying lengths, has handles on each side, has velcro and two quick-release buckles for closure, is made of cotton-canvas type material, fits snugly around lower abdomen and is grasped at handles.

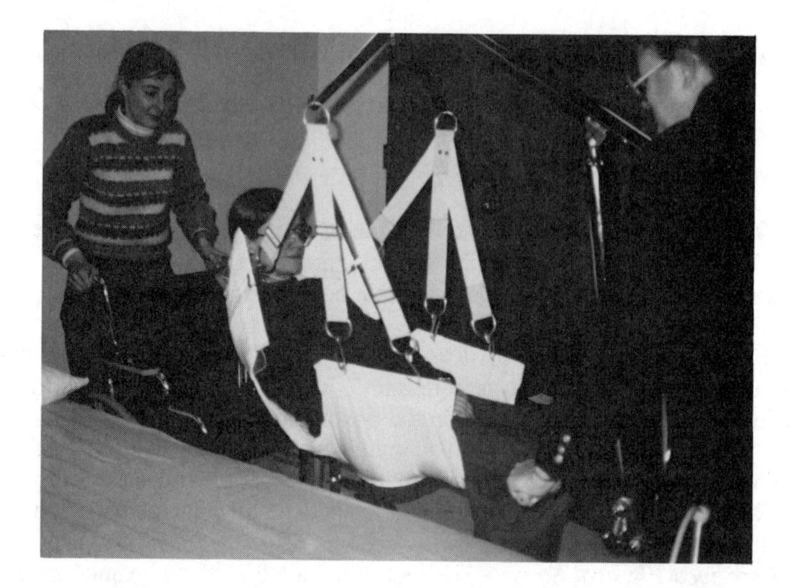

Figure 5. Hoyer Lift is a hydraulic lift that has an adjustable base; a pump handle for raising and lowering the patient and a variety of slings that attach through hooks, chains or web straps.

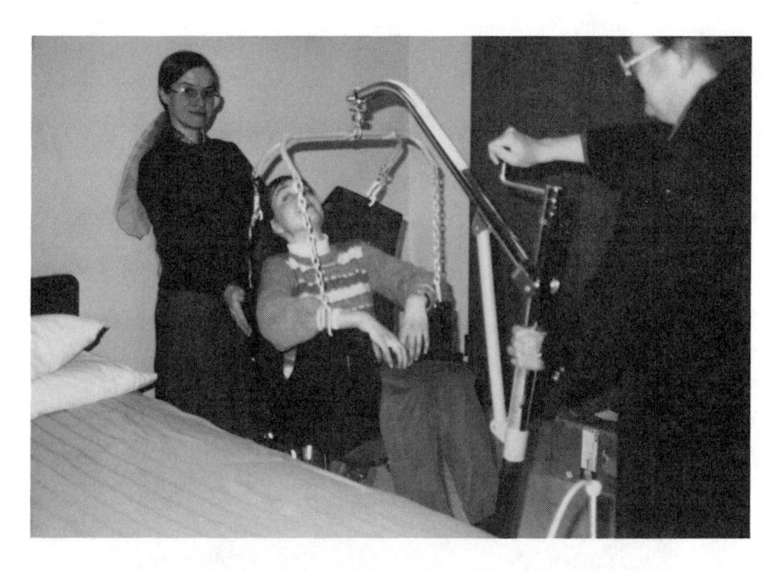

Figure 6. Trans-Aid Lift has a non-adjustable "C" base, a ball-bearing screw lifting mechanism with crank in horizontal plane and has a variety of slings that attach by hooks and dangling color-coded chains.

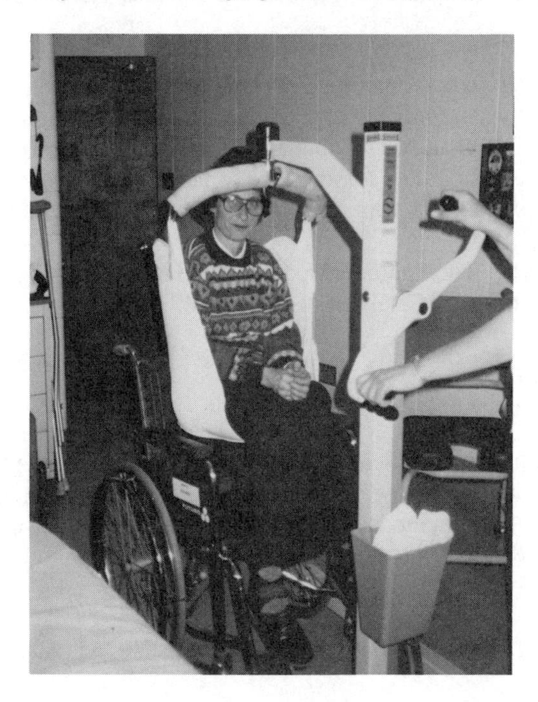

Figure 7. Ambulift (C_3) has a semi-adjustable base, a mechanical chain-winding mechanism for lifting/lowering with crank in vertical plane and the sling attaches by loops and hooks.

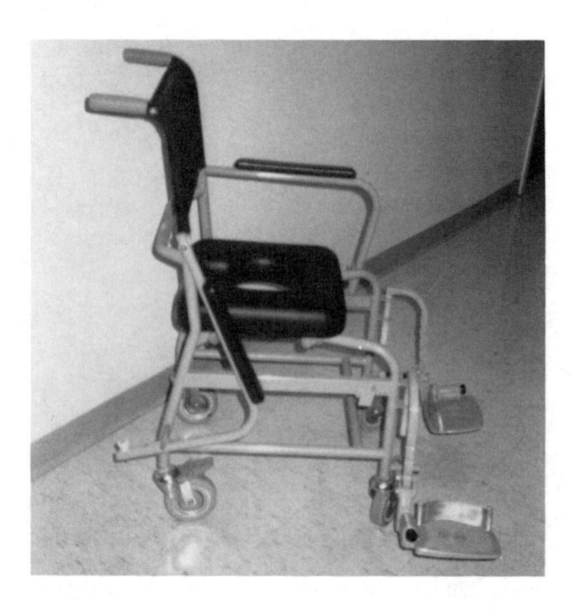

Figure 8. Shower/toileting chair is a heavy chair with padded removable seat and adjustable/removable foot rests and arm rests that can be lowered.

Biomechanical Data. Compressive force at L_5S_1 was estimated to be about two times greater when transferring a patient using the traditional manual lifting method ($\bar{x}$ = 4757N), than when using any of the other methods for transfer (see Table 4). The least amount of compressive force was experienced while carrying out the two person transfers with gait belt ($\bar{x}$ = 2080N) and with the walking belt ($\bar{x}$ = 2044N).

"Patient" Data. The traditional axilla lift and transfers using the gait belt and Lift H were uncomfortable and felt insecure (Table 5).

Summary

The present method of manually transferring the patient was the most stressful, least comfortable, most insecure and provided the greatest amount of compressive force to L_5S_1. The methods taken into the clinical area to be used with patient care in Goal 5 were: the Posey Walking Belt and Lift A.

GOAL 5: APPLICATION OF FINDINGS TO CLINICAL AREAS

The purpose of this part of the study was to take the positive findings from the laboratory and apply them to clinical patient care (Garg and Owen, 1992). The

Table 4. Summary of stress ratings and biomechanical data by method of transferring the "patient."

Stress[1] Body Part	Method of Transfer															
	Manual Lifting (2)[2]		Gait Belt (2)		Walking Belt (2)		Walking Belt (1)		Patient Handling Sling (1)		Lift H (2)		Lift T (2)		Lift A (2)	
	$\bar{x}$	SD	$\bar{x}$	SD	$\bar{x}$	SD	$\bar{x}$	SD	$\bar{x}$	SD	$\bar{x}$	SD	$\bar{x}$	SD	$\bar{x}$	SD
Shoulder	6.5	1.2	5.1	1.3	3.4	1.7	3.2	0.9	4.1	1.5	5.2	1.8	4.1	1.5	2.5	1.1
Upper back	6.0	1.6	4.8	1.6	3.2	1.1	3.4	1.0	4.0	1.5	4.5	1.5	3.2	1.2	1.9	1.1
Lower back	6.0	1.5	4.9	1.8	3.1	1.4	3.3	1.6	4.4	1.9	3.7	1.9	2.8	1.7	1.3	1.0
Whole	6.2	2.3	4.9	1.4	3.1	1.0	3.3	1.1	4.1	1.5	3.5	1.6	3.1	1.6	1.2	0.8
Compressive force (N)	4757	244	2080	210	2044	176	2391	362	2446	430						

1. Scale: 0 = no stress; 9 = extremely stressful
2. Indicates number of nurses making the transfer

Table 5. Summary of "Patient" Data.

Variables	Method of Transfer															
	Manual Lifting (2)[1]		Gait Belt (2)		Walking Belt (2)		Walking Belt (1)		Patient Handling Sling (1)		Lift H (2)		Lift T (2)		Lift A (2)	
	$\bar{x}$	SD	$\bar{x}$	SD	$\bar{x}$	SD	$\bar{x}$	SD	$\bar{x}$	SD	$\bar{x}$	SD	$\bar{x}$	SD	$\bar{x}$	SD
Comfort[2]	5.7	0.9	5.7	0.8	1.6	0.5	1.7	0.8	2.6	1.3	5.2	1.1	3.0	1.7	0.5	0.6
Security[3]	5.1	1.2	4.0	1.4	1.1	0.8	1.1	1.0	3.2	1.4	5.4	1.4	3.0	1.7	0.5	0.6

1. Indicates number of nurses making the transfer
2. Scale: 0 = extremely comfortable; 7 = extremely uncomfortable
3. Scale: 0 = extremely secure; 7 = extremely insecure

focus was education of nursing personnel on use of the Posey walking belt and Ambulift C_3 when transferring patients in and out of bed, on and off the toilet and on and off the shower chairs. Post intervention data were collected for eight months on floor one and four months on floor two.

Subjects. All of the NAs were involved in this part of the study as were the 140 patients on floors one and two of the nursing home.

Procedure. The patients were categorized by amount of care needed. The environmental changes were made for worksite redesign, and all nursing staff on both floors received education and practice with transfer techniques. Two nurse observers were trained in the use of data collection instruments. The Borg (1962) Rating of Perceived Exertion scale was used after completing a transfer. The two nurse observers randomly observed nursing personnel to collect the above data and to determine acceptability rates for the use of belts and Ambulifts. They also timed each patient handling task as it was being performed.

The Accident Investigative Reports and the OSHA 200 logs were reviewed to determine the number of back injuries that occurred on floors one and two for four years prior to intervention and eight months post intervention.

Intervention. The 140 patients were grouped into three categories according to ability to assist with the transfer: independent (n = 42), dependent but weight-bearing (n = 49) and dependent non-weightbearing (n = 49). The Ambulift C_3 was used with dependent non-weightbearing and heavy patients. The walking belt was recommended for dependent weightbearing patients weighing less than 150 pounds. Each patient needing a walking belt was provided one at the bedside. An Ambulift C_3 was stationed on each wing (20 patients per wing).

Environmental changes were made such as toilet seat risings were placed on toilets so transfer surfaces were flush and showers were modified so shower chairs could be used.

All nursing personnel were trained in use of walking belt, Ambulift and shower chairs until they felt comfortable and demonstrated competency. The technique taught for using the walking belt was: two nursing personnel stood facing the patient, each had one hand gripping the belt handle; placement of feet was one foot facing the patient and other foot in direction of the move; with flexed knees and backs straight, they used a synchronized, gentle rocking motion to create momentum and then pulled the patient toward themselves, shifted their weight to the foot facing the direction of the move, pivoted and transferred the patient. Instruction on use of the hoist followed the printed instructions provided by the manufacturer.

The following summarizes the results of intervention.

Perceived Exertion. In general, the average stressful ratings were about 14 (between "hard" and "somewhat hard") before intervention and 9 ("very light" for the walking belt) and 8 (between "very, very light" and "very light" for the Ambulift) after intervention (Table 6).

Acceptability Rates. Most of the time the nursing personnel used the walking belt or hoist as directed according to the categorization of the patient as either

Table 6. Ratings of perceived exertion for lower back from pre- and post-intervention phases for selected tasks.

	Ratings of Perceived Exertion		
	Pre-Intervention	Post-Intervention	
Task	Manual Lifting	Walking Belt	Ambulift
Bed to WC	14.1	8.6**	8.2**
WC to bed	14.2	9.3**	7.6**
WC toilet	14.1	9.5**	7.2**
Toilet to WC	14.3	8.8**	8.6**
Chairlift to WC	13.4	9.5**	6.0**
Weighing patients	13.8	—	6.3**

** Significant at $p \leq 0.01$

dependent and weightbearing or non-weightbearing. Sometimes, however, the condition of the patient changed so the device used also changed. Ninety-five percent of the time the nursing personnel used either a hoist or walking belt when transferring a dependent, non-weightbearing patient from bed to WC or WC to bed. This was true 92% of the time for transferring patients on and off the toilet. It is probable that the hoist should have been used rather than the walking belt in some instances; however, the manual, under-the-axilla technique was not used!

Time. As expected it took less time to use the devices as personnel became more skilled in their use. During preintervention it took an average of 8 to 18 seconds to transfer a patient using the under-the-axilla manual method. It took an average of 25 seconds to transfer using the walking belt and 73 seconds for the hoist.

Injuries. There were 83 injuries per 200,000 work hours in the four years prior to the intervention stage of this study; these were back injuries that occurred on floors one and two. During the eight months post intervention there were 47 back injuries per 200,000 work hours. Most of the back injuries that occurred post intervention were during non-patient transfer activities.

Summary

The results of this study indicate that an ergonomic intervention is appropriate for reducing physical and biomechanical stresses while transferring patients from one location to another. The most physically challenging tasks were redesigned or eliminated (e.g., through use of shower chairs for toileting and showering) and the acceptance rate of these new approaches was high.

ERGONOMIC APPROACH WITHIN A HOSPITAL SETTING

A similar study design has been implemented in the hospital setting. The PHTs ranked as most stressful by hospital personnel were very similar to those found in the nursing home study:

- Lift patient up from the floor
- Transfer patient from WC to bed
- Transfer patient from bed to cardiac chair
- Transfer patient from bed to commode
- Transfer patient from commode to bed
- Transfer patient from bed to WC
- Transfer patient from cardiac chair to bed
- Transfer patient from stretcher/cart to bed
- Transfer patient from bed to stretcher/cart

Criteria for selection of *assistive devices* were applied and the following devices were studied:

transfer on and off stretcher:

*Slipp　　Air Pal　　Dixie Smooth Mover

transfer in and out of bed, on and off commode/toilet:

*MediMan Hoist	*Sara Lift
*Posey Walking Belt	Total Lift
Maxi Lift	Dextra Lift
MediMaid	

lift up in bed:

Air Pal　　*Magic Sheet

toileting in bed:

* Kimbro Pelvic Lift

transfer from floor to WC or bed:

* Mediman Lift　　Maxi Lift

Those devices with an asterisk (*) were perceived to be least stressful, most comfortable and secure, hand force was reduced and the estimated compressive force to L_5S_1 disk was lower than the NIOSH action limit (U.S. Department of Health and Human Services, 1981); these assistive devices were used within the intervention phase of the study (analysis of data is continuing).

Implementing an Ergonomic Program

A very important element in implementing any program is to have commitment from management and from the individuals who must carry out the program. Change from routine is difficult so definitive strategies must be planned to reinforce the driving forces and decrease the restraining forces.

Some driving forces may be: the need to decrease the number of personnel with back problems in order to have a safer work environment and adequate staffing levels, a need to decrease workers' compensation insurance premiums, an opportunity to expend resources for primary prevention instead of at the tertiary or illness/injury level and a need to exhibit to employees personal concerns for their health and safety.

Some restraining forces may include time and financial resources. As shown in the nursing home study, the use of assistive devices does take more time and the equipment does demand financial investment.

A consultant with knowledge in ergonomics would be helpful in order to evaluate the interaction between equipment, environment, task, personnel and overall work organization.

Someone on the staff should assume leadership within this back injury prevention program. Usually individuals in top level management cannot take on the responsibilities because of ongoing commitments to other elements of the organization. The individual selected should be interested in the program, capable of providing leadership, have the respect of management and staff, be willing to test out various approaches to reducing back stresses and have the ability to keep management informed of progress/problems in the program.

In some institutions it may be possible for a staff person from each unit to be assigned to this injury prevention team. Some advantages to this model are: staff can relate to an individual who is already an integral part of the unit, her/his presence and work with this project will be a reminder of preventive efforts, this person can receive and funnel to the teams and the coordinator of the program ideas, concerns, etc. and one person can see that maintenance of equipment is up to date on that unit. This model of staff involvement incorporates important principles such as permitting the people who are at risk for the back problems to be involved in problem solving and even decision making in relation to this problem. This sharing in decision making also facilitates a feeling of ownership of and commitment to the program.

Tasks that are most stressful to carry out should be determined by those who normally carry out the work of the unit. The technique similar to the one described in the nursing home study can be used. Owen, Garg and Jensen (1992) used four methods to determine stressfulness of tasks and found no statistically significant difference among them. The methods were: asking the individuals to list stressful tasks and then to rank these tasks by stressfulness felt, asking them to rate the amount of stress felt to various parts of the body while carrying out those tasks cited as stressful (using Borg scale), estimating compressive force to L_5S_1 using the three-dimensional static biomechanical model of Garg and Chaffin and estimating tensile force on the erector spinae muscles using a biomechanical model. Based on the results of the comparison of these four methods, it seems feasible that it is adequate to ask the individuals to rank and rate the stressfulness of those tasks perceived to be most stressful.

The task(s) found to be most stressful should be studied. Questions such as the following may help to crystallize the problem. How is the task performed? How much weight is lifted? Are good body mechanics being used? (Certain types of lifting carry a particularly high level of risk, e.g., lifting at a distance, lifting involving asymmetry and twisting). Can manual lifting be eliminated? Are there environmental aspects that contribute to the stress such as unequal lifting surfaces or confined work space? Can a technique of pull or push be used instead of lift? Can the task be redesigned, can it be eliminated? How can the workplace be changed?

Answers to some of these questions should be helpful in proposing alternative methods for carrying out the task. *One alternative* may be to redesign the task by introducing an assistive device for transfer of patients. It is helpful for staff who are involved in the transfer task to be involved in testing out various assistive devices and even to contribute to decision making for which devices to purchase. An adequate number of devices must be purchased so they are readily available to staff.

Training in the alternative method is essential. Owen (1988) and Bell (1984) found personnel were reluctant to use assistive devices because they lacked knowledge and skill. Training to competency is important which includes demonstration and hands-on return demonstration in the classroom and at the bedside. A systematic schedule for retraining and also education of new staff should be instituted and maintained.

Management policies may be necessary to help assure/encourage staff to fulfill the goals of the project. It is important for personnel to see and experience management involvement and support throughout the program.

Maintenance of equipment is imperative also as this is often a reason for non-use. Delegation of this responsibility to one person on the unit may be helpful.

A health management program including an element directed at back pain prevention should be built into every prevention program. This is often lacking in settings where health care professionals work such as hospitals and nursing homes. Personnel within the health management program should encourage the staff to participate in the health program, report symptoms early relating to back problems, conduct prompt follow-up, conduct worksite tours to study work practices and add their expertise to the ergonomic approach to prevention of occupationally related back problems.

Evaluation of outcomes of the project are important. Tools such as the Borg Rating of Perceived Exertion Scale or 7 to 10 point Likert Scales are adequate to determine if there has been reduction in perceived physical stress. An on-going monitoring of the program is essential to assure staff are using the devices and using them correctly. A study of the incident reports and OSHA 200 logs also contribute to the evaluation process.

SUMMARY

Many approaches to the prevention of occupationally related back pain have been attempted. However, nursing personnel continue to rank fifth nationally in workers' compensation claims filed for this problem. The ergonomic approach has been successful in decreasing the physical stress involved in the lifting and transferring patients.

REFERENCES

Andersson GBJ: Epidemiological aspects on low-back pain in industry. *Spine,* No. 1, 53–60, 1981.

Baty D, Stubbs DA: Postural stress in geriatric nursing, *Int. J. Nursing Studies,* 24, 339–344, 1987.

Bell F: *Patient lifting devices in hospitals.* London: Groom Helm, 1984.

Borg G: *Physical performance and perceived exertion.* Thesis, Gleerups, Lund, 1962.

Christensen B, Kockrow E: Foundations of nursing. St. Louis: Mosby Year Book, 1991.

Cole G; *Basic Nursing Skills and Concepts.* St. Louis: Mosby Year Book, 1991.

Cust G, Pearson J, Mair A: The prevalence of low back pain in nurses. *International Nursing Review,* 19, 169–178, 1972.

Earnest V: Clinical Skills in Nursing Practice (2nd Ed.). Philadelphia: J.B. Lippincott Co., 1993.

Frymoyer JW, Pope MH, Costanza MC: Epidemiological studies of low-back pain. *Spine,* 5, 419–423, 1980.

Gagnon M, Sicard C, Sirois J: Evaluation of forces on the lumbo-sacral joint and assessment of work and energy transfers in nursing aides lifting patients. *Ergonomics,* 29, 407–421, 1986.

Garg A, Chaffin D: A biomechanical computerized simulation of human strength. *AIIE Transactions,* 7, 1–15, 1975.

Garg A, Owen BD, Beller D, Banaag J: A biomechanical and ergonomic evaluation of patient transferring tasks: Wheelchair to bed and bed to wheelchair. *Ergonomics,* 34(3): 289–312, 1991.

Garg A, Owen BD: Reducing back stress to nursing personnel: An ergonomic intervention in a nursing home. *Ergonomics,* 35(11):1353–1375, 1992.

Garg A, Owen BD, Carlson B: An ergonomic evaluation of nursing assistant's job in a nursing home. *Ergonomics,* 35(9):979–995, 1992.

Harper P, Billet E, Gutowski M, Soo Hoo K, Lew M, Roman A: Occupational low-back pain in hospital nurses. *Journal of Occupational Medicine,* 27(7): 518–524, 1985.

Jensen R: Events that trigger disabling back pain among nurses. *Proceedings of the 29th Annual Meeting of the Human Factor Society.* Santa Monica, CA: Human Factors Society, 1985.

Klein B, Jensen R, Sanderson L: Assessment of workers' compensation claims for back strains/sprains. *Journal of Occupational Medicine,* 26(6):443–448, 1984.

Kozier B, Erb G, Blais K, Johnson J, Temple J: Techniques in Clinical Nursing (4th Ed.). Redwood City, CA: Addison-Wesley, pp. 460–463, 1993.

Magora A: Investigation of the relation between low back pain and occupation. *Industrial Medicine and Surgery,* 39(11):31–37, 1970.

Nordin M, Ortengran R, Andersson G: Measurements of Trunk Movements During Work. *Spine*, 5, 465–469, 1984.

Owen BD, Damron C: Personal characteristics and back injury among hospital nursing personnel. *Research in Nursing and Health*, 7, 305–313, 1984.

Owen BD: The need for application of ergonomic principles in nursing. In S. Asfour (Ed.), *Trends in ergonomics/human factors IV*, North Holland: Elsevier Science Publishers, pp. 831–838, 1987.

Owen BD: Patient handling devices: an ergonomic approach to lifting patients. In F. Aghazadeh (Ed.), *Trends in Ergonomics/human Factors V.* North-Holland: Elsevier Science Publishers B.V., 1988.

Owen BD: The magnitude of low-back problems in nursing. *Western Journal of Nursing Research*, 11(2):234–242, 1989.

Owen BD, Garg A: Patient handling tasks perceived to be most stressful by nursing assistants. In A. Mital (Ed.), *Advances in Industrial Ergonomics and Safety I*, Philadelphia: Taylor & Francis Ltd., pp. 775–781, 1989.

Owen BD, Garg A: Assistive Devices for use with patient handling tasks. In B. Das (Ed.), *Advances in Industrial Ergonomics and Safety II*. London: Taylor & Francis, Ltd., pp. 585–592, 1990.

Owen BD, Garg A: Reducing risk for back pain in nursing personnel. *AAOHN Journal*, 39(1):24–33, 1991.

Owen BD, Garg A, Jensen RC: Four methods for identification of most back-stressing tasks performed by nursing assistants in nursing homes. *International Journal of Industrial Ergonomics*, 9, 213–220, 1992.

Pheasant S: Ergonomics, Work and Health. Gaithersburg, MD: Aspen Publishers, Inc., 1991.

Smith S, Duell D: Clinical Nursing Skills: Nursing Process Model Basic to Advanced Skills (3rd Ed.). Norwalk, CT: Appleton & Lange, 1992.

Snook S, Campanelli R, Hart J: A study of three preventive approaches to low back injury. *Journal of Occupational Medicine*, 20, 478–481, 1978.

Stobbe T, Plummer R, Jensen R, Attfield M: Incidence of low back injuries among nursing personnel as a function of patient lifting frequency. *Journal of Safety Research*, 19(2): 21–28, 1988.

Stubbs D: Back pain in nurses: Summary and Recommendations. University of Surrey, 1986.

Stubbs D, Buckle P, Hudson M, Rivers P, Baty D: Backing out: Nurse wastage associated with back pain. *International Journal of Nursing Studies*, 23, 325–336, 1986.

Stubbs D, Baty D, Buckle P, Fernandes H, Hudson M, Rivers P, Worringham C, Barlow C: Back Pain in Nurses: Summary and Recommendations. Guildford, England. University of Surrey, 1986.

Stubbs D, Rivers P, Hudson M, Worringham C: Back pain research. *Nursing Times*, 77, 857–858, 1981.

Stubbs D, Buckle P, Hudson M, Rivers P, Worringham C: Back pain in the nursing profession I. Epidemiology and pilot methodology. *Ergonomics*, 26, 755–765, 1983.

Torma Krajewski J: Occupational Hazards to Health Care Workers. Northwest Center for Occupational Health and Safety, September, 1986.

U.S. Department of Health and Human Services. *Work Practices Guide for Manual Lifting* (DHHS (NIOSH) Publication No. 81-122), Cincinnati, OH, 1981.

Valles-Pankratz S: What's in back of nursing home injuries. *Ohio Monitor*, 62(2): 4–8, 1989.

Venning P, Walter S, Stitt L: Personal and job-related factors as determinants of incidence of back injuries among nursing personnel. *Journal of Occupational Medicine*, 28(10): 820–825, 1987.

PART 2

The Lifting Team: A Method to Reduce Lost-Time Back Injury in Nursing—a Four Hospital Study

William Charney

Occupational back injury continues to be one of the most costly injuries in hospitals. Studies of back-related compensation claims show nurses among the highest claim rates in any occupation or industry.[1-6] Harper reported that over a six-month period in a large tertiary hospital, 52% of nurses missed at least one day of work due to lower back pain.[7] In addition to causing substantial absenteeism and injury, back pain contributes to the decision of some nurses to leave the profession. Stubbs investigated the reasons registered nurses left current positions. Of 1000 nurses who left the profession, 35% listed back pain as a major reason for leaving. Of the nurses in this study who decided to leave nursing permanently, 12% did so because of back pain.[8]

Harper found that 48% of nurses with back pain related patient lifting as the cause.[9] Jensen found that although numerous factors affect risk, a widely recognized but poorly documented factor is exposure to patient lifting.[10] Klein and Stellman both point out that a leading cause of lost work time for nurses is back injury and the trigger event for the majority of these back injuries is moving patients.[11-12]

Given the epidemic of back injury in nursing, intervention approaches were appraised for effectiveness and cost. Venning showed that education alone will not solve this problem.[13] Stubbs et al. pointed to the paucity of specific data to support systematic instruction of lifting techniques.[14] In view of this climate of uncertain data for training results vs. cost for training, a new method was tested.

METHOD

The lifting team method incorporates the philosophy of removing nurses from the everyday task of moving patients in a facility. Within the method three assumptions are made:

1. The lifting of patients should be considered a skill rather than a random task. It should be performed by skilled professional client movers trained in the latest techniques, equipped with the most "user friendly" mechanical equipment and protected with essential personal protective equipment.

2. A risk assessment guideline should place actual risk where it can be controlled: in a team of two rather than a whole nursing department.

1-56670-083-3/94/$0.00+$.50

3. A linear relationship exists between the percentage of removal of nursing from lifting to a decrease in compensation dollars spent.

Data from four hospitals that have implemented a lifting team are reported.

Hospital A: Data of three years implementation. Two years of data published in peer review.[15–16]
Hospital B: Data of one year reported.
Hospital C: Data of one quarter of implementation
Hospital D: Two years of data. 1993 is actually three quarters of fiscal year.

All four hospitals have been running the lifting team on the day shifts where all four reported that the day shift had the majority of accidents. The data from each hospital are reported in somewhat different ways, yet give an overall impression of the effectiveness of the method and approach. The mandate of the teams in all four hospitals was the same; respond to the majority of all total body lifting needs of the facility and remove nursing from the overall responsibility of lifting patients during the attended shift. All four hospitals reported starting with a scheduled and paging system for team response. Hospital A and B have moved over to almost an entire paging system, eliminating the schedule approach.

PROCEDURE

The median census for the four hospitals is approximately 300 beds. During the course of the study Hospital B's census expanded to 450 beds with the opening of new units. Hospital A averaged 5500 lifts per year over three years, Hospital B had 9877 lifts for the year reported and there are no statistics available for number of lifts for Hospital C. Hospital D did not report number of lifts for the year. Lifting teams were chosen from employee pools and given training by the physical therapy departments. Hospital D is running the lifting team during the day shift, five days per week. The hospital initiated a weekend lifting team effective May, 1993.

RESULTS

Hospital A went from 16 lost time accidents to 1 lost-time accident on the day shift during the first year of implementation. Hospital A had lost-work days decrease from 215 days lost to 6 days lost during the first year of the study, October, 1989 to October, 1990. In the subsequent two years of data collection Hospital A has maintained a statistically significant decrease in lost-time injury and lost days as reported in Table 1.

Table 1. Lost-time back injuries in nursing. Day shift, October, 1988 to September 1993. San Francisco General Hospital.

Year	Hours	No. of Accidents	No. Lost Days
10/88 - 9/89	8AM - 4PM	16	215
10/89 - 9/90	8AM - 4PM	1	6
10/90 - 9/91	8AM - 4PM	0	0
10/91 - 9/92	9AM - 5PM	2	7
10/92 - 9/93	9AM - 5PM	4	18+

Hospital B for the one-year day shift implementation reported lost-time injuries decreased from 23 to 13 (43.4% reduction). Lost restricted workdays from this type of injury decreased from 405 to 268 (33.8% reduction). Severity rates and frequency rates are reported in the table and all show a dramatic downward trend. Of the 13 lost-time cases in Hospital A that did occur during times the lift team was available, 9 (69.2%) occurred during the first six months of the program and 4 (30.7%) occurred during the remaining six months. This suggests a downward trend due to familiarity with the program. When Hospital A compares by quarters there is a dramatic decrease in the first third quarter as the team began as compared to the last computed quarter; lost-time back injuries in quarter 3, 1991 (9 injuries) fell to 1 lost-time injury in third quarter, 1993 and lost restricted workdays fell from 159, third quarter 1991 to 15 lost days, third quarter 1993. This downward trend is also predictive as the team and hospital staff acquire experience in the method.

Hospital C has reported the following data for the first quarter that the team has been implemented: 140 lost days to zero lost days.

Hospital D showed statistical decreases in all areas: monetary loss category, total claims, indemnity claims, frequency rate and loss cost rate. Hospital D data also show a progressive decline in rates for the two years of lifting team implementation: 1992, the first year of implementation, shows significant decreases from the previous year where no lifting team was implemented and 1993 shows even a greater decrease than 1992.

MONETARY IMPLICATIONS

Hospital A shows a decrease in compensable dollars for back injury in nursing of approximately $144,000. Deduction of cost of team of $70,000 leaves a total savings of $70,000 per year.

Hospital B shows a decrease of approximately $30,000 in the fiscal year the lifting has operated, after all costs have been deducted.

Hospital C shows a reduction of $41,000 in the first quarter of operation after all costs have been deducted.

Hospital D had compensation losses in 1991 (year before implementation) of $414,235. The first year of implementation of the team the dollar losses were decreased to $239,085, a 42% decrease, and in the third year the dollars spent on compensation dropped to $138,829, a decrease of 66%.

DISCUSSION OF RESULTS

All four hospitals have reported statistically significant reductions in major categories for back injury. Lifting teams in all four hospitals were able to reliably meet the needs of patient transfers while removing the major percentage of nursing responsibility for patient lifting, thus allowing nurses to utilize more time for clinical bedside care. Hospital A quantified that by having the lifting team ward nurses save approximately 1 hour per shift. Lifting team members were not injured in the four hospitals studied due to an important provision in the policy and procedure that for every total body transfer use of a mechanical lifting device was mandatory (Hoyer lift, slide board, etc.).

The success of these four hospitals merits continuing study in the lifting team approach to reduce lost-time injury due to lifting patients. However, the lifting team is only one of many variables in a hospital's overall ergonomic program.

REFERENCES

1. Howells RA, Kurgut TC, Jr: Safety Considerations for Hospital Populations. In *Handbook of Hospital Safety,* Stanley PE (Ed.). CRC Press, Inc., pp 13–42, 1981.
2. Cust G, Pearson JCG, Mam A: The Prevalence of Low Back Pain in Nurses. *Int Nursing Rev* 19:169–179, 1972.
3. Harper P, Billet E, Gutouski M, Soo Hoo K, Lew M, Romon A: Occupational Low Back Pain in Hospital Nurses. *J Occup Med* 27:518–524, 1984.
4. Magon A, Tausten I: An Investigation of the Problem of Sick Leave in the Patient Suffering from Low Back Pain. *Ind Med Surg* 38:398–408, 1969.
5. Owen BD, Damron CF: Personnel Characteristics and Back Injury Among Hospital Nursing Personnel. *Res Nursing Health,* 7:305–313, 1984.
6. Klein BP, Jensen RC, Samelson LM: Assessment of Worker's Compensation Claims for Back Strains. *J Occup Med* 26:443–448, 1984.
7. Harper P, et al.: Low Back Pain in Hospital Nurses, 27:518–524, 1984.
8. Stubbs DA, Buckle PW, Hudson MP, Rivers PM, Worthington CJ: Back Pain in the Nursing Profession. *Ergonomics,* 26:755–765, 1983.
9. Harper P, Billet B, et al.: Occupational Low Back Pain in Hospital Nurses. *J Occup Med* 27:518–525, 1985.
10. Jensen RC: Low Back Pain in Nursing. *J Safety Research,* Vol 19, pp. 21–25.
11. Klein BP, Jensen RC, Svadenson LM: Assessment of Worker's Compensation Claims for Back Sprains. *J Occup Med* 26:443–448, 1984.
12. Stellman, JM: Safety in the Health Care Industry. *Occup Health Nursing,* 30:17–21, 1982.
13. Venning PJ: Back Injury Prevention Among Nursing Personnel: The Role of Education. *AAOHN J,* 36(8):327–333, 988.

14. Stubbs DA, Buckle PW, Hudson MP, Rivers OM, Washington CJ: Back Pain in the Nursing Profession. *Ergonomics* 26:767–779, 1983.
15. Charney W, Zimmerman K, Walera E: The Lifting Team: A Design Method to Reduce Lost Time Back Injury in Nursing. *AAOHN*, Vol. 35, No. 5.
16. Charney W: The Lifting Team: Second Year Data. *AAOHN*, Vol. 40, No. 10.

Presented to the American Health Association Conference, October, 1993.

Table 2. Lifting team total lifts, October, 1989 through September, 1993. San Francisco General Hospital (Hospital A).

Year	Scheduled	On-Call	Total
10/89 - 9/90	4288	1598	5886
10/90 - 9/91	3730	2129	5859
10/91 - 9/92	2355	3289	5644
10/92 - 9/93	2198	3383	5581

Table 3. Lifting team statistics. October, 1992 through September, 1993. San Francisco General Hospital (Hospital A).

Types of lifts	# of Lifts
Out of bed; bed to bed	2274
To/from commode	101
To/from gurney	459
Back to bed	2124
To/from tub/shower	43
Pulled/lifted up in bed	147
Turned over in bed; scale	72
Pulled up in chair; pulled up on gurney	32
From chair to chair/walker; floor to chair; chair to floor	99
To/from x-ray/exam table	97
Hoyer lift	133
# of Lifts	
Scheduled	2198
On-Call	3383
Total	5581

Table 4. Frequency Rate. Lost-time back injuries—lifting/moving patients. Hospital B.

	July 1, 1991–30 June 30, 1992	July 1, 1992–June 30, 1993
ICU	7.76	1.48
CCU	7.15	3.60
2SO	10.43	6.36
3SO	7.04	0.0
4SO	2.29	4.48
PED	4.09	0.0
FCMC	1.99	0.0
L&D	8.70	2.18
SUR	2.81	1.34
X-RAY	6.09	3.74
DIAL	0.0	30.37
PT	6.20	27.79
HOME	0.0	10.85
ER	0.0	0.0

$$\text{formula: } \frac{\text{\# lost-time back injuries} \times 200{,}000}{\text{productive hours}}$$

rate = number of injuries per 100 FTE's

Note: Other cost centers were not calculated due to geographical relocation, temporary closures and/or changes in patient population/unit utilization.

Table 5. Severity Rate. Lost/restricted days from lifting/moving patients. Hospital B.

	July 1, 1991–30 June 30, 1992	July 1, 1992–June 30, 1993
ICU	130.70	32.65
CCU	53.68	7.20
2SO	33.39	16.98
3SO	79.20	0.0
4SO	75.77	13.45
PED	98.20	0.0
FCMC	95.69	0.0
L&D	232.78	13.13
SUR	177.14	9.41
X-RAY	294.67	24.31
DIAL	0.0	60.75
PT	49.66	1063.16
HOME	0.0	50.63
ER	0.0	0.0

formula: $\frac{\text{\# lost/restricted days} \times 200{,}000}{\text{productive hours}}$

rate = number of lost days per 100 FTE's

Note: Other cost centers were not calculated due to geographical relocation, temporary closures and/or changes in patient population/unit utilization.

Table 6. Report on the lift team: performance. July 1, 1992 through June 30, 1993. Hospital B.

Type of lift	Number	% of total
Back to bed/over in bed	3355	33.9
Chair in bed	1641	16.6
Out of bed	1538	15.5
Up in bed/dressed	1078	10.9
Transport	1062	10.7
Gurney transfer	581	5.8
Up in chair	137	1.3
Chair to chair	128	1.2
Missed lift	99	1.0
Move wheeled object	94	0.9
From toilet	63	0.6
Standby/not needed	33	0.3
Mechanical transfer	18	0.1
Chair to scale/walker	17	0.1
Other	16	0.1
X-ray table	8	0.08
Tub/shower	8	0.08
Restrain pt.	1	0.01
Total	9877	

Table 7. Report on the lift team: utilization. July 1, 1992 through June 30, 1993. Hospital B.

Unit	Number	% of total
3 South	1954	19.6
ICU (CVICU & SICU)	1257	12.7
4 North	1219	12.3
CCU	980	9.9
NCCU	913	9.2
3 North	578	5.8
3 West	526	5.3
2 South	505	5.1
4 South	494	5.0
4 West	479	4.8
Dialysis	184	1.8
PACU	182	1.8
ER	128	1.2
Endoscopy	102	1.0
Orthopedics	76	0.7
Radiology	69	0.6
1 Southwest	53	0.5
EEG	50	0.5
To/from automobile	44	0.4
AM admit	25	0.2
Other	22	0.2
Pediatrics	17	0.1
Rehab	8	0.08
Surgery	7	0.07
Cardiology	6	0.06
Bronchoscopy	3	0.03
FCMC	3	0.03
Labor and delivery	2	0.02
Outpatient	1	0.01
Total	9877	

Table 8. Report on the lift team: distribution. July 1, 1992 through June 30, 1993. Hospital B.

Hour of day	Number	% of total
0800–0900	882	8.9
0900–1000	1121	11.3
1000–1100	1352	13.6
1100–1200	1293	13.0
1200–1300	662	6.7
1300–1400	1571	15.9
1400–1500	1218	12.3
1500–1600	1247	12.6
1600–1630	531	5.3
Total	9877	

Table 9. Report on the lift team: effectiveness. July 1, 1992 through June 30, 1993. Hospital B.

All back injuries lifting/moving patients	3rd CY quarter 1991	1992	1993
a. Monday-Friday 0800–1630 at DMC	11	8	4
b. All other hours at DMC	10	7	6
c. Home health	—	2	1
Total	21	17	11
Lost-time back injuries lifting/moving patients			
a. Monday-Friday 0800–1630 at DMC	9	5	1
b. All other hours at DMC	4	3	2
c. Home health	0	2	1
Total	13	10	4
Lost/restricted days from lifting/moving patient back injuries			
a. Monday-Friday 0800–1630 at DMC	159	233	15
b. All other hours at DMC	252	25	13
c. Home health	0	12	23
Total	411	270	51

Table 10. Report on the lift team: effectiveness (composite). July 1, 1992 through June 30, 1993. Hospital B.

All back injuries lifting/moving patients	**July 1, 1991 through June 30, 1992**	**July 1, 1992 through June 30, 1993**	**% change**
a. Monday-Friday 0800–1630 at DMC	37	24	- 35.1
b. All other hours at DMC	38	26	- 31.5
c. Home health	0	3	+ 100.0
Total	75	53	- 29.3
Lost-time back injuries lifting/moving patients			
a. Monday-Friday 0800–1630 at DMC	23	13	- 43.4
b. All other hours at DMC	21	14	- 33.3
c. Home health	0	3	+ 100.0
Total	44	30	- 31.8
Lost/restricted days from lifting/moving patient back injuries			
a. Monday-Friday 0800–1630 at DMC	405	268	- 33.8
b. All other hours at DMC	626	120	- 80.8
c. Home health	0	14	+ 100.0
Total	1031	402	- 61.0

Table 11. Report on the lift team: effectiveness (unit specific). All back injuries lifting/moving patients. July 1, 1992 through June 30, 1993. Hospital B.

Unit	July 1, 1991 through June 30, 1992			July 1, 1992 through June 30, 1993			Overall % change
	M-F 8–4:30	all other		M-F 8–4:30	all other		
3 South	3	7	(10)	1	1	(2)	- 80.0%
4 North	2	2	(4)	0	1	(1)	- 75.0*
ICU (CVICU and SICU)	6	6	(12)	4	0	(4)	- 66.6
1 Southwest	1	2	(3)	1	0	(1)	- 66.6*
Surgery	2	1	(3)	1	0	(1)	- 66.6
Labor and delivery	2	2	(4)	1	1	(2)	- 50.0
Rehab	2	2	(4)	0	2	(2)	- 50.0*
Pediatrics	2	0	(2)	0	1	(1)	- 50.0
Float pool	3	6	(9)	1	4	(5)	- 44.4
2 South	5	2	(7)	2	2	(4)	- 42.8
4 South	2	1	(3)	1	1	(2)	- 33.3
3 North	0	4	(4)	1	2	(3)	- 25.0*
CCU	1	1	(2)	1	1	(2)	0.0
FCMC	0	1	(1)	0	1	(1)	0.0
ER	1	0	(1)	1	0	(1)	0.0
Radiology	3	1	(4)	4	2	(6)	+ 50.0
Dialysis	0	0	(0)	0	1	(1)	+100.0
Physical therapy	2	0	(2)	3	1	(4)	+100.0
Lift team	—	—		1	0	(1)	— *
4 West	—	—		0	2	(2)	— *
3 West	—	—	—	1	3	(4)	— *
Totals	37	38	(75)	24	26	(50)	
Home health	0	0	(0)	2	1	(3)	
	37	38	(75)	26	27	(53)	- 29.3

* incomplete data

Table 12. Report on the lift team: effectiveness (unit specific). Lost-time back injuries from lifting/moving patients. July 1, 1992 through June 30, 1993. Hospital B.

Unit	July 1, 1991 through June 30, 1992			July 1, 1992 through June 30, 1993			Overall % change
	M-F 8–4:30	all other		M-F 8–4:30	all other		
3 South	2	2	(4)	0	0	(0)	-100.0%
FCMC	0	1	(1)	0	0	(0)	-100.0
Pediatrics	1	0	(1)	0	0	(0)	-100.0
Rehab	1	1	(2)	0	0	(0)	-100.0*
1 Southwest	1	0	(1)	0	0	(0)	-100.0*
ICU	4	2	(6)	1	0	(1)	- 83.3
Labor and delivery	2	2	(4)	1	0	(1)	- 75.0
4 North	2	2	(4)	0	1	(1)	- 75.0*
Surgery	2	0	(2)	1	0	(1)	- 50.0
CCU	1	1	(2)	1	0	(1)	- 50.0
2 South	3	2	(5)	2	1	(3)	- 40.0
Radiology	2	1	(3)	1	1	(2)	- 33.3
3 North	0	3	(3)	1	2	(3)	0.0*
ER	0	0	(0)	0	0	(0)	0.0
Float pool	1	3	(4)	1	3	(4)	0.0
4 South	0	1	(1)	1	1	(2)	+100.0
Physical therapy	1	0	(1)	3	1	(4)	+300.0
Dialysis	0	0	(0)	0	1	(1)	+100.0
Lift team	—	—		0	0	(0)	—*
3 West	—	—		0	1	(1)	—*
4 West	—	—		0	2	(2)	—*
Totals	23	21	(44)	13	14	(27)	
Home health	0	0	(0)	2	1	(3)	+100.0
	23	21	(44)	15	15	(30)	- 31.8

* incomplete data

Table 13. **Report on the lift team: effectiveness (unit specific). Lost or restricted days from lifting/moving patients. July 1, 1992 through June 30, 1993. Hospital B.**

Unit	July 1, 1991 through June 30, 1992			July 1, 1992 through June 30, 1993			Overall % change
	M-F 8–4:30	all other		M-F 8–4:30	all other		
3 South	11	34	(45)	0	0	(0)	- 100.0%
FCMC	0	48	(48)	0	0	(0)	- 100.0
Pediatrics	24	0	(24)	0	0	(0)	- 100.0
1 Southwest	149	0	(149)	0	0	(0)	- 100.0*
Rehab	4	2	(6)	0	0	(0)	- 100.0*
Surgery	126	0	(126)	7	0	(7)	- 94.4
Labor and delivery	12	95	(107)	6	0	(6)	- 94.3
Radiology	4	141	(145)	4	9	(13)	- 91.0
4 North	8	92	(100)	0	12	(12)	- 88.0*
CCU	13	2	(15)	2	0	(2)	- 86.6
4 South	0	33	(33)	4	2	(6)	- 81.8
ICU (CVICU and SICU)	29	72	(101)	22	0	(22)	- 78.2
3 North	0	53	(53)	12	10	(22)	58.4*
2 South	10	6	(16)	7	1	(8)	50.0
ER	0	0	(0)	0	0	(0)	0.0
Float pool	7	48	(55)	62	32	(94)	+ 70.9
Physical therapy	8	0	(8)	142	11	(153)	+1812.5
Dialysis	0	0	(0)	0	2	(2)	+ 100.0
Lift team	—	—		0	0	(0)	— *
3 West	—	—		0	3	(3)	— *
4 West	—	—	—	0	38	(38)	— *
Totals	405	626	(1031)	13	14	(388)	
Home health	0	0	(0)	2	1	(14)	+ 100.0
	405	626	(1031)	15	15	(402)	- 61.0

* incomplete data

Table 14. ABMC, Lift team followup. 4 months operations. Hospital C.

	1991–1992 (Ashby only)		1992–1993 (Ashby only)		Savings	Comments
	Lost days	Hours	Lost days	Hours		
RN	140	1,120		0	1,120	65+ hrs of 1991–1992 are for case which occurred during uncovered period
PCA	43	344	5	40	304	
Funds spent		31,676		2,205	29,471	for 1992–1993 two cases are too early to estimate potential cost
Sick time		23,726		168	23,558	
Replacement		51,680		800	50,880	
		107,082		1,173	103,909	
Lift team costs					62,713	5.6 FTEs with an annual salary of 28k plus benefits
					41,196	
						Herrick site has remained comparable between years for the number of reported injuries 84 lost days in 1991–1992 vs. 70+ lost days in 1992–1993
						There were two reported injuries in '92–'93 which did not result in any loss time from the job

Table 15. Multi-year loss rate analysis nursing cost centers—back injuries only. Hospital D.

	1991	1992	1993*	1993**
Data				
Losses	$414,235	$239,085	$138,829	$170,895
Valuation date	12/31/91	12/31/92	9/30/93	9/30/93
Prod. P/R hrs	619,852	625,331	457,578	519,397
Total claims	38	28	18	19
Indemnity claims	27	12	10	11
TD days paid	N/A	N/A	375	382
Loss Rates				
● Loss cost rate	$.67	$.38	$.30	$.33
O Loss cost rate	$133,656	$76,467	$60,080	$65,805
O Frequency rate	12.3	9.0	7.9	7.3
O Indemnity rate	8.7	3.8	4.4	4.2
O TD days paid rate	N/A	N/A	164	147

● Per productive payroll hours
O Per 200,000 productive payroll hours

Comments: 1. 1993* data excludes STCC experience; 2. 1993** data includes STCC experience; 3. Lifting team initiated on day shift only, effective 4/92; 4. Weekend lifting team initiated effective 5/93.

Table 16. **Frequency and indemnity rates. Nursing cost centers only. Hospital D.**

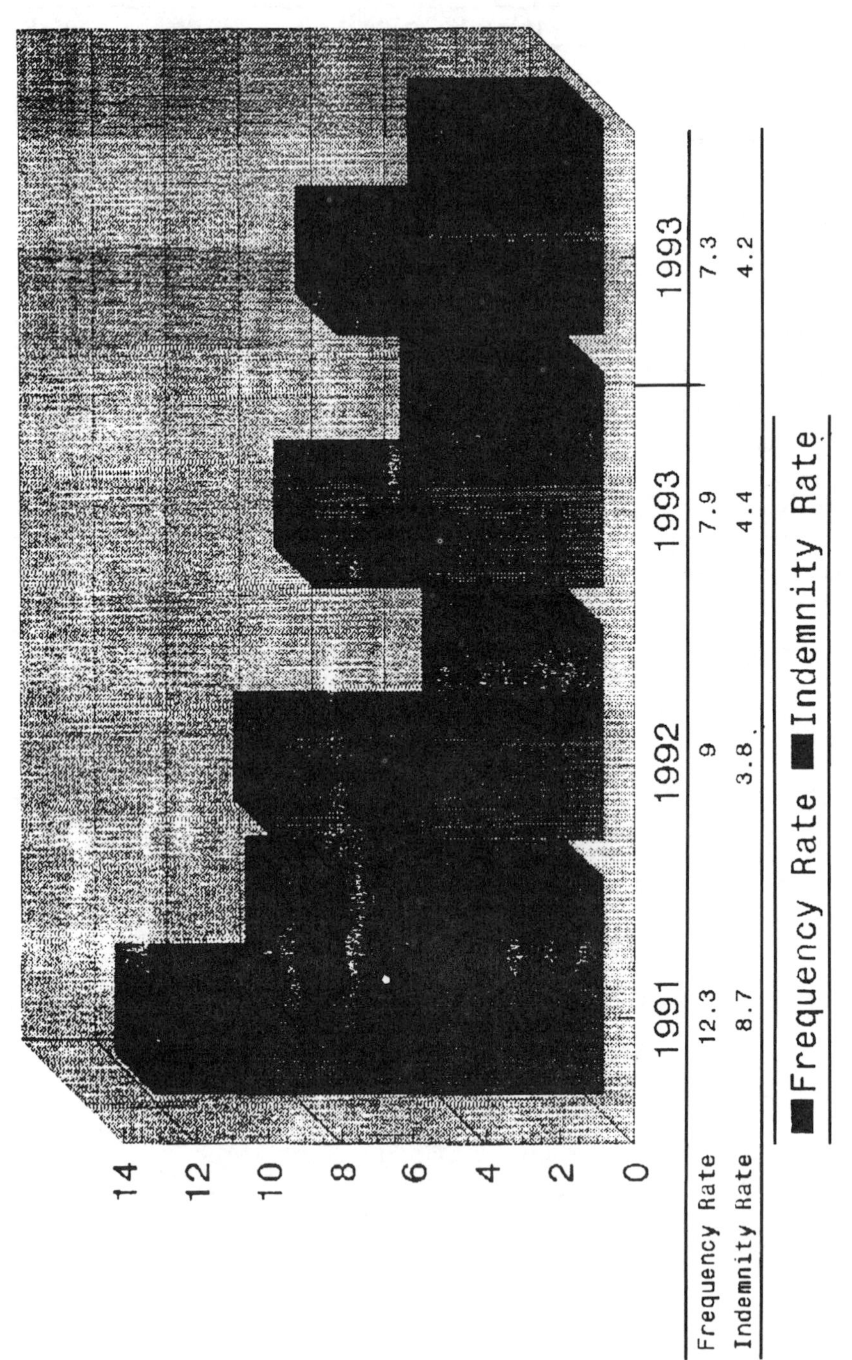

Table 17. Loss cost rate analysis. Hospital D.

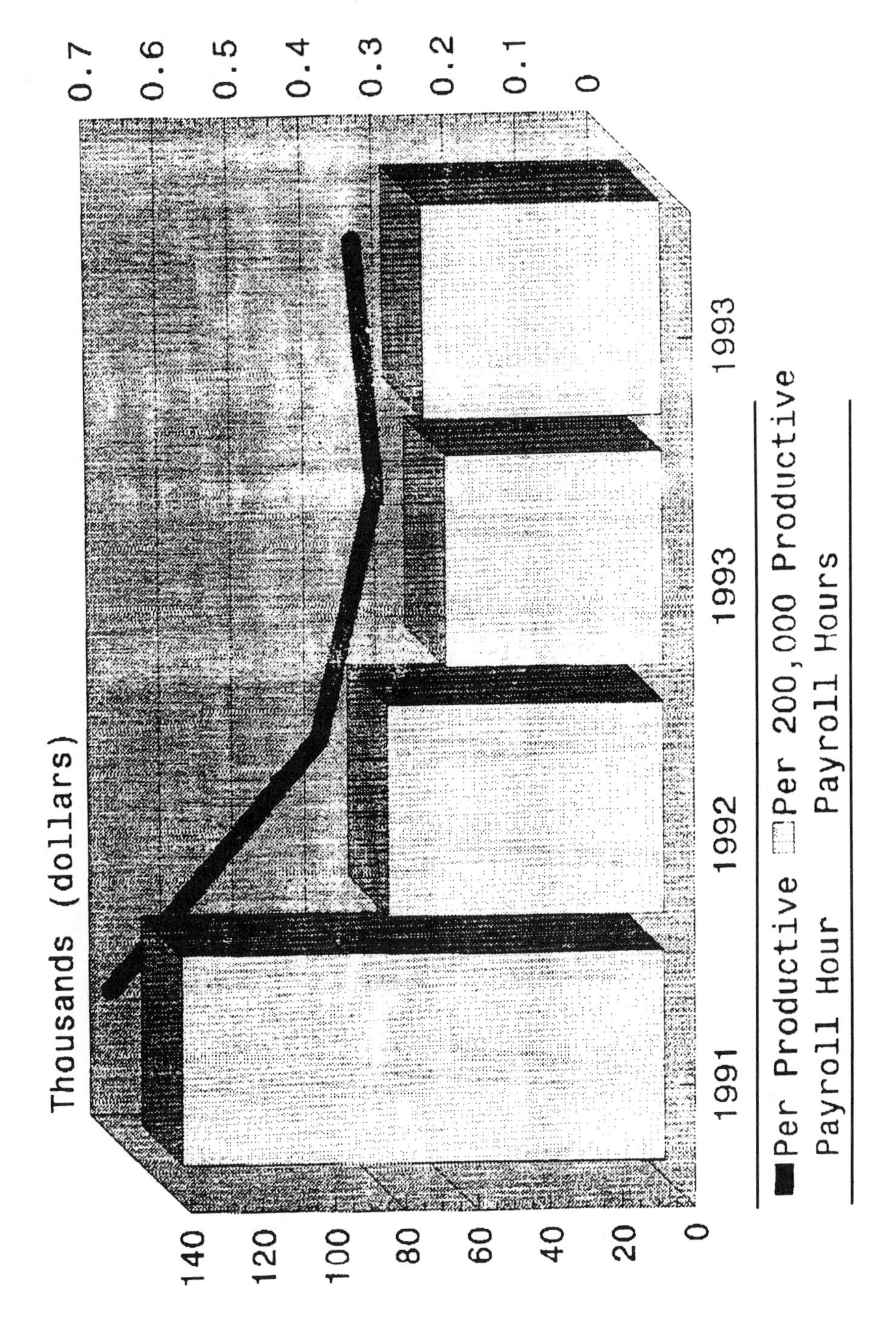

Table 18. Lifts per shift—SGH/STCC.

	January 1993	May 1993	June 1993	July 1993	August 1993
Average census*	148.1	162.2	199.7	200.7	185.7
Weekday team	11	15	11	14	18
Weekend team	0	15	21	19	22
Total average	11	15	14	15	19

* per business service

Weekday lifting team initiated April 1992; Weekend lifting team initiated May 1993; 5 North, sub-acute unit in operation, May 1993; 5 East and 5 West/STCC in full operation, June 1993; A comparison: at San Francisco General Hospital—average census = 353, average number of lifts per shift = 30.

APPENDIX 1

Sample Facility Ergonomic Program to Address Back Injuries Caused by Lifting

Ergonomics can simply be defined as the study of work. Ergonomics help adapt the job to fit the person, rather than force the person to fit the job. Adapting the job to fit the worker can help reduce ergonomic stress and eliminate many potential ergonomic disorders. The objective of ergonomics is to adapt the job and workplace to the worker by designing tasks, work stations, tools and equipment that are within the worker's physical capabilities and limitations.

Below is a list of key elements essential to the success of any ergonomics program. While the key elements specifically relate to back injuries, the methodology can be used to address any cumulative trauma disorders, which are injuries to the body's tendons, nerves and muscles caused by overuse.

KEY ELEMENTS OF IMPLEMENTING AN ERGONOMICS PROGRAM

1. Management and Labor Involvement

Commitment and involvement are essential elements of a sound safety and health program.

- Commitment by management provides organizational resources and motivating force necessary to deal effectively with ergonomic hazards.
- Management demonstrates involvement by placing a priority on eliminating ergonomic hazards.
- Management commitment is demonstrated by assigning and communicating responsibility for the ergonomic program to all managers, supervisors and employees. Accountability mechanisms are in place.
- An effective written program for job safety, health and ergonomics is in place, outlining the facility's goals, and is communicated to all employees. All managers, supervisors and employees involved know what is expected of them.
- Labor and employee involvement in the ergonomics program is encouraged through health and safety committees, a complaint/suggestion program, prompt and accurate reporting of injuries and training of employees in the skills necessary to analyze jobs for ergonomic stress.
- Procedures are implemented to regularly evaluate the effectiveness of the ergonomics program, to monitor stress in meeting goals and objectives. This process includes review of injuries recorded in the Occupational Injury Reporting System (OIRS), workers' compensation reports, DOSH 900 reports and reports from employees of unsafe working conditions. Also included are regular walk-around inspections, safety and health committee meetings and employee surveys regarding worksite changes.

2. Worksite Analysis

Worksite analysis identifies existing hazards and conditions, work habits which create hazards and areas where hazards may develop. Also included are close scrutiny and tracking of injury and illness records to identify patterns of traumas or strains. The objective of worksite analysis is to recognize, identify and correct ergonomic hazards.

- The facility gathers relevant information solutions to back injury problems.
- Baseline screening surveys using a checklist are conducted to evaluate ergonomic risk factors and determine which tasks are most stressful and need improvement.
- Job analysis is performed by persons skilled in evaluating ergonomic risk factors. Evaluation is conducted during peak activity times.
- Changes are implemented to avoid the most stressful tasks.
- Periodic surveys and followups are conducted to evaluate changes.
- Screening surveys/checklists are utilized to evaluate workplace stressors.
- After each back injury, a determination is made if a task can be modified to reduce future risk.

3. Hazard Prevention and Control

Ergonomic hazards are prevented primarily by proper selection of equipment, effective use of assistive devices and by implementation of proper work practices. The equipment must be available in sufficient quantities, convenient for use and properly maintained.

4. Work Practices/Administrative Controls

- Supervisors are familiar with ergonomic guidelines and enforce facility policies.
- Injuries are accurately reported.
- An effective program for facility and equipment maintenance is implemented which minimizes ergonomic hazards and includes:

  a. A preventive maintenance program for devices
  b. Maintenance whenever employees report problems; sufficient spares exist for out-of-service equipment
  c. Implementation of housekeeping programs which minimize slippery work surfaces and slip/fall hazards

5. Medical Management

Proper medical management is necessary both to eliminate or reduce the risk of lifting injuries through early identification and treatment, and to prevent future problems through rehabilitation and training. Health care providers must be part of the injury prevention team, and after on-site review interact and exchange information with management.

- The program is supervised by a person trained in the prevention of cumulative trauma disorders and is a consultant to the Safety and Health Committee. The program is periodically evaluated and includes:

  a. Accurate injury and illness recording
  b. Early recognition and reporting
  c. Conservative treatment with specific restrictions during recovery
  d. Systematic monitoring—return to work only after skills have been reassessed
  e. Baseline health assessment for comparing changes in health status, and preplacement evaluation of skills

6. Training and Education

The purpose of training and education is to ensure that managers, supervisors and employees are sufficiently informed about the ergonomic hazards to which they may be exposed and thus are able to participate actively in their own protection.

- Training programs should be designed and implemented by qualified persons. The program should include an overview of the potential risk of back and other musculoskeletal injuries, their causes and early symptoms and means of prevention and treatment.
- Training should be presented in a language and at a level of understanding, appropriate for the individuals being trained, and should include opportunities for interactive questions and answers with the trainer.
- Training should include an opportunity to "practice" with equipment and devices used at each facility.

- Refresher training is provided and addresses specific needs. Provisions are made to train absent employees.
- Staff physical fitness is encouraged.
- Training on early identification, reporting and conservative treatment of back injuries is provided to all workers, supervisors and management on an annual basis.

CHAPTER 8

Operating Theater Safety

PART 1

Inhalation Anesthesia Agents

Edward W. Finucane

INTRODUCTION AND BACKGROUND

Considerable valid concern exists for the potential harm—particularly to certain categories of exposed individuals—that could occur as a result of an exposure to volatile inhalation anesthesia agents. Included in this general broad category of chemicals are the following seven, tabulated below, along with some of their more important physical properties:

Nitrous Oxide

Synonyms: Laughing Gas, Hyponitrous Acid Anhydride, Factitious Air, and/or Hippie Crack

Name & Formula: Dinitrogen Monoxide or Nitrous Oxide: N_2O

Structure: $N \equiv O = N$

Physical Properties: Boiling Point = – 88.5°C (gas at room temperature)
Molecular Weight = 44.01 amu
Gaseous Nitrous Oxide Specific Gravity = 1.53; (the Specific Gravity of Air = 1.00)
Colorless gas with a slightly sweetish odor and taste
Normally supplied in metal cylinders containing a vapor phase above a liquid phase, at an approximate internal pressure of 800 psig

1-56670-083-3/94/$0.00+$.50

Halothane

Synonyms: Fluothane and/or Rhodialothan

Name & Formula: 2-bromo-2-chloro-1,1,1-trifluoroethane: $C_2HBrC1F_3$

Structure:

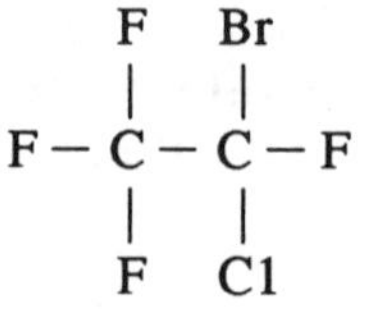

```
     F   Br
     |   |
F -  C - C - F
     |   |
     F   C1
```

Physical Properties: Boiling Point = 50.2°C (liquid at room temperature)
Molecular Weight = 197.39 amu
Liquid Density = 1.871 gms/cm^3
Non-flammable and highly volatile liquid, with a sweetish but not wholly unpleasant odor
Normally supplied in specially keyed glass bottles

Enflurane

Synonyms: Ethrane, Efrane, Alyrane, NSC-115944, Compound 347, and/or Methylflurether

Name & Formula: 2-chloro-1-(difluoromethoxy)-1,1,2-trifluoroethane

2-chloro-1,1,2-trifluoroethyl difluoromethyl ether
$C_3H_2C1F_5O$

Structure:

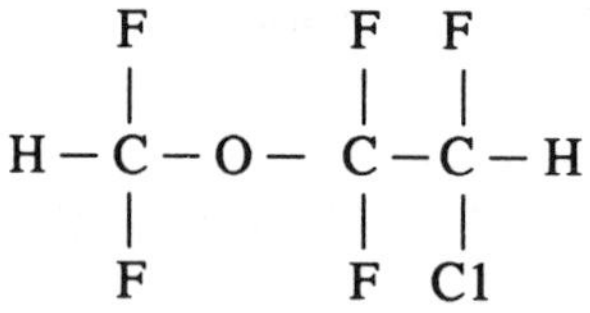

```
    F         F   F
    |         |   |
H - C - O -   C - C - H
    |         |   |
    F         F   C1
```

Physical Properties: Boiling Point = 56.5°C (liquid at room temperature)
Molecular Weight = 184.50 amu
Liquid Density = 1.517 gms/cm^3
Stable, volatile and non-flammable colorless liquid, with a faint but characteristic ethereal odor
Normally supplied in specially keyed glass bottles

Isoflurane

Synonyms: Forane, Forene, Aerrane, and/or Compound 469

Name & Formula: 2-chloro-2-(difluoromethoxy)-1,1,1-trifluoroethane
or
1-chloro-2,2,2-trifluoroethyl difluoromethyl ether
$C_3H_2ClF_5O$

Structure:

```
     F   H       F
     |   |       |
 F — C — C — O — C — H
     |   |       |
     F   Cl      F
```

Physical Properties: Boiling Point = 48.5°C (liquid at room temperature)
Molecular Weight = 184.50 amu
Liquid Density = 1.450 gms/cm^3
Stable, highly volatile and non-flammable colorless liquid, with a mild but characteristic ethereal odor
Normally supplied in specially keyed glass bottles

Methoxyflurane

Synonyms: Metofane, Penthrane, Pentrane, and/or DA 759

Name & Formula: 2,2-dichloro-1,1-difluoro-1-methoxyethane
or
2,2-dichloro-1,1-difluoroethyl methyl ether
$C_3H_4Cl_2F_2O$

Structure:

```
     H           F   Cl
     |           |   |
 H — C — O —     C — C — H
     |           |   |
     H           F   Cl
```

Physical Properties: Boiling Point = 105°C (liquid at room temperature)
Molecular Weight = 164.97 amu
Liquid Density = 1.425 gms/cm^3
Stable, moderately volatile and non-flammable liquid with a moderately strong and pungent ethereal odor
Normally supplied unkeyed glass bottles

Comments: Limited to veterinary anesthesia

Suprane

Synonym: Desflurane

Name & Formula: 1-(difluoromethoxy)-1,2,2,2-tetrafluoroethane
or
1,2,2,2-tetrafluoroethyl difluoromethyl ether
$C_3H_2F_6O$

Structure:

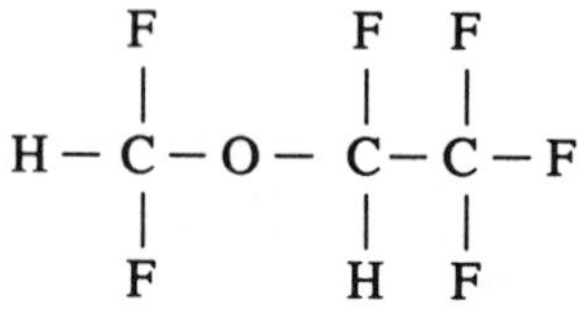

```
        F       F   F
        |       |   |
    H — C — O — C — C — F
        |       |   |
        F       H   F
```

Physical Properties: Boiling Point = 22.8°C (liquid, when in a check valve sealed bottle at room temperature)
Molecular Weight = 168.04 amu
Liquid Density = 1.467 gms/cm^3
Stable, extremely volatile and non-flammable colorless liquid with a moderately strong and pungent ethereal odor
Normally supplied in unkeyed glass bottles equipped with a pressure check valve

Comments: Approved by the U.S. Food and Drug Administration for human inhalation anesthesia use in the United States in mid-1993

Sevoflurane

Synonyms: Sevofrane

Name and Formula: 2-fluoromethoxy-1,1,1,3,3,3-hexafluoropropane
or
1,1,1,3,3,3-hexafluoro-2-propyl fluoromethyl ether
$C_4H_3F_7O$

Structure:

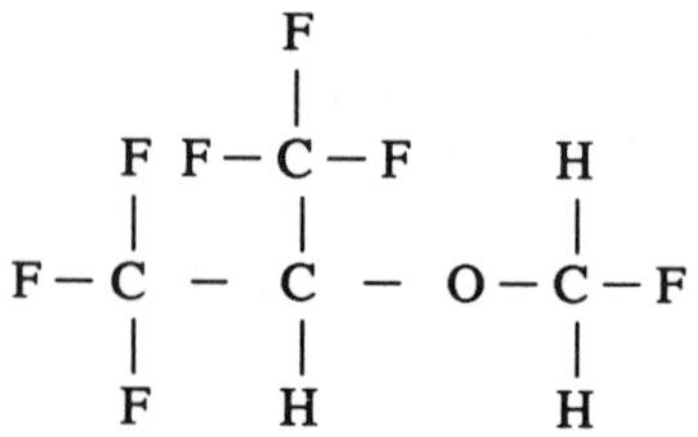

```
                F
                |
      F     F — C — F       H
      |         |           |
  F — C   —     C   —   O — C — F
      |         |           |
      F         H           H
```

Physical Properties: Boiling Point = 58.6°C (liquid at room temperature)
Molecular Weight = 200.06 amu
Liquid Density = 1.525 gms/cm^3
Stable, volatile and non-flammable colorless liquid, with a fairly pungent ethereal odor
Normally supplied in unkeyed glass bottles

Comments: Not yet approved by the U.S. Food and Drug Administration for human inhalation anesthesia use in the United States

EXISTING EXPOSURE STANDARDS[1]

The currently established Government and Professional Society Exposure Limits and Standards for these seven volatile agents are tabulated below; these listings will use appropriate portions of the following sets of abbreviations:

Applications applicable to all areas, organizations and agencies:

TWA: Time Weighted Average (usually evaluated over 8 hours)
STEL: Short Term Exposure Limit (usually evaluated over a 15-minute period, as a Time Weighted Average)
C: Ceiling Value (a "Never-to-be-exceeded" Concentration)

Abbreviations applicable to the American Conference of Government Industrial Hygienists (ACGIH):

TLV: Threshold Limit Value (the concentration level above which an exposed individual might reasonably anticipate the onset of adverse effects—as determined by the ACGIH in specific evaluation efforts covering these effects)

Abbreviations applicable to the U.S. Department of Labor, Occupational Safety and Health Administration (OSHA):

PEL: Permissible Exposure Limit (the concentration level above which an employer can be cited for failure to maintain a "safe" workplace)

Abbreviations applicable to the National Institute for Occupational Safety and Health (NIOSH):

REL: Recommended Exposure Limit: (the concentration level above which NIOSH data and experience have indicated the possible onset of adverse effects in exposed individuals)

Abbreviations applicable to the Deutsche Forschungsgemeinschaft Agency of Germany:

MAK: Maximum Arbeitsplatz Konzentration (Maximum Concentration Value for the Workplace) (the concentration level above which German employers may be cited for failure to maintain a "safe" workplace)

CURRENTLY ESTABLISHED LIMITS AND STANDARDS

Nitrous Oxide: ACGIH 8-hour TLV-TWA = 50 ppm
NIOSH 8-hour REL-TWA = 25 ppm

Halothane: ACGIH 8-hour TLV-TWA = 50 ppm
NIOSH 60-minute REL-C = 2 ppm
DFG 8-hour MAK-TWA = 5 ppm

Enflurane: ACGIH 8-hour TLV-TWA = 75 ppm
NIOSH 60-minute REL-C = 2 ppm

Isoflurane: No established exposure limits or standards of any type from any agency or organization

Methoxyflurane: NIOSH 15-minute REL-C = 2 ppm

Suprane: No established exposure limits or standards of any type from any agency or organization

Sevoflurane: No established exposure limits or standards of any type from any agency or organization

ANESTHESIA AGENT TOXICOLOGY

Much of the data on the toxicology of the seven agents have been developed as a result of observing and cataloging the complaints, illnesses and various other complications that have been reported by individuals who have worked in locations where there are non-zero ambient concentration levels of these same agents. In such cases, it is invariably difficult to establish direct, irrefutable "cause and effect" relationships between the reported ailment and the suspect causal exposure to the particular volatile anesthesia vapor. Comparative statistical studies on exposed vs. unexposed populations have tended to confirm several of these hypothetical "cause and effect" relationships; however, several important unanswered questions still remain.

Much of the actual systematic work on the toxicology of these chemicals has involved the use of laboratory animals; and much of this work has involved exposing these animals to concentration levels that are *far higher* than currently recognized exposure standards—as tabulated above. The evident primary goal of most of these laboratory investigations has been to identify and document the actual health risks of relatively large scale exposures to these agents. This fact notwithstanding, it is largely these data—and particularly, the extrapolation of these data back to lower and lower concentration ranges—that have provided the most reliable basis for identifying the health hazards associated with volatile anesthesia agents.

The data in this area, so far as it is currently known and recognized, will next be listed: (1) by each specific volatile agent and (2) by the various health difficulties for which that agent is thought to be responsible.

NITROUS OXIDE

There is a steadily growing body of data that supports the contention that exposures to nitrous oxide (at 8-hour TWA concentration levels in excess of 25 ppm), by second and third trimester pregnant women can produce significant increases in their rates of: (1) spontaneous abortion, (2) stillbirth and (3) congenital abnormalities among their children.[2–7] Although there is not yet total agreement on these "cause and effect" relationships, it is safe to say that the general attitude of health professionals is that nitrous oxide exposures—particularly to second and third trimester pregnant women—must be carefully monitored and controlled.

In addition to the foregoing, there have been several epidemiological studies and a lesser number of laboratory investigations, the results of which have suggested that exposures to trace levels of nitrous oxide might also be responsible for increases in hepatic as well as renal disease (excluding renal pyelonephritis and cystitis).[2–4,7] These studies have indicated greater adverse impacts on exposed women than on men. Considerably more investigative work needs to be done in these areas before specific toxicological conclusions can be definitively reached.

There have also been scattered reports of ambient nitrous oxide exposures causing certain genetic abnormalities—as measured by Sister-Chromatid Exchange Analyses.[8] As is true in the case of the previously mentioned situation involving hepatic and renal disease, considerably more investigative work needs to be done in these areas before specific genetic toxicological conclusions can be confirmed.

HALOGENATED ANESTHESIA AGENTS—GENERALIZED

Exposures to the halogenated inhalation anesthesia agents will be treated, first, by identifying any of the similar broadly based negative effects that can be or have been caused by *any* of these six compounds (a not unreasonable approach, since these six are relatively similar, chemically), and second, by identifying health problems that appear to be uniquely caused by any single member of this group—to the extent that data specific to any of these individual agents has been developed.

The most commonly reported health risk associated with the halogenated anesthesia agents are similar to those reported and documented for many of the other halogenated organic compounds (i.e., refrigerant, vapor degreasing solvents, etc.). These include various chronic difficulties in the areas of psychomotor, hematopoietic, central nervous, hepatic and renal system diseases and dysfunctions.[3–4,7] Clearly, all six of these agents function as acute central nervous system depressants. The chronic ailment category most commonly reported among individuals exposed to these volatile agents involves the liver, and includes a number of actual cases of chemical hepatitis hypothetically projected to have been caused by exposures to these vapors. Considerable experimental effort has been devoted to evaluating the relative liver toxicity of four of these anesthesia agents, as determined by monitoring the intracellular K^+ content (this, as an indicator of the time, concentration, and oxygen dependent cytotoxicity of the biotransformation processing of these agents in the liver). The result of these undertakings has been a ranking—from the most to the least harmful—as potential hepatotoxins—this ranking is: Halothane, Isoflurane, Enflurane and Sevoflurane.[9]

A small number of reports indicate that repeated low level exposures to these agents may also be responsible for increased susceptibility to infections and neoplastic disease.[4]

Finally, there have been a few reports of ambient halogenated anesthesia agent exposures causing certain genetic abnormalities—again, as measured by Sister-Chromatid Exchanges—however, the clear determination that one, or even any, of these agents was the principal causal factor in these cases, as contrasted to nitrous oxide being the causal agent, has not been conclusively made.[8]

HALOTHANE

Halothane is the member of this group that: (1) is not an ether and (2) contains bromine. As such, it would not be surprising if there were at least some differences between it and the other halogenated agents, from the perspective of their effects on exposed individuals, and indeed there are some differences. Chemical (or clinical) hepatitis is far more closely associated with Halothane exposures than is the case with any other of these agents.[10–13] The most likely mechanism for this chemical induced hepatitis is the hypersensitivity reaction to liver neo-antigens that are produced by the Halothane metabolite, 2-chloro-1,1,1-trifluoroethane.[10]

Various other studies have implied that Halothane may be teratogenic and/or embryotoxic. These conclusions have been developed through studies of the resultant ultrastructural effects in rats exposed to Halothane at 10 ppm(vol) or more vs. the same effects observed in unexposed rats.[13] Finally, ambient occupational Halothane exposures have been linked to increased and potentially unhealthy plasma bromide concentrations. In contrast, when Enflurane is the inhalation anesthesia agent being used in surgical procedure, this plasma bromide concentration buildup in exposed individuals is completely absent.[15]

METHOXYFLURANE

Methoxyflurane—administered to male Fisher rats, at a concentration of 50 ppm(vol), over a 14-week period—has been found to be both growth depressing and hepatotoxic. At the end of the exposure period, the livers of all of the exposed rats were examined, and all showed focal hepatocellular degeneration and necrosis, as well as evidence of liver cell regeneration.[16]

SOURCES OF WASTE OR FUGITIVE ANESTHESIA AGENTS

In the overall consideration of the important and/or significant sources of potentially harmful ambient occupational concentration levels of these anesthetic agents, we must consider two principal sources; namely: (1) waste and/or (2) fugitive sources. A *waste anesthetic gas* is any gas or vapor that has been, or is being, exhaled by an anesthetized person, either during or after a procedure that has involved the use of inhalation anesthesia agents. A *fugitive anesthetic gas* is any gas or vapor that has escaped from any system that is functionally upstream of the patient (i.e., the anesthetic gas vaporizer/ventilator, the high pressure nitrous oxide supply system outlets, etc.). By definition, fugitive anesthetic gases exist only in surgical suites, operating rooms, and/or the piping systems that extend between the master gas supply manifold and the outlet fittings wherever they may be located. These shall each be considered in order.

WASTE ANESTHETIC GASES

Typically, in the modern hospital operating room, there will be two separate and distinct systems that have been engineered, in whole or in part, to mitigate problems associated with the potential buildup of waste anesthetic gases. The first of these, a vacuum scavenging system, is designed to function solely as a local exhaust hood, situated directly at the anesthesiologist's (or anesthetist's) station. Its purpose simply is to capture and remove the patient's exhaled breath during the surgical procedure. The second of these systems is the overall ventilation system, the main function of which is to circulate and replenish the room's air.

Vacuum scavenging systems should be (and are) used as the principal mechanism for removing waste anesthetic gases, whether the anesthesia mixture is administered to the patient by open, mask or endotracheal tube induction. When such a system is of sufficient design capacity (exhaust flow rate volume ≥25 liters/minute, typically), and is utilized properly, it will be effective—or even very effective—in removing waste anesthetic gases from the surgical suite, thereby minimizing the exposure of the surgical team.[17] Alternatively, for a scavenging system that has less than this level of exhaust capacity, or—even worse—for an operating room that is not equipped with any type of waste gas scavenging system, the buildup of potentially unsatisfactorily high concentration levels of waste anesthetic gases will be almost a certainty.

In the most general sense, the scavenging of waste anesthetic gases will always be more difficult in the cases of either open-circuit or mask induction—either, in marked contrast to induction via endotracheal tube. In these difficult types of situations, the exhaust connection between the anesthesia induction circuit and the waste gas scavenging system will be both more tenuous and much more difficult to maintain; thus the potential for unscavenged anesthetic gases collecting in unsatisfactorily high concentration levels will be much greater.

As a fairly typical example, in 1990, ambient air measurements were made in seven different hospitals in Vienna, Austria. In one of these measurement situations—during an otolaryngological surgical procedure that necessitated anesthesia induction by open-circuit, and in a facility without a waste gas scavenging system (a textbook worst possible case scenario)—peak concentration levels of nitrous oxide and Halothane were found to be 2,600+ ppm and 150+ ppm, respectively. These astonishingly high concentration levels contrasted markedly to their counterparts in operating rooms at other Viennese hospitals, all of which had properly designed and operating waste gas scavenging systems. For this second group of operating rooms, and in situations where the anesthetic gases were administered to the patient via endotracheal tube, the concentrations, expressed as 8-hour TWAs, were found to be in the range: 8–15 ppm (mean = 11 ± 3 ppm) for nitrous oxide; and 0.1–0.6 ppm (mean = 0.3 ± 0.2 ppm) for Halothane. For the same or similar operating rooms, when the anesthetic gases were mask induced, these concentrations, also expressed as 8-hour TWAs, were found to be in the range: 24–211 ppm (mean = 83 ± 49 ppm) for nitrous oxide, and 0.3–1.3 ppm (mean = 0.8 ±0.3 ppm) for Halothane.[17]

Clearly, in the operating room, the occupational waste anesthetic gas exposure of every individual involved can be significantly mitigated through the correct use of an adequately designed and properly functioning exhaust scavenging system.[18–25]

General operating room ventilation systems, too, contribute to the removal of waste anesthetic gases; however, the primary function of these systems is to circulate and replenish the air in the room. This replenishment function must always be emphasized, since this aspect of the operation of a general ventilation system is always its prime function. In accomplishing this, however, the ventilation system will clearly supplement the vacuum scavenging system in the removal of waste anesthetic gases. Replenishment here should be understood to imply that *all* of the air that is removed from the operating room should ultimately be exhausted from the building by that operating room's ventilation system. None of the air should ever be recirculated.

In addition, for any operating room, the ventilation system's capacity—expressed in terms of its overall room air exchange rate—should be at least 15 room volumes/hour, and preferably 18–20 room volumes/hour.[18–19,26] An operating room with a ventilation system having this sort of overall air handling capacity (assuming also that the overall ventilation flow patterns in the room include no short circuits) will generally be an occupationally safe place in which to work. Such a ventilation system can be regarded as sufficient, both in and of itself, and as a functional backup to the room's waste anesthetic gas scavenging system. For reference, a short-circuiting air flow pattern is one in which inlet air will flow directly to an exhaust register, without having swept out every volume or space within the room itself—such upswept spaces are called "dead volumes." For an operating room that has "dead volumes," the potential for harmful ambient occupational anesthetic gas buildups may be very great, and this is true even if the overall room air exchange rate were high enough to appear to guarantee against such situations.

General surgical recovery areas, too, are potential sources for dangerously high concentration levels of waste anesthetic gases. Recovering surgery patients will tend to exhale exponentially decreasing concentrations of those gases they have inhaled during their surgery. For any patient whose surgical procedure has required an extended period of time (i.e., more than four hours), the initial concentration of waste anesthetic gases in their exhaled breath can be *very* high—nitrous oxide concentrations in excess of 2000 ppm have been measured in the breathing zone of patients immediately after their arrival in the recovery area. Additionally, the time period during which these patients must be monitored—in order to *fully* complete their "offgasing" process—can be quite prolonged. Nurses, and any other medical practitioner, who must tend to these patients or who must work in areas where they are recovering are, themselves, very good candidates for receiving an unsatisfactorily high level exposure to waste anesthetic gases.

In general, the ventilation requirements for a surgical recovery area will be the same in terms of the necessary minimum air exchange rate as was the case for an operating room; and the same comments with respect to short circuiting and "dead volumes" also apply. An even more effective way to ensure against the "dead volume," that may, or actually does, contain high concentration levels of waste anesthetic gases, involves the use of individual slot exhaust hoods at each patient location. Such an approach will always provide the best possible engineering solution, insofar as worker exposures to waste anesthetic gases are concerned.[27]

A brief comment with respect to veterinary surgical suites is in order at this point. The risks of waste anesthetic gas exposures in these operatories are probably considerably greater than those that would be expected to occur in their human operating room counterparts. The reasons for this are several and include all of the following: (1) veterinary clinics frequently employ portable gas delivery carts that are not designed to capture waste anesthetic gases; (2) the costs of specific waste gas scavenging systems may be prohibitively high, and therefore beyond the reach of the relatively smaller veterinary clinic; and (3) the overall effectiveness of such scavenging systems has yet to be fully verified in the veterinary area, and there are, therefore, neither current OSHA recommendations nor standards for veterinary scavenging systems (to say nothing of the fact that there are no OSHA PELs for any of the anesthetic agents).[28]

In a typical survey of fourteen different veterinary operatories, each performing surgeries in which anesthetic mixtures of nitrous oxide and Halothane were used, the following geometric mean TWA concentrations (as well as the overall concentration ranges measured) were as follows: 100 ppm (range: 14–1700 ppm) and 2.6 ppm (range: 0.5–119 ppm), respectively, for nitrous oxide and Halothane. Since a typical veterinary surgery rarely lasts more than four hours, the same geometric mean, TWA concentrations, now extrapolated out to cover an 8-hour period, and again for nitrous oxide and Halothane were as follows: 34 ppm (range: 5–530 ppm) and 1.3 ppm (range: 0.5–34 ppm), respectively.[29]

FUGITIVE ANESTHETIC GASES

Anesthetic gas leaks, as implied earlier, will typically originate from one of the following three sources: (1) from the high pressure piping systems that deliver nitrous oxide to the operating suites (including the low pressure connections to the anesthesia circuit); (2) from leakage on or around the anesthetic gas vaporizer/ventilator machine—in particular, from any and all of the low-pressure connections whose many seals and joints should always be suspect (these machines must have components of this type in order to facilitate the required frequent cleanings and replacements); or (3) from the poor work practices and/or habits of the anesthesiologist/anesthetist who utilizes this equipment. In the event that the source of the unsatisfactorily high ambient concentration levels of anesthetic gases turns out to have been attributable to the poor work practices of the anesthesiologist/anesthetist, then the ambient anesthetic gas vapors present can be in either the waste or the fugitive category. Again, as stipulated earlier, an overall air exchange ventilation rate of 18–20 room volumes/hour can help to mitigate these types of problems; however, elimination of the specific leaks would be the preferred solution.[30]

MONITORING ANESTHESIA AGENTS

Monitoring ambient and occupationally significant concentration levels of the various anesthesia agents can be accomplished in several ways; the choice of the method or approach involved will usually be dictated by the type of information required (i.e., real-time concentrations, individual exposure dosimetry, etc.); and/or the nature, type and number of potential interferants that might be present.

BASIC ANALYTICAL PARAMETERS

These functional requirements of *any* gas analysis can be understood, most beneficially, in terms of the four characteristic factors that apply to *every* ambient analytical application, namely: (1) the Sensitivity, (2) the Selectivity, (3) the Reproducibility and (4) the Timeliness of the method. The following four descriptions will hopefully adequately define and describe these functionality parameters:

1. A method's **Sensitivity** is usually expressed in terms of its **M**inimum **D**etection **L**imit—or **MDL**—to the Material being analyzed. Ideally, a method's **MDL** should be equal to, or less than, 10% of some "Significant Concentration" (usually, the **PEL-TLV**) of the material to be monitored.

2. A method's **Selectivity** is the measure of its ability to distinguish the material to be monitored from *everything else* that might be in the matrix to be monitored (i.e., interfering chemicals). Interferences can be either positive or negative. Positive ones increase the analytical reading; negative ones, on the other hand, decrease or "quench" it.

3. A method's **Reproducibility** is a measure of its ability to provide consistent readings for the same concentration of the material to be measured: (a) at different times, (b) under different ambient conditions, or (c) by different analysts. It is also a measure of the method's ability to remain in calibration for extended periods of time.

4. A method's **Timeliness** is a measure of the time interval required for it to provide its assessment of the unknown concentration answer for the material being analyzed. This time interval can be quite broad, ranging from instantaneous (real-time) to very extended periods—as, for example, when a passive dosimeter must be sent to some distant external location for evaluation.

ANALYTICAL METHODS

There are a number of viable analytical methods that purport to monitor ambient concentration levels of the anesthetic gases. In general, analyzers in each of the method categories listed below—in Tables 1 and 2—have been advertised as being both well suited to and available for these tasks. Certain units employing these analytical methods can be provided as portables, others are available in fixed or installed arrangements. In addition, some of the fixed units can be provided in single or multiple point monitors. In a few, systems are available in both of these configurations.

The following two listings cover those analytical methods that appear to the author to offer the greatest potential for successfully analyzing the anesthetic gases. Table 1 focuses on the analysis of nitrous oxide, while Table 2 addresses the six halogenated anesthesia agents.

This following Functional Score Sheet represents the structure within which the author's assessment of the relative merits and/or capabilities of each of the several different analytical methods (analyzer types) listed above will be presented, with separate and distinct assessments for each of the two different analyses—namely, the one for nitrous oxide and the one for the halogenated anesthetic gases.

Functional Score Sheet

Rating No.	(1) Sensitivity	(2) Selectivity	(3) Reproducibility	(4) Timeliness
5	< 0.2 ppm	Excellent	Excellent	Real-Time
4	< 0.5 ppm	Very Good	Very Good	< 2 min. delay
3	< 1.0 ppm	Good	Good	≤ 10 min. delay
2	≤ 5.0 ppm	Fair	Fair	> 10 min. delay
1	> 5.0 ppm	Poor	Poor	> 1 day delay

Table 1. Nitrous oxide.

IR-n	Infrared spectrophotometric absorbance analyzers
PAS-n	Photoacoustic infrared spectrophotometric analyzers
GC-n	Gas chromatographic analyzers
PD-n	Passive dosimetric systems
AD-n	Active dosimetric systems

Table 2. Halogenated anesthesia agents.

IR-h	Infrared spectrophotometric absorbance analyzers
PAS-h	Photoacoustic infrared spectrophotometric analyzers
AEC/SS-h	Pumped or active electrochemical/solid state analyzers
PEC/SS-h	Passive electrochemical/solid state analyzers
GC-h	Gas chromatographic analyzers
PD-h	Passive dosimetric systems
AD-h	Active dosimetric systems

NITROUS OXIDE ANALYSIS

From Table 1, it can be seen that there are five potentially viable analytical methods available for this task. That there will be additional methods available in the future is a virtual certainty. That there may be even additional ones available today cannot be denied; however, the author is unaware of any other current candidates. There are, to be sure, well-established methods that analyze nitrous oxide at concentrations in the percent range; however, for ambient analyses focused on occupational health risks, the five-member listing is quite complete.

The following two tabulations list the author's evaluation of the relative merits of each analytical method in successfully completing ambient, occupationally significant nitrous oxide analyses. Table 3 focuses on "portable" systems, while Table 4 provides the same assessment for their installed or fixed counterparts.

Clearly, if the real-time or near real-time concentration data must be obtained, the best currently available methods are either: (1) long pathlength infrared spectrophotometry (IR-n) or (2) photoacoustic infrared spectrophotometry (PAS-n).

If it is necessary to actually monitor real-time for ambient levels of nitrous oxide, it should be noted that there is an existing NIOSH Standard (Method No. 6600) that specifies the use of long pathlength infrared for this task. This approach, which accomplishes N_2O analyses extremely well, calls for making absorbance measurements at a mid-infrared wavelength of 4.48 micrometers (a wavenumber of 2,232 cm^{-1}), and an appropriately long pathlength, usually—for ambient situations in which the N_2O concentrations are less than 200 ppm—at a pathlength of 20.25 meters.[31]

There are both portable and fixed long pathlength infrared analyzers that are very well suited to analyzing ambient levels of nitrous oxide. Fixed systems have the capability of being multiplexed so that a single analyzer can be used to monitor as many as 24 separate locations, simultaneously datalogging and summarizing the analysis at each location on any time or sequence basis that may be required. Multiplexing a single unit to analyze at many separate points will offer both advantages and disadvantages. Among the former are the following two: (1) lower purchase costs, since only a single analyzer must be obtained and (2) lower operating costs that arise from having to calibrate and service only a single analyzer. On the negative side, this approach will always extend the time interval between subsequent analyses at each point. Finally, fixed systems can also be set up to compensate for potential interferants, thereby improving the selectivity of the system.

Near real-time measurements (i.e., delays of less than 2 minutes for each discrete N_2O concentration readout) can be very adequately accomplished by photoacoustic infrared spectrophotometric analyzers. Although NIOSH Method No. 6600 neither specifies nor identifies photoacoustic infrared methods as suitable for nitrous oxide analyses, the author feels that the methods are sufficiently similar, that this technology could be regarded as having been included in this method. A photoacoustic infrared spectrophotometric analyzer will employ the same mid-infrared wavelength in its analytical setup as does its long pathlength infrared cousin, namely, 4.48 micrometers (or a wavenumber of 2,232 cm^{-1}).

As was the case with the long pathlength infrared systems, there are both portable and fixed photoacoustic infrared analyzers that are very well suited to analyzing ambient levels of nitrous oxide. Fixed systems of this type also have the capability of being multiplexed, with the same simultaneous datalogging and summarizing capability as was the case for the long pathlength infrared systems. In addition, multiplexing a single photoacoustic infrared analyzer to work at many points will offer the same generic advantages as was the case for its long pathlength infrared system cousin.

Table 3. Ratings of the various portable nitrous oxide analyzers and/or systems.

Analytical Method	1 Sensitivity	2 Selectivity	3 Reproducibility	4 Timeliness
Ir-n	5	4	4	5
PAS-n	5	5	5	4
PD-n	3	4	4	1–2
AD-n	4	3	3	1–2

Table 4. Ratings of various fixed nitrous oxide analyzers and/or systems.

Analytical Method	1 Sensitivity	2 Selectivity	3 Reproducibility	4 Timeliness
IR-n	5	5	5	4
PAS-n	5	5	5	4
GC-n	5	5	4	2

An active dosimetric system (AD-n) is one that utilizes a battery-powered personal air sampling pump—on which the pumping rate can be very accurately set *and* maintained for a specific time period. Connected in series with this pump will be a sorbent filled tube. This combination system will then be clipped to an individual for whom nitrous oxide exposure data are being sought. This active dosimeter, with its pump operating, will be worn by this person for a prescribed time period, at the end of which, it will be retrieved and forwarded to an analytical laboratory. In this lab, the sorbent material in the tube will be desorbed and its contents analyzed chromatographically.

A passive dosimeter (PD-n) is simply a card or tag, on which will be mounted a porous pad impregnated with some material that will react chemically with nitrous oxide. An initially unexposed card or tag will be clipped to the collar or the shirt pocket of the individual for whom nitrous oxide exposure data are being sought. During the time period when this tag is being worn, nitrous oxide will diffuse into the pad and undergo the desired chemical reaction. This chemical reaction will cause a change of some type (color, reflectivity, etc.) that can be measured accurately so as to quantify the nitrous oxide exposure.

Finally, both active and passive dosimetry systems are available for nitrous oxide measurements. In general, these two approaches are clearly the very best methods for documenting a single individual's exposure to nitrous oxide, over either a full 8-hour workday, or any shorter, but specific and well-documented time interval. Dosimetry systems that will permit the user to obtain an individual's dose directly on site and at the conclusion of any identified time interval are currently under development. Such systems will eliminate the requirement for sending exposed dosimeters to some remote location or lab for evaluation.

Fixed gas chromatographic systems (GC-n) are also capable of completing the analysis for nitrous oxide. The simplest and least expensive of these systems—one that would employ a Thermal Conductivity Detector (TCD)—will be limited to ambient concentrations equal to or greater than 20 ppm. More sophisticated systems that use a Discharge Ionization Detector (DID) can be set up to work at ambient concentrations down to less than 1.0 ppm. For this application, one would choose a chromatographic column packed with one of the readily available, common, rigid structured, uniform pore-sized, porous polymers such as Poropak or Chromosorb. Capillary columns coated with dimethyl polysiloxane will likely also be effective for this separation. Successful analytical results could be obtained by operating the chromatograph's oven either isothermally or under temperature programming. In all of the cases listed above, one would probably choose a helium carrier gas for ambient nitrous oxide analyses. Although this analytical method will provide excellent selectivity, gas chromatographic systems will provide excellent selectivity, gas chromatographic systems will almost always be more difficult and expensive to calibrate, maintain, and operate.

HALOGENATED ANESTHETIC GAS ANALYSES

From Table 2, it can be seen that there are seven potentially viable analytical methods available for this task. Again, it is certain that there will be additional methods available in the future, and there may even be other current ones available to perform these analyses. For the ambient analysis of the halogenated anesthetic gases, however, and with a focus on occupational health risk significant concentration levels, the seven-member listing of Table 2 will be both very satisfactory and relatively complete. The following two tabulations list the author's evaluation of the relative merits of each of these halogenated anesthetic gas analytical methods. Table 5 focuses on portable analyzers, while Table 6 provides the same assessment for installed or fixed systems.

Although there are no current NIOSH (or other) Standard Methods for measuring any of the halogenated anesthesia agents, clearly both the long pathlength infrared (IR-h) and the photoacoustic infrared (PAS-h) approaches offer fine solutions to obtaining these types of data. For reference, the following tabulation, Table 7, identifies both the "preferred" and a "second choice" wavelength (wavenumber) applicable to either of the infrared analytical approaches listed above, for any of the six halogenated anesthesia agents. For the long pathlength system, assuming occupationally important ambient concentration levels of any of these six chemicals (i.e., in the range of 0–25 ppm), the pathlength should be set at 20.25 meters. There is no pathlength consideration for the photoacoustic infrared system.

Table 5. Ratings of the various portable halogenated anesthetic gas analyzers and/or systems.

Analytical Method	1 Sensitivity	2 Selectivity	3 Reproducibility	4 Timeliness
IR-h	5	3	4	5
PAS-h	5	4	5	4
AEC/SS-h	2	2	3	4
PEC/SS-h	2	2	3	3
PD-h	3	2	3	1–2
AD-h	3	5	3	1–2

Table 6. Ratings of the various fixed halogenated anesthetic gas analyzers and/or systems.

Analytical Method	1 Sensitivity	2 Selectivity	3 Reproducibility	4 Timeliness
IR-n	5	4	5	2–4
PAS-n	5	5	5	2–4
AEC/SS-n	2	2	3	3–4
PEC/SS-n	1	2	3	4
GC-n	5	5	4	1–3

There are both portable and fixed long pathlength infrared spectrophotometric analyzers (IR-h) that are very well suited to analyzing ambient levels of the six halogenated anesthetic gases. The fixed systems all have the capability of being multiplexed so that a single analyzer can be used to monitor many locations, and these systems can even do this for up to four of the six different halogenated anesthetic gases, simultaneously. These systems will also simultaneously datalog as well as generate periodic data reports that summarize the measurements at each location. As was the case for the nitrous oxide analytical situation, multiplexing a single unit to analyze at many separate points will offer both advantages and disadvantages. There appear to be two advantages, namely: (1) lower purchase costs, since only a single analyzer must be obtained, and (2) lower operating costs that arise from having to calibrate and service only a single analyzer. Since fixed systems can be set up to compensate for potential interferants, they can monitor for up to four of these halogenated anesthetic gases simultaneously. They accomplish this by adjusting their response to each of these six different gases by the

Table 7. Analytical wavelength choices for the halogenated anesthetic gases.

Anesthesia Agent	Preferred		Second Choice	
	Wavelength	Wavenumber	Wavelength	Wavenumber
Halothane	7.85 µ	1,274 cm^{-1}	8.83 µm	1,132 cm^{-1}
Enflurane	8.67 µm	1,153 cm^{-1}	12.05 µm	830 cm^{-1}
Isoflurane	8.57 µm	1,167 cm^{-1}	8.17 µm	1,224 cm^{-1}
Methoxyflurane	9.20 µm	1,087 cm^{-1}	7.61 µm	1,314 cm^{-1}
Suprane	8.19 µm	1,222 cm^{-1}	8.46 µm	1,182 cm^{-1}
Sevoflurane	8.09	1,236 cm^{-1}	9.73 µm	1,027 cm^{-1}

effect that each of the others has on it. Because of this, the significantly improved; however, the broad similarity in the infrared absorptive fingerprints of these six halogenated anesthetic agents means that the analytical selectivity will never be perfect. On the negative side, this multipoint and/or multicomponent analytical approach will always extend the time interval between subsequent analyses at each point.

Near real-time measurements (i.e., delays of less than 2 minutes for each discrete halogenated anesthetic gas concentration readout) can be very adequately accomplished by photoacoustic infrared spectrophotometric analyzers (PAS-h). This type of analyzer will employ the same preferred and second choice mid-infrared wavelengths (wavenumbers), as listed in Table 7, as will its long path-length infrared spectrophotometric cousin.

As was the case with the long pathlength infrared systems, there are both portable and fixed photoacoustic infrared analyzers that are very well suited to analyzing ambient levels of the six halogenated anesthetic gases. Fixed systems of this type, also, can be multiplexed to many different points, while providing highly useful and functional datalogging and summarizing capabilities. These analyzers can also simultaneously analyze for up to four different halogenated anesthetic gases, and in doing so, achieve the same general benefits (and dis-advantages) as do their long pathlength infrared cousins.

The pumped or active electrochemical/solid state analyzer (AEC/SS-h) is a unit that uses either fairly analyte specific electrochemical cells or solid state detectors to provide their concentration readouts. These units also employ internal sampling pumps to aspirate air through their respective detector sections. Their passive electrochemical/solid state analyzer counterparts (PEC/SS-h) are identical in every respect *except* that they do not have internal sampling pumps—rather, they rely on the ambient diffusion of the target analyte to, and ultimately into, their detectors. This type of analyzer will usually also be equipped with some sort of a selective membrane that is positioned in front of its detector. This membrane will have been selected, because it is permeable to only a limited number of chemicals, and its placement in the analytical path will have the effect of improving the analyzer's selectivity.

Active and passive electrochemical/solid state analyzers are both available in virtually any desired configuration—portable or fixed, single or multipoint. These units will not, however, be able to analyze simultaneously for different halogenated anesthetic gases. The two principal advantages of these types of analyzers over any of their alternatives is their very low original cost and their previously mentioned wide configurational flexibility. They are particularly well suited to determining the existence and/or the location of any significant leak of any of the halogenated anesthetic gases; however, their relatively poor selectivity and, particularly, their very poor sensitivity usually disqualify them from active consideration for any monitor task at ambient and occupationally important concentration levels.

Both active (AD-h) and passive dosimetry systems (PD-h) are available for the measurement of the halogenated anesthetic gases. As was the case with nitrous oxide dosimetry, a dosimeter of either type will clearly be the very best method available for documenting a single individual's exposure to these materials. Dosimetry systems that will permit the use to obtain an individual's dose directly on site and at the conclusion of any identified time interval are currently under development. Such systems will eliminate the requirement for sending exposed dosimeters to some remote location or lab for evaluation.

Fixed gas chromatographic systems (GC-h) are also very capable of completing the analysis for any, or all, of these halogenated anesthesia agents. Gas chromatographs designed to analyze for these materials could do so successfully using either a Flame Ionization Detector (FID) or a Photoionization Detector (PID). Minimum Detection Limits (MDLs) for any of these gases would be well below 1.0 ppm. For this application, one would choose a chromatographic column packed with a low polarity, porous polymer such as Graphitized Carbon, Gas Chrom 220, Poropak Q, or Chromosorb 102. Successful analytical results could be obtained by operating the chromatograph's oven either isothermally or under temperature programming. Although this analytical method will almost certainly provide the very highest levels of analytical sensitivity and selectivity, gas chromatographic systems will always be more difficult and expensive to calibrate, maintain, and operate.

REFERENCES

1. Guide to Occupational Exposure Values—1993, compiled by the American Conference of Government Industrial Hygienists, 1993.
2. AANA Journal Course: Update for Nurse Anesthetists—Occupational Exposure to Trace Anesthesias: Quantifying the Risk; Foley, Kevin, MD; *Journal of the American Association of Nurse Anesthetists,* 61(4):405–412, August, 1993.
3. Buring JE, Hennekens CH, Mayrent SL, Rosner B, Greenberg ER, Colton T: Health Experiences of Operating Room Personnel. *Anesthesiology,* 62(3): 325–330, March, 1985.

4. Green CJ: Anesthetic Gases and Health Risks to Laboratory Personnel: A Review. *Laboratory Animals,* 15(4):397–403, October, 1981.
5. Gardner RJ: Inhalation Anesthetics - An Update. *Toxic Substances Bulletin,* Issue 9, page 11, June, 1988.
6. Gardner RJ: Inhalation Anesthetics - An Update. *Toxic Substances Bulletin,* Issue 17, page 3, December, 1991.
7. Rogers B: A Review of the Toxic Effects of Waste Anesthetic Gases. *Journal of the American Association of Occupational Health Nurses,* 34(12):574–579, December, 1986.
8. Sardas S, Cuhruk H, Karakaya AE, Atakurt Y: Sister-Chromatid Exchanges in Operating Room Personnel. *Mutation Research,* 279(2):117–120, May, 1992.
9. Ghantous HW, Fernando J, Gandolfi AJ, Brendel K: The Toxicity of the Halogenated Anesthetic Gases in Guinea Pig Liver Slices. *Toxicology,* 62(1): 59–69, May, 1990.
10. Sutherland DE, Smith WA: Chemical Hepatitis Associated with Occupational Exposures to Halothane in a Research Laboratory. *Journal of Veterinary and Human Toxicology,* 34(5):423–424, October, 1992.
11. Lind RC, Gandolfi AJ, Hall PD: Hepatotoxicity of Subanesthesia Halothane in the Guinea Pig. *Journal of Anesthesia and Analgesia,* 74(4):559–563, April, 1992.
12. Lings S: Halothane Related Liver Affections in an Anaesthetist. *British Journal of Industrial Medicine,* 45(10):716–717, 1988.
13. Toker K, Ozer NK, Yalcin AS, Tuzuner S, Gogus FY, Emerk K: The Effect of Chronic Halothane Exposure on Lipid Peroxidation, Osmotic Fragility and Morphology of Rat Erythrocytes. *Journal of Applied Toxicology,* 10(6): 407–409, December, 1990.
14. Baeder C, Albrecht M: The Embryotoxic/Teratogenic Potential of Halothane. *International Archives of Occupational and Environmental Health,* 62(4): 263–271, June, 1990.
15. Carlson P, Ekstrand J, Hallen B: Plasma Fluoride and Bromide Concentrations During Occupational Exposure to Enflurane and Halothane. *Acta Anaesthesiologica Scandinavica,* 29(7):669–673, 1985.
16. Plummer JL, Hall PD, Jenner MA, Ilsley AH, Cousins MJ: Hepatic and Renal Effects of Prolonged Exposure to Rats to 50 ppm Methoxyflurane. *Acta Pharmacologica et Toxicologica,* 57(3):176–183, September, 1985.
17. Gilly H, Lex C, Steinbereithner K: Anesthetic Gas Contamination in the Operating Room - An Unsolved Problem? *Anaesthesist,* 40(11):629–637, November, 1991.
18. Girompaire D, Landais A, Allegrini D: Pollution of Operating Rooms. *Cahiers D'Anesthesiologie,* 39(4):253–256, 1991.
19. Gardner RJ: Inhalation Anaesthetics - An Update. *Toxic Substances Bulletin,* Issue 17, page 3, December, 1991.
20. Snef L, Stapf F, Mueller W, Schimmel G: Investigations on Operating Theatre Staff with Long-Term Exposure to Halothane. *Zeitschrift für de Gesamte Hygiene und ihre Grenzgebeit,* 35(8):480–483, September, 1989.
21. Breum NO, Kann T: Elimination of Waste Anaesthetic Gases from Operating Theatres. *Acta Anaesthesiologica Scandinavica,* 32(5):388–390, July, 1988.
22. Sonander H, Stenqvist O, Nilsson K: Nitrous Oxide Exposure During Routine Anaesthetic Work. *Acta Anaesthesiologica Scandinavica,* 29(2):203–208, February, 1985.
23. Carlsson P, Ljungqvist B, Hallen B: The Effect of Local Scavenging on Occupational Exposure to Nitrous Oxide. *Acta Anaesthesiologica Scandinavica,* 27(6):470–475, December, 1983.
24. Dula DJ, Skiendzielewski JJ, Snover SW: The Scavenger Device for Nitrous Oxide Administration. *Annals of Emergency Medicine,* 12(12):759–761, December, 1983.

25. Thompson JM, Sithamparanadarajah R, Hutton P, Robinson JS, Stephen WI: Evaluation of the Efficacy of an Active Scavenger for Controlling Air Contamination in an Operating Theatre. *British Journal of Anaesthesia,* 53(3): 235–240, March, 1981.
26. Witner CG, Hamm G, Hueck U, Apel G, Lamprecht E: Toxicity Risk in the Exposure of Surgical Personnel to Gaseous Anaesthetics. *Zeitschrift für de Gesamte Hygiene und ihre Grenzgebeit,* 33(12):622–627, December, 1987.
27. Gardner RJ: Inhalation Anaesthetics - Exposure and Control. *Annals of Occupational Hygiene,* 33(2):159–173, 1989.
28. Burkhart JE, Stobbe TJ: Real-Time Measurement and Control of Waste Anesthetic Gases During Veterinary Surgeries. *American Industrial Hygiene Association Journal,* 51(12):640–645, December 1990.
29. Gardner RJ, Hampton J, Causton JS: Inhalation Anaesthetics - Exposure and Control During Veterinary Surgery. *Annals of Occupational Hygiene,* 35(4): 377–388, 1991.
30. Pothmann W, Shimada K, Goerig M, Fuhlrott M, Schulte am Esch J: Pollution of the Workplace by Anesthetic Gases, Causes and Prevention. *Anaesthesist,* 40(6):339–346, June, 1991.
31. Burroughs GE, Woebkenberg ML: NIOSH/DPSE. Nitrous Oxide Analyses - Criteria for a Recommended Standard, Occupational Exposure to Waste Anesthetic Gases and Vapors. *NIOSH Method No. 6600,* February 15, 1984.

PART 2

An Anesthesia Mask Gas-Scavenging System

Anthony Schapera and William Charney

ABSTRACT

Problem: Reducing operating room contamination with N_2O during delivery of anesthesia by mask.

Method: The level of N_2O contamination was measured in the breathing zone of anesthesiologists while they administered inhalation anesthesia by mask to five patients. A mask gas-scavenging attachment was used for 30 minutes and then removed while anesthesia continued for a further 30 minutes. The levels of N_2O with and without the scavenging attachment were computed.

Results: Using the scavenging attachment, N_2O contamination was reduced from greater than 150 PPM to less than 5 PPM, a level well below the 25-PPM limit recommended by the National Institute of Occupational Safety and Health (NIOSH).

Conclusions: The scavenging device is a simple and effective way to reduce operating room contamination with N_2O during delivery of anesthesia by mask.

INTRODUCTION

The significance of contamination of the operating room by anesthesia gas is controversial. Several studies and reviews have described an increased risk to the health of operating room staff in association with exposure to waste anesthetic gases.[1–4] It is reasonable to assume that reducing environmental contamination will lessen the health risk to operating room staff. The National Institute of Occupational Safety and Health (NIOSH)[5] recommends that levels of N_2O in the operating room do not exceed 25 PPM over the course of a general anesthetic. In our operating rooms, levels of N_2O measured in the immediate area or "breathing zone" of anesthesiologists often ranged from 80 PPM to greater than 200 PPM during delivery of inhalation anesthesia by an anesthesia mask. We therefore sought a way in which to reduce this level of contamination.

Leaks from a semi-closed breathing circuit will reduce the effectiveness of any anesthesia gas-scavenging system. Common sources of leaks are at the CO_2 absorber gaskets, in the anesthesia ventilator bellows, at the pressure release valve ("pop-off valve"), at the reservoir bag or at the high pressure gas connections at the central gas manifold. These leaks can be eliminated by routine maintenance checks and pre-anesthesia equipment checks. An additional major leak, however, occurs from beneath the edge of an anesthesia mask during administration of inhalation anesthesia. Both the contour of the patient's face and the design of the

1-56670-083-3/94/$0.00+$.50

mask may contribute to an imperfect seal between the mask and face, allowing leakage of gas from beneath the edge of the anesthesia mask. Anesthesia masks have a soft inflatable rim which improves the seal but the fit is often imperfect, especially in patients with thick beards or those with loss of subcutaneous tissue around the cheek area.

The purpose of this study was to test the effectiveness of a device designed in order to prevent anesthesia gases that had escaped from beneath the edge of the anesthesia mask from contaminating the operating room environment.

METHODS

The scavenging device (Figures 1 and 2) consists of a conventional anesthesia mask with a soft inflatable rim (King Systems, Model #1065, size 6, large adult mask). A 5-mm internal diameter 6-foot length of suction tubing is attached with one open end adherent to the right side of the anesthesia mask. A distal inch of the suction tubing has several side-holes in order to allow continued suction flow if the tip were to be occluded. A clear polyethylene plastic drape of approximately 12 in. diameter and 8 in. tall is attached to the anesthetic mask over the suction tubing. The free rim of this drape is tailored to conform to the contours of a normal adult face, and the edges are elasticized to provide a more effective seal where it contacts the skin. The suction tube exits through an opening in the plastic drape and is connected to a source for vacuum (300–500 mmHg). The anesthesia face mask is applied in the usual fashion during induction and maintenance of anesthesia. The suction tubing is connected to the source of suction and as anesthesia is induced, the clear plastic drape is brought over the patient's face and adjusted so as to provide the best beneath the edge of the face mask will be scavenged through the suction tubing. In the event that the anesthesiologist needs to gain rapid access to the patient's face, the drape may be easily removed.

We studied the effectiveness of the mask gas-scavenging system in five healthy patients undergoing elective surgery with inhalation anesthesia that included N_2O. The study was approved by the University of California San Francisco Committee on Human Research and informed consent was obtained from all study subjects.

The operating rooms used ranged in size from 8000–10,000 ft^3 and all rooms underwent ten complete air changes per hour.

A Miran 1A™ infrared spectrophotometer (calibrated to manufacturer's specifications) was used to measure N_2O levels. This has a detection limit of 1 PPM. We first sampled the operating room air for background levels of N_2O and next tested the breathing circuit for leaks by pressurizing the anesthesia circle system with N_2O. We sampled for N_2O at various locations in the breathing circuit and at scavenger reservoir bag. If obvious leaks were found, we corrected the leak or, if not correctable, we did not conduct the study in that operating room.

Anesthesia was then induced in the test patients with a combination of intravenous agents and volatile inhaled agents (halothane or isoflurane as determined by

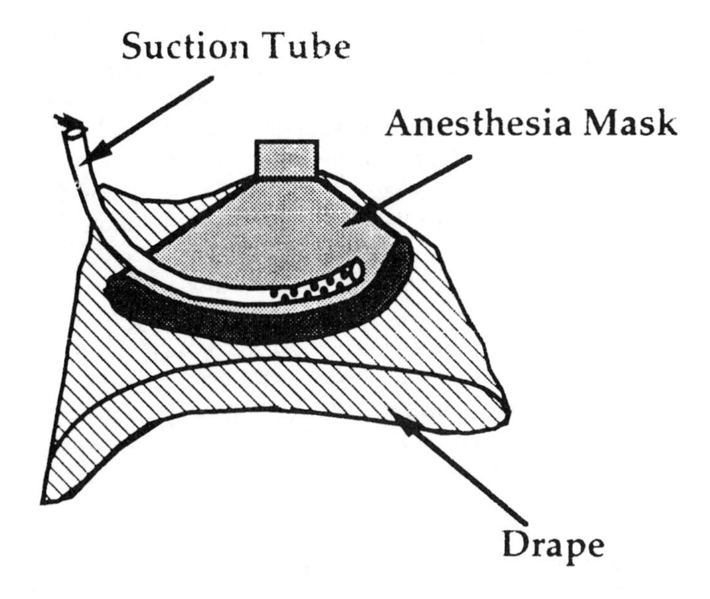

Figure 1. Anesthesia mask with scavenging attachment.

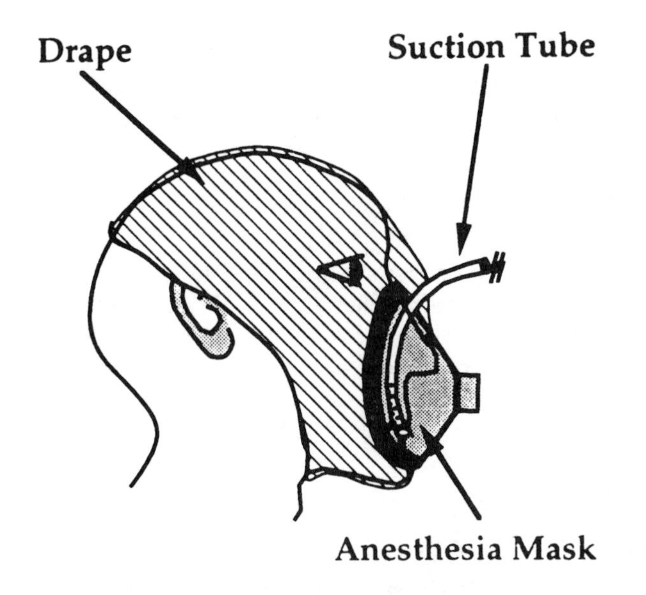

Figure 2. Method of applying scavenging drape over patient's face.

the anesthesiologist) in oxygen. The test period began when a steady state of anesthesia had been achieved and the anesthesiologist was satisfied that the airway was secured. An oral or nasal airway was used at the discretion of the anesthesiologist, but if used was left in place throughout the study period.

For the study period, patients breathed spontaneously while anesthesia was maintained with either halothane or isoflurane and a N_2O (3 L/minute)/O_2 (2 L/minute) gas mixture. For the first half hour of the study, the anesthesia mask gas scavenging system was used to administer anesthesia.

The anesthesiologist held the anesthesia mask with the left hand and applied the mask to the patient's face in the manner preferred by the anesthesiologist. We did not advise the anesthesiologist in the method of application of the mask itself, but did ensure that the plastic shroud was draped over the patient's face in the most effective way to collect escaping anesthetic gasses. Head straps were not used. In no cases were oral or nasal esophageal stethoscopes or temperature probes used. The breathing circuit safety-release ("pop-off") valve was adjusted so that the reservoir bag on the circuit completely refilled during exhalation and partially emptied during inhalation.

Samples of air were obtained continuously from the environment adjacent to the anesthesia mask (approximately 2 feet from the anesthesia mask), and the highest N_2O specific light absorbance in each minute time frame was recorded for 30 minutes. The anesthesiologist was unaware of the level of N_2O being measured. After 30 minutes, we disconnected the scavenging device and carefully removed it from the anesthesia mask while the anesthesiologist continued to apply the mask to the patient's face as before. We continued to sample room air for N_2O over the next 30 minutes.

Data were described as the mean values ± 1 standard deviation for N_2O concentration over each study period.

RESULTS

Patient characteristics are shown in Table 1. The scavenging system markedly reduced the average N_2O concentration over the course of 30 minutes of inhalation anesthesia in each case (Figure 3). The N_2O concentration was maintained at consistently low levels throughout the test period as illustrated in the data from patient number 5 (Figure 4).

DISCUSSION

The most significant result of this study is the demonstration that the newly designed anesthesia mask-scavenging attachment reduces contamination with N_2O within the anesthesiologist's breathing zone to a concentration well below the 25 PPM level recommended by NIOSH. Without the attachment, levels of N_2O

Table 1. Patient characteristics.

Patient	Age	Sex	Bearded	Dentition
1	33	male	no	full
2	63	female	no	full
3	49	male	no	full
4	38	female	no	edentulous
5	48	male	no	edentulous

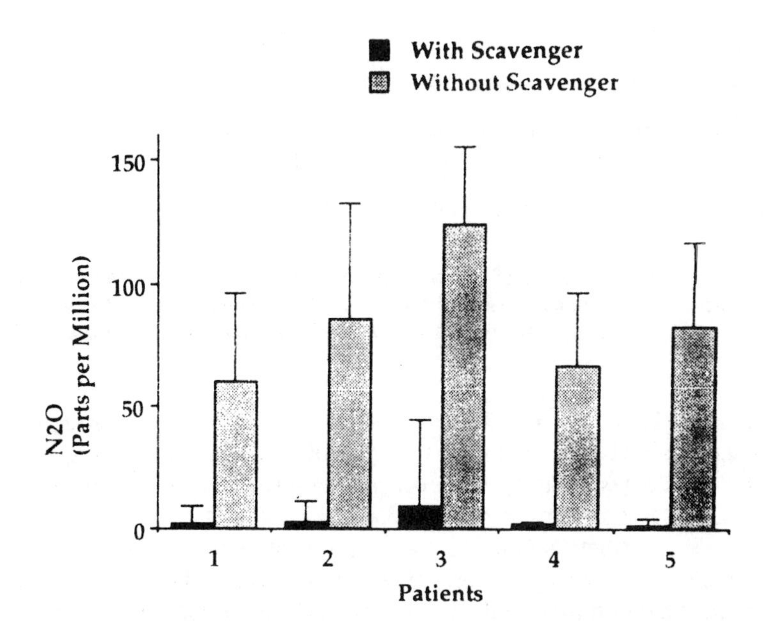

Figure 3. N_2O concentrations (mean ± standard deviations) in breathing zone of anesthesiologist while delivering inhalation anesthesia by mask to five patients.

greatly exceeded the recommended limit. Using a double mask scavenging device, (essentially an inner mask surrounded by an outer rigid shell into which escaping gas is evacuated) N_2O levels were reduced from 145 to 15 PPM in one study.[6] Use of this device results in a further seven fold reduction of N_2O in the anesthesiologist's breathing zone.

The scavenging attachment has several design points that deserve emphasis. We used the main-line hospital vacuum system as a source of suction for the evacuation of escaping gases. The static suction pressure in this system is not

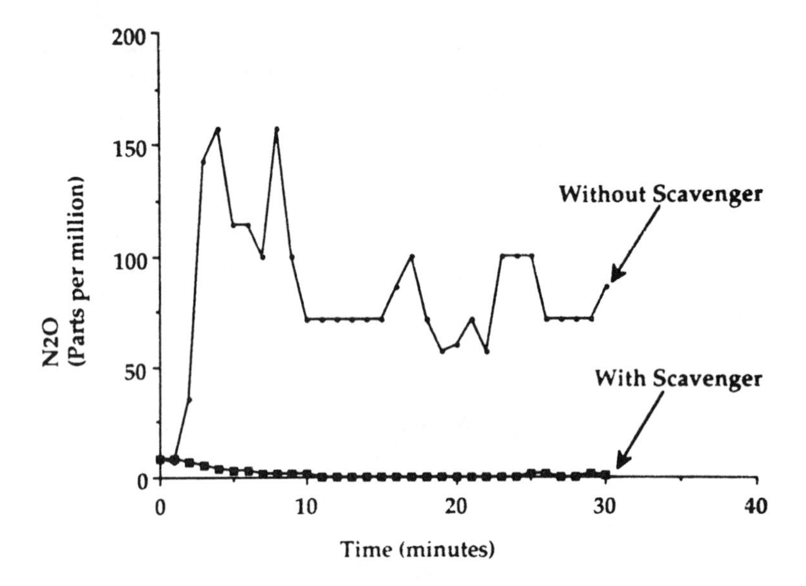

Figure 4. Plot of N_2O concentration versus time in the breathing zone of anesthesiologist delivering inhalation anesthesia by mask to patient number five. Values shown are the highest value sampled during each minute of the test period.

constant, but varies between 300–500 mmHg and results in an air flow through 5-mm ID tubing of about 60 L/minute. This rate of flow should be more than adequate to evacuate gas escaping from beneath an anesthesia mask (fresh gas flow in most anesthesia systems seldom exceeds 10 L/minute).

Because most anesthetists hold the anesthesia mask with the left hand, the right side of the mask is usually less effectively sealed against the patient's face. We attached the suction tubing to the right hand side of the anesthesia mask in order to more effectively contain and evacuate gas escaping from the right edge of the mask. The distal inch of the suction tubing has several side-holes in place in order to minimize occlusion of the suction by the plastic drape being sucked up against a single orifice.

The plastic drape itself is of clear polyethylene plastic. The contour is designed so that the edges fit well over the occiput and chin reducing the chance of slippage of the edges during use. A loose elasticized edge assists the fit of the drape. Although the drape fits well during use, its overall fit is fairly loose so that it can be instantly removed if necessary. The clear plastic enables the anesthetist to see the patients face clearly during the procedure. The patient's eyes should be taped shut during use of the device in order to prevent corneal abrasion by the plastic. In contrast to the double mask device,[6] the attachment of the drape does not add to the bulk of the anesthesia mask, an important factor in minimizing hand fatigue when delivering inhalation anesthesia by mask.

CONCLUSIONS

We have demonstrated a new scavenging device that effectively reduces environmental contamination with anesthetic gases during inhalation anesthesia delivered by a standard anesthesia mask.

REFERENCES

1. Cohen EN, Bellville JW, Brown BW: Anesthesia, pregnancy and miscarriage: A study of operating room nurses and anesthetists. *Anesthesiology.* 35:343–347, 1991.
2. Report of an ad hoc committee of the effect of trace anesthetics of the health of operating room personnel, American Society of Anesthesiologists. Occupational disease among operating room personnel: a national study. *Anesthesiology.* 41: 321–340, 1974.
3. Corbett TH, Cornell RG, Endres JL, Lieding K: Birth defects among children of nurse anesthetist. *Anesthesiology.* 1974;41:341–344.
4. Brodsky JB, Cohen EN, Brown BW, Wu ML, Whitcher CE. Exposure to nitrous oxide and neurologic disease among dental professionals. *Anesth Analg.* 60: 297–301, 1981.
5. National Institute for Occupational Safety and Health: Criteria for a Recommended Standard—Occupational Exposure to Waste Anesthetic Gases and Vapors. *DHEW Pub. No (NIOSH)* 77–140. Cleveland, US Department of Health, Education and Welfare, National Institute for Occupational Safety and Health; page 3, 1977.
6. Reiz S, Gustavsson A-S, Häggmark S, et. al.: The double mask—a new local scavenging system for anæsthetic gases and volatile agents. *Acta Anæsthesiol Scand.* 30:260–265, 1986.

CHAPTER 9

Taking Care of Ourselves: Ensuring Safety in the Handling of Antineoplastic Drugs

Bernadette Stringer

INTRODUCTION

There is an irony in a nurse developing cancer because of handling anti-cancer drugs that makes a "great story" for the news media. For a health and safety representative who has been trying to get members of her union to take seriously the potential hazards of working with antineoplastics, the sudden interest of the media in the story can only be beneficial, but while attending a funeral with a grieving husband and two newly motherless children the overwhelming emotion is simply one of sorrow. Moments like those make you wish you had another job, but they also bring home the human reality of statistics.

We'll never be able to "prove" Sally Giles' cancer was caused by her work with antineoplastic drugs over 20 years in hospitals in eastern Canada and on Vancouver Island, but based on the evidence there is a chance it was. Her death and a family's grief are not the only sources of sorrow. For someone with a knowledge of hospital health and safety there is the added sorrow of knowing that something could have been done to prevent exposure.

"Back then" work practices almost ensured exposure. But what about today? Unfortunately, our knowledge of potential risks and our ability to engineer safer equipment and work practices do not guarantee health care workers a safe environment. In fact, despite guidelines that should ensure little or no worker exposure to antineoplastics there is evidence that exposure remains a problem. Is this a problem even if the existing industrial hygiene guidelines are in place or is it a problem because of poor compliance? So far, we don't really know.

Canadian hospitals, where the majority of these drugs are handled, are rarely inspected for adherence to whatever guidelines exist (provincial jurisdiction)!

HAZARDS

Some hazards associated with handling antineoplastic drugs have been recognized by various health professionals for more than a dozen years (Perry, 1992).

1-56670-083-3/94/$0.00+$.50

Potential health problems associated with handling antineoplastics that have been studied include toxic effects on the skin, eye injuries, systemic problems, allergic reactions, carcinogenicity, mutagenicity, teratogenicity and menstrual abnormalities.

NATURE OF ANTINEOPLASTICS

Since 1942 cancer treatment has included the use of drugs, beginning with the nitrogen mustards (Perry, 1982). More than 35 drugs are currently available (Perry, 1992) and "development of new agents for chemotherapy is, in fact, one of the top four categories of research endeavor worldwide" (Schardien, 1993).

Antineoplastics are grouped into several major categories depending on their different modes of action against cancerous cells. Their effects are not limited to those cells—they also damage normal cells, especially those rapidly proliferating in the intestinal tract, bone marrow, hair follicles and reproductive organs. Because of this, antineoplastics are hazardous to a developing embryo or fetus (Jochimsen, 1992).

Anti-cancer drugs fall into two principal groups:

- Those that interact directly with DNA
- Those that inhibit nucleic acid synthesis

The modes of action of most antineoplastic drugs fall within the following categories.

Alkylating Agents

Alkylating agents introduce alkyl groups into the molecules. This interferes with the cancer cells' ability to divide and multiply—leading to their destruction. Cyclophosphamide is one of the most commonly used alkylating agents.

Antimetabolites

Antimetabolites block the cancer cells' use of normal nutrients. Methotrexate, from this category, does not permit folic acid normally found in our diet to be transformed and used for the biosynthesis of nucleic acids.

Antibiotics

Antibiotics interfere with DNA to block RNA production. Each antibiotic has a specific target or site within the genetic structure. Actinomycin is a commonly used antibiotic-antineoplastic.

Miscellaneous

This category consists of hormones such as steroids, the enzyme L-asparaginase, cisplatin and inorganic metallic salt, the vinca alkyloids derived from plants such as vincristine and others (Rogers, 1986; Plowman, 1991).

PATIENT SIDE EFFECTS

Acute and chronic side effects in patients receiving antineoplastic drugs have been documented since the beginning of their use. These range from allergic reactions, hair loss, nausea and vomiting and effects on a developing fetus to long-term effects on all organs, especially the gonads, liver and endocrine system. Secondary cancers in patients are also known to arise, especially leukemias, because of antineoplastic therapy (Plowman, 1991).

Carcinogenicity tests in animals reveal a reasonable concordance between the tumors developed during the animal experiments and the tumors found in human data. (See Table 1.)

Antineoplastic drugs' variable effects on fertility and gonadal function in patients receiving treatment for cancer range from irregular to absent menses, as well as a range of symptoms that resemble those experienced by women in menopause. When treatment with an antineoplastic is terminated some women return to a normal menstrual pattern while others do not, remaining amenorrheic. Whether or not the symptoms are permanent seems to be influenced by the dose received and the age when the woman was treated (Chapman, 1992).

OCCUPATIONAL EXPOSURE

The most common exposures to antineoplastic drugs are via the respiratory route or from skin contact, although exposure as a result of ingestion is possible (Valanis et al. 1993). Antineoplastic drugs are usually in the powder form, which must be mixed with a sterile fluid for administration, or in the sterile liquid state, which must be diluted. In Canada, patients receiving chemotherapy are most often hospitalized, sometimes for the duration of treatment only (several hours in outpatient clinics) and the drugs are most often given intravenously.

AIR CONTAMINATION

The earliest reported findings of workplace air contamination are from studies of air in pharmacy rooms where antineoplastics were being mixed under hoods that forced filtered air over the workers hands (over the drug) and toward the worker (horizontal laminar flow) to ensure drug sterility, and of air in rooms where workers were mixing antineoplastic drugs without any type of ventilation hood.

Table 1. Principal sites of tumors induced by antitumor alkylating agents in laboratory animals and in patients. (Plowman et al. 1991.)

Agent	Laboratory Animals	Patients
Busulfan	Leukemia, lymphoma	Acute leukemia
Chlorambucil	Lung; leukemia, lymphoma	Lung; acute leukemia
Cyclophosphamide	Bladder	Bladder; acute leukemia
Melfalan	Lung; lymphoma	Acute leukemia
Nitrogen mustard	Lung; lymphoma	Acute leukemia(?)

Kleinberg et al. (1981) measured levels of 5-fluorouracil and cefazolin in an experiment simulating normal mixing under *horizontal laminar flow,* and detected up to 0.07 $\mu g/m^3$ in a number of the air samples from within the flow hood.

Neal et al. (1983) carried out air monitoring for 5-fluorouracil, methotrexate, doxorubicin and cyclophosphamide in a room where drugs were *mixed without any form of local exhaust ventilation.* Methotrexate and doxorubicin could not be detected, but it was estimated that airborne particulate concentration from 0.12–82.26 ng/m^3 of 5-fluorouracil was present during 200–320 hours of monitoring and 370 ng/m^3 of cyclophosphamide was present during 80 hours.

A U.S. study (1986) by McDiarmid et al., found no airborne 5-fluorouracil during a total of 152 hours of monitoring in a hospital pharmacy where *vertical laminar flow* was used in a biological safety cabinet. In a Finnish study by Pyy et al. (1988) only the HEPA (High Efficiency Particulate Air) filter used inside a hospital pharmacy's vertical laminar flow hood was contaminated with cyclophosphamide.

More recently Lees et al. (1993) found 3 of 34 air samples collected in a hospital pharmacy positive for cyclophosphamide. Two were taken inside the biological safety cabinet—(which uses vertical laminar flow), where mixing of the drugs took place and therefore not unexpected, but one air sample was taken outside the cabinet, in the breathing zone of the operator. The authors speculated that because the three positive samples occurred with the largest air volumes and were only marginally above the minimum detectable level, similar low levels of cyclophosphamide in the air could go undetected on a regular basis.

ENVIRONMENTAL CONTAMINATION

The Lees et al. study also included environmental wipe tests to assess cyclophosphamide surface exposure and found 18% of wipe tests in a university hospital pharmacy and 14% of wipe tests in the oncology clinic contaminated. The positive wipe tests in the pharmacy samples were found inside the biological safety cabinet, on the countertop and on the floor; the positive wipe tests in the oncology clinic were taken from countertops, sinks and from a video display terminal.

Sessink et al. (1992), in the Netherlands, carried out the same type of study and detected environmental contamination by 5-fluorouracil, cyclophosphamide and methotrexate. The positive wipe tests in this study were from floor, tables (in the rooms where the drugs were administered) and the outside of cleaned chamber pots and urinals. Unexpectedly, they found that at the beginning of the workday, the floor of the room where the drugs were administered was contaminated, even though samples taken at the end of the previous workday, were negative. The authors speculated that cleaning procedures carried out between times may have been responsible.

BIOLOGICAL FLUID CONTAMINATION

Reports of urine and blood tests to assess antineoplastic exposure of healthcare personnel have been found in the literature for the last 15 years.

A study by Hirst et al. (1984) found that two oncology nurses handling cyclophosphamide were retaining some of the unconverted drug in their urine (cyclophosphamide cannot be completely converted to its byproducts). The nurses were followed for a total of 57 days. In all, 7 of 30 afternoon and 1 of 57 morning urine samples were positive for cyclophosphamide.

Other biological fluid studies of workers have focused on either: i) the mutagenic ability of the fluid (urine) or ii) indicators that indicate exposure such as chromosome breaks and sister chromatid exchanges (blood).

Urine mutagenicity studies in workers exposed to antineoplastics began with Falck et al. (1979) in Finland, in which seven non-smoking oncology nurses were compared to non-exposed controls. The oncology nurses were found to have more highly mutagenic urine.

Many of these studies followed and Harrison (1992) summarized the results of 22 studies in a table. He concluded that increased urine mutagenicity in exposed workers did occur in some studies but that a reduction in exposure could be achieved through the use of engineering controls, improved work practices and personal protective equipment.

Sister chromatid exchanges in the peripheral lymphocyte cells were first assessed by Waksvik et al. (1981) in a study where 10 nurses were compared to 10 female hospital clerks. The two groups found comparable for factors such as smoking, diagnostic and therapeutic irradiation and drug intake were found to differ in rate of exchanges. The exposed nurses experienced a greater number of chromosomal gaps and only a slightly greater number of sister chromatid exchanges.

Chromosomal abnormalities were also assessed in 11 exposed nurses compared to 16 unexposed laboratory workers and hospital clerks by Nikula et al. (1983), and a statistically significant number of breaks and aberrations in the nurses' chromosomes were found. The researchers took the same risk factors into consideration as Waksvik et al.

Kolmodin-Hedman et al. (1983) though, when comparing sister chromatid exchanges in exposed and non-exposed nurses, found no difference; this was also the case for Barale et al. (1985) who compared 21 antineoplastic nurses to office and kitchen employees and found no difference between the two. In this study there was a wide variation in exposure, but even when the 5 least exposed nurses were excluded from the analysis, there was no difference between the two.

Harrison has also summarized these types of tests in a table. Of 12 studies, half are positive, and half are negative.

Like many biological tests though, *even if* differences between exposed and non-exposed workers exist, it is not an indication of disease but a red flag indicating that exposure may be occurring.

ACUTE AND LONG-TERM EFFECTS

McDiarmid and Egan (1988) reported on two cases where workers experienced "accidental" heavy exposure to vincristine and carmustin in a hospital. Symptoms included an allergic-type reaction with respiratory sequelae in the first case and severe gastric upset lasting one day in the second.

In a recent survey of pharmacy personnel, Valanis et al. (1993) report that skin contact was the best predictor of 27 acute symptoms such as nausea and diarrhea and that exposed personnel had more symptoms than non-exposed pharmacy personnel.

Reports of chronic effects were provided by Sotaniemi et al. (1983), when each of three consecutive Finnish head nurses in an oncology department suffered chronic liver damage.

In a 1990 case-control study of physicians, Skov et al. found 20 cases of leukemia and 25 cases of non-Hodgkins lymphoma in 21,781 physicians, members of the Danish medical association from 1965–1988. Each case was matched by sex and age to 4 controls from among other physicians who were not cases. This gave a Relative Risk (RR) of 2.85 (C.I. 0.51–16.02) for developing leukemia and an RR of 0.74 (C.I. 0.13–4.26) for developing lymphoma in physicians who reported having been exposed, or currently exposed. Although the difference in the risk of leukemia between the exposed and non-exposed physicians was not statistically significant, and based on relatively few cases, it warrants further observation.

In 1992, Skov et al. assessed 794 Danish nurses from 5 oncology wards for the same cancers. The nurses contributed 5636 person years at risk from the early 1970s to 1987. A total increase in cancers was not found but when the lymphatic and hemapoietic cancers were looked at separately, it was found that the RR of leukemia was 10.65 (C.I. 1.29–38.5). Although the RR is statistically significant, it is based on 2 cases and the authors emphasize the need to carry out further analyses in a few years to see if the trend continues.

REPRODUCTIVE EFFECTS

Over the last decade, several studies have suggested that there is a link between exposure to antineoplastic drugs and adverse reproductive effects.

Two appeared in 1985. Hemminki et al. reported the results from a Finnish case control study in which 46 malformation cases in the children of nurses (occurring between 1973–1979) were each compared to three nurse controls/case. They found that nurses who handled antineoplastics less frequently than once a week had an odds ratio (OR) of 2.0 for producing a malformed child while those with exposure greater than once a week had an OR for producing a malformed child of 4.7, which was statistically significant (p=0.02).

This was also the finding in McDonald et al.'s (1988) study, based on 56,067 interviews of Montréal women who had delivered or had a spontaneous abortion during 1982–1984. They found 8 defects in the children of women who had administered antineoplastics when 4.07 were expected. The authors stressed that this was based on small numbers and that their exposure information was very crude.

Selevan et al., in the other 1985 Finnish case control study, evaluated spontaneous abortions in nurses registered in a central registry between 1973 and 1980. The spontaneous abortions were considered cases if the mother worked on one of the wards where antineoplastics were handled as per the Finnish Institute of Occupational Safety and Health's criteria.

Each case had 3 controls who did not handle antineoplastic drugs from the same hospital. They found that nurses who had spontaneously aborted in the first trimester were more than twice as likely to have had first-trimester exposure to antineoplastic drugs than nurses who had given birth. The OR was statistically significant at 2.30 (C.I. 1.20–4.39). Later exposure during the pregnancy and cumulative exposure was not found to have increased the risk.

A cohort study of spontaneous abortion in Danish pharmacy assistants who reported monthly exposure to antineoplastic drugs was associated with an OR of 1.6, but it was not statistically significant and based on a small number (Schaumburg and Olsen, 1990).

Five years after Selevan et al.'s study, Stucker et al. (1990) in France carried out a cross-sectional study to compare 139 exposed pregnancies (the hospital pharmacy distributed at least 10 vials of antineoplastic/nurse/week in the unit where they were exposed) to unexposed pregnancies and found after adjusting for a number of confounders (age, smoking, pregnancy order), that the exposed had a rate of 26% for spontaneous abortion versus 15% in the unexposed.

Using the same data, the authors assessed the birth weight of exposed nurses' children and found exposed newborns slightly smaller than non-exposed, but the difference were not statistically significant (Stucker et al., 1993).

Skov et al. (1992), in the same report that compared cancer rates in Danish nurses handling antineoplastic drugs to cancer rates in nurses not handling drugs, looked at low birth weight, spontaneous abortions, congenital abnormalities and pre-term births among those handling antineoplastic drugs from 1970–1987. They found no difference in the rate of these outcomes in 1282 oncology nurses compared to 2572 nurses from other departments who had not been exposed.

The most recent report on reproductive effects of antineoplastic drugs comes from France, from a survey of nurses primarily looking for an association between reproductive risks and anesthetic gases. They also found an OR of 10.0 (95% C.I. 2.1–56.2) for ectopic pregnancy in those handling antineoplastic drugs. This was based on only 6 ectopic pregnancies in 734 pregnancies and adjusted for one variable—gravity (number of pregnancies). (Saurel-Cubizolles et al., 1993.)

OTHER REPRODUCTIVE EFFECTS

A relationship between antineoplastic drug exposure and menstrual abnormality may exist. It is known that antineoplastic drugs have detrimental effects on the ovaries, the endocrine system and the conceptus. Many studies of women treated with antineoplastics in therapeutic doses have found disturbances in these patients' menstrual cycles especially in older patients who often did not ever resume normal menstruation after termination of treatment.

Shortridge surveyed U.S. oncology nurses in 1987, to assess menstrual function. Although exposure in oncology nurses would be low-level, it could occur for many years. She found that the prevalence of menstrual irregularity and amenorrhea (not menstruating for at least 3 months) increased in nurses who handled antineoplastic drugs, but that this was more pronounced in nurses aged 30–45 years versus nurses who were less than 30 years old.

In her study the prevalence of amenorrhea in non-exposed nurses was 1%, in the under 30-year-old *exposed* nurses it was 3.5%, and in the 30–45-year-old *exposed* nurses it was 5.7%.

Some studies described in previous sections were not adequate in terms of design. In order to have confidence in a study's results it is important that other factors which might explain the outcome, besides exposure to antineoplastic drugs, be accounted. Nevertheless, an incomplete review of the literature indicates that there has been *and continues to be* concern about the effects that exposure to antineoplastic drugs may have on workers' health.

WORKER ADHERENCE TO GUIDELINES

Since the implementation of safe-handling guidelines (about 1985) several authors have evaluated adherence. The main emphasis has been placed on use of proper training, ventilation and personal protective equipment. None of the studies have been satisfied that workers handling antineoplastics are fully complying with the regulations.

A national U.S. survey of 4800 eligible personnel, carried out in 1988–1989, but reported in 1992, by Valanis et al. found:

- 92% of staff who mix drugs wear gloves, and 63% wear gowns
- 82% of staff who administer wear gloves and 23% wear gowns
- 75% of staff who clean contaminated excreta wear gloves, and 6% wear gowns
- Between 65–69% of pharmacists used vertical laminar flow hoods to mix in outpatient departments in small and large hospitals
- 72% of nurses in outpatient departments, 66% of nurses in small hospitals and 44% of nurses in large hospitals used vertical laminar flow hoods to mix antineoplastics

The authors concluded that staff mixing drugs were the most compliant with OSHA guidelines but not sufficiently, that non-compliance tended to be the worst for handling cancer patient excreta and similar for large and small hospitals.

CONCLUSION

Developing procedures that follow Harrison's guidelines (see Appendix A to this section) should ensure a good level of safety. Ensuring *continued adherence* to those guidelines is *critical.*

During the last 15 years, some studies, including recent ones, have demonstrated that workers handling antineoplastic drugs have been exposed or continue to work in contaminated environments.

Workers should not be exposed to any amount of antineoplastic drug.

REFERENCES

Barale R, Sozzi G, Toniolo P, Borghi O, Reali D, Loprieno N, Porta GD: Sister-chromatid exchanges in lymphocytes and mutagenicity in urine of nurses handling cytostatic drugs. *Mutation Research* 157:235–240, 1985.

Chapman RC: Gonadal toxicity and teratogenicity. In: *The Chemotherapy Sourcebook.* Baltimore: Williams and Wilkins Press, 710–753, 1992.

Harrison BR: Safe handling of cytotoxic drugs—a review. In: *The Chemotherapy Sourcebook.* Baltimore: Williams and Wilkins Press, 799–832, 1992.

Hemminki K, Kyyronen P: Spontaneous abortions and malformations in the offspring of nurses exposed to anesthetic gases, cytostatic drugs and other potential hazards in hospitals. *J Epi Comm Health* 39:141–147, 1985.

Hirst M, Mills DG, Tse S, Levin L, White DF: Occupational exposure to cyclophosphamide. *Lancet* 1:86–188, 1984.

Jochimsen PR: Handing cytotoxic drugs by healthcare workers—A review of the risks of exposure. *Drug Safety* 7:374–380, 1992.

Kleinberg ML, Quinn MJ: Airborne drug levels in a laminar flow hood. *Am J Hosp Pharm* 38:1301–1303, 1981.

Kolmodin-Hedman B, Hartvig P, Sorsa M, Falck K: Occupational handling of cytostatic drugs. *Arch Tox* 54:25–33, 1983.

McDevitt JJ, Lees PJ, McDiarmid MA: Exposure of hospital pharmacists and nurses to antineoplastic agents. *J Occup Med* 35:57–60, 1993.

McDiarmid MA, Egan T: Acute antineoplastic exposure to antineoplastic agents. *J Occup Med* 30:984–987, 1988.

McDiarmid MA, Egan T, Furio M: Sampling for airborne fluorouracil in a hospital drug preparation area. *Am J Hosp Pharm* 43:1942–1945, 1986.

McDonald AD, McDonald JC, Armstrong B, Cherry NM, Cote R, Lavoie J, Nolin AD, Robert D: Congenital defects and work in pregnancy. *Br J Ind Med* 45:581–588, 1988.

Neal AD, de Wadden RA, Chiou NL: Exposure of hospital workers to airborne antineoplastic agents. *Am J Hosp Pharm* 40:597–601, 1983.

Nikula E, Kiviniity, Leisti J: Chromosome aberrations in lymphocytes of nurses handling cytostatic agents. *Scan J Work Environ Health* 10:71–74, 1984.

Perry MC: Chemotherapy, toxicity and the clinician. *Seminars in Oncology* 9:1–4, 1982.

Plowman PN, McElwain T, Meadows A: Complications of cancer management. Cambridge: Butterworth-Heinemann, 36, 1991.

Pyy L, Sors M, Hakala E: Ambient monitoring of cyclophosphamide in manufacturing and hospitals. *Am Ind Hyg Assoc J* 49:314–317, 1988.

Rogers B: Antineoplastic agents: actions and toxicities. *AAOHN Journal* 34:530–538, 1986.

Saurel-Cubizolles MJ, Job-Spira N, Estryn-Behar M: Ectopic pregnancy and occupational exposure to antineoplastic drugs. *Lancet* 341:1169–1171, 1993.

Selevan S, Lindbohm ML, Hornung RW, Hemmink K: A study of occupational exposure to antineoplastic drugs and fetal loss in nurses. *N Eng J Med* 313:1173–1178, 1985.

Sessink PJM, Boer KA, Scheefhals APH, Anzion RBM, Box RP: Occupational exposure to antineoplastic agents at several departments in a hospital. *Int Arch Occup Environ Health* 64:105–112, 1992.

Schaumburg I, Olsen J: Risk of spontaneous abortion among Danish pharmacy assistants. *Scand J Work Environ Health* 16:169–174, 1990.

Schardien JL: Cancer chemotherapeutic agents. In: *Chemically induced birth defects*. New York: Marcel Dekker, 457–508, 1993.

Shortridge L: Assessment of menstrual variability in working populations. *Rep Tox* 2:171–176, 1988.

Skov T, Maarup B, Olsen J, Rorth M, Winthereik H, Lynge: Leukemia and reproductive outcome among nurses handing antineoplastic drugs. *Br J Ind Med* 49:855–861, 1992.

Sotaniemi EA, Sutinen S: Liver injury in subjects occupationally exposed to chemicals in low doses. *Acta Med Scand* 212:207–215, 1982.

Stellman JM, Zoloth SR: Cancer chemotherapeutic agents as occupational hazards: a literature review. *Cancer Investigation* 4:127–135, 1986.

Stucker I, Caillar J-F, Collin R, Gout M, Poyen D, Hemon D: Risk of spontaneous abortion amount nurses handling antineoplastic drugs. *Scand J Work Environ Health* 16: 102–107, 1990.

Valanis B, Vollmer WM, Labuhn KT, Glass AG: Association of antineoplastic drug handling with acute adverse effects in pharmacy personnel. *Am J Hosp Pharm* 50:455–462, 1993.

Valanis B, Vollmer WM, Labuhn K, Glass AG, Corelle C: Antineoplastic drug handling protection after OSHA guidelines—comparison by profession, handling activity, and work site. *J Occup Med* 149–155, February 1992.

Waksvik H, Klepp O, Brogger: Chromosome analyses of nurses handling cytostatic agents. *Cancer Treat Rep* 65:607–610, 1981.

APPENDIX 1

Handling of Antineoplastic Drugs

*1. **Purpose:*** To establish a policy for handling antineoplastic drugs within the Medical Center.

2. Policy: All services in the Medical Center which handle antineoplastic drugs will establish a written policy outlining procedures for the safe transport, storage preparation, administration and disposal of these agents. Each policy will be submitted to the Chief of Staff for review. Each service will establish a training program which will inform the employee of the identity of the potentially hazardous materials used in the service, of emergency procedures to be followed in the case of accidental exposure, of potential health hazards and symptoms expected from exposure to antineoplastic drugs, of procedures for cleaning spills, who to notify in the event of an exposure or spill, who to notify for additional information on these chemicals and where to seek medical attention. Each service chief will submit to the Safety Manager and the Employee Health Unit a permanent record listing the names of employees preparing, administering and/or disposing of antineoplastic drugs (high-risk employees), and a record of all areas in these respective services where antineoplastic agents are stored, prepared, administered or destroyed. These records will be updated as necessary.

3. Delegation of Authority:

a. Chief, Supply Service: Will ensure the safe storage, transport and disposal of all antineoplastic drugs which are received by this Medical Center through normal supply channels, that all antineoplastic drugs are identified as potentially hazardous materials during their movement through the Supply Service and that all Supply Service employees are aware of their hazardous properties and procedures for proper handling.

b. Chief, Pharmacy Service: Will ensure the safe storage, preparation, dispensing and disposal of all antineoplastic drugs which are received from Supply Service or from other sources, that all pharmacy personnel required to handle or dispense these drugs are trained in the proper handling techniques and that all antineoplastic drugs leaving the Pharmacy are properly labeled.

c. Chief, Nursing Service: Will ensure the safe storage, administration and disposal of antineoplastic drugs which are received from Pharmacy Service, and that all nursing personnel required to administer these drugs or care for patients receiving these drugs are trained in the proper handling techniques.

d. Chief, Building Management Service: Will ensure that antineoplastic drug wastes are safety transported to storage and disposal areas, that all Building Management employees working in areas where antineoplastic drugs are handled are aware of their potential hazard and are trained in the safe handling of these wastes and will assist all using services in cleanup of spills of antineoplastic drugs.

e. Chief, Engineering Service: Will ensure that all Engineering Service employees involved with disposal of antineoplastic drug wastes are aware of their potential hazard and are trained in the safe handling of these wastes; that antineoplastic drug wastes are properly stored prior to on-site disposal; that these wastes are properly and safely incinerated in accordance with local, state and Environmental Protection Agency (EPA) guidelines; and that all antineoplastic drug wastes to be disposed of by an outside firm are safely stored, properly labeled, manifested and transported in accordance with state, EPA and Department of Transportation regulations.

f. Safety Management: Will monitor all affected services for compliance with the requirements of this memorandum; will establish contact with the responsible person in each service and establish a mechanism of reporting acute exposures, spills and unsafe working conditions; and will investigate the circumstances surrounding each incident that results in employee exposure or in personal injury. The Safety Manager will report all accidents and records of service compliance to the Medical Center Safety, Occupational Health and Fire Protection Committee.

4. Procedures:

a. Medical Surveillance:

(1) Every employee determined to be at high risk will be given an appropriate physical examination by the Employee Health Unit. The purpose of this examination is to set a baseline against which changes can be measured.
(2) All employees determined to be at high risk will be given an annual physical examination. These examinations will be designed to detect changes in general health and in specific areas vulnerable to exposure to chemicals such as the skin, buccal and nasal mucosal membranes and the eyes.
(3) All employees acutely exposed to an antineoplastic drug will be examined and treated by the Employee Health Unit.
(4) Service chiefs will ensure that all high-risk employees and other exposed employees report for the appropriate examination.
(5) Records, when established, will be maintained as a part of the employee health record.

b. Personnel Practices:

(1) All employees handling antineoplastic drugs will wear personal protective clothing to include disposable gloves, disposable gown and eye protection. Gowns will be closed in the front and will have long sleeves and tight-fitting cuffs at the wrists. Potentially contaminated clothing will be removed prior to leaving a work area.
(2) An OSHA-approved dust/mist respirator or face mask will be worn when airborne particles or aerosols of antineoplastic drugs are generated during handling, unless proper ventilation (i.e., Class II Biological Safety Cabinet) is available.
(3) There will be no eating, drinking, smoking, chewing of gum or tobacco, application of cosmetics or storage of food in areas where antineoplastic drugs are used.
(4) All personnel will wash their hands immediately after completion of any procedures in which antineoplastic drugs have been used.
(5) Mechanical pipetting aids will be used for all pipetting procedures. Oral pipetting is prohibited.

(6) Only employees who have received proper training will prepare or administer antineoplastic drugs. All services will document this training.
(7) Antineoplastic drug work load will be distributed among the trained personnel in order to minimize daily exposure.
(8) Employees will report all spills, exposures or unsafe conditions to their supervisors.
(9) Employees who are pregnant, planning a pregnancy (male or female), breast feeding or who have a written statement from a physician which provides medical reasons why they should not be exposed to antineoplastic drugs will not prepare or administer these agents or care for patients during their treatment (up to 24 hours after completion of therapy).

c. Operational Practices:

(1) All areas where antineoplastic drugs are stored, prepared or disposed of will be posted with signs bearing the "BIOHAZARD" or "CANCER HAZARD" symbol.
(2) Supervisors of all areas where antineoplastic drugs are stored, prepared or disposed of will designate personnel authorized entry to the area and post the area as "off limits" to patients and general employees.
(3) Preparation of antineoplastic drugs for clinical use will be conducted only in a Class II Biological Safety Cabinet by properly trained personnel. These hoods will be marked as "BIOHAZARD" or "CANCER HAZARD" preparation areas. These hoods will meet the specifications of the National Sanitation Foundation Standard No. 49 and will be certified at least every 6 months. Antineoplastic drugs will not be prepared in open environments such as clinics and nursing units. Clinical services requiring these drugs for their patients should contact Pharmacy Service.
(4) Only minimum working quantities of antineoplastic drugs will be stored in patient care and preparation areas.
(5) Syringes containing antineoplastic agents and fluid bags to which they have ben added will be labeled "Chemotherapy Agents: Dispose of Properly" prior to dispensing.
(6) General housekeeping procedures will suppress the formation of aerosols by the use of a wet mop or a vacuum cleaner equipped with a HEPA filter or water trap. Dry sweeping or mopping is prohibited in areas where spills have occurred or in preparation or administration areas.
(7) Materials contaminated with antineoplastic drugs (i.e., syringes, gauze pads, gowns, gloves, etc.) will be discarded only in closed metal containers lined with 4-mil. plastic bags. These will be labeled as "CAUTION: CHEMOTHERAPY WASTES." Needles, syringes and needles or i.v. bags with sets and needles will be discarded in special "Chemotherapy Disposal Containers" which are puncture-proof and have "tamper-proof" lids.
(8) Spills

a. In the event of breakage of containers of powder or liquid antineoplastic drugs, steps should be taken to prevent spread of the spill and to prevent other employees from coming into the contaminated area. Absorbent materials such as spill pillows or towels should be placed over the spill to prevent it from moving under cabinets or shelving units.

b. Immediately put on personal protective clothing including heavy duty gloves, respirator, gown and eye protection.

c. Absorb spill with chemical-spill pillows, gauze pads, paper towels or plastic-backed absorbent pads. Wash contaminated surfaces with water and reabsorb. Place all contaminated materials in 4-mil. plastic bags. Decontaminate area if a neutralizing agent is available. Wash all contaminated surfaces at least three times.

d. Seal and double bag all contaminated materials. Mark as hazardous wastes. Contact housekeeping to remove waste and to terminally clean area.
e. Personnel involved in the cleanup should wash all potentially exposed skin surfaces with soap and water.
f. Notify the Medical Center Safety Manger of the spill.
g. If the spill is a dry powder, cover with a generous supply of water-dampened absorbent towel or gauze.
h. Broken glass should be handled only with heavy-duty gloves, placed in a cardboard or plastic container prior to disposal in hazardous waste bags or placed in a Chemotherapy Disposal Container.

(9) Accidental Exposure
a. In the event of an accidental acute exposure to an antineoplastic drug, all exposed surfaces should be rinsed thoroughly with copious amounts of water. An emergency shower or eyewash will be available.
b. If applicable, wash exposed surfaces with appropriate neutralizing solution, then wash thoroughly with soap and water.
c. For eye exposure, immediately flood the affected eye with water or isotonic eyewash designated for that purpose for at least 15 minutes. Seek medical attention immediately.
d. Report exposure to supervisor.
e. Report to the Employee Health physician for assessment and local treatment.

(10) Building Management employees will collect chemotherapy disposal containers when full and transport them to the secure holding area for incineration. Chemotherapy disposal bags or containers will not be left unattended but will be taken directly to the secure holding area and placed in the approved closed containers. Building Management personnel will wear gloves when handling these containers. Isolation gowns, respirators and eye protection will be worn when handling open bags or improperly sealed or broken containers.

(11) Antineoplastic drug wastes will be considered hazardous wastes and destroyed by incineration. Engineering Service personnel receiving these wastes for incineration will wear gloves when handling bags or disposal containers. Disposable gowns, respirators and eye protection will also be worn when handling open containers or improperly sealed or broken containers.

(12) Employees will wear personal protective clothing when handling urine, feces, or soiled linens or clothing, or patients who have received antineoplastic drugs within the previous 24 hours. They will dispose of urine and feces carefully but in the usual manner. Contaminated linens and clothing will be placed and sealed in a water-soluble laundry bag (same as for contaminated linens) and sent to Laundry Service for cleaning. Potentially contaminated garments will not be worn or transported outside the work area but disposed of in the appropriate containers.

(13) If a patient expires within 24 hours of receiving an antineoplastic drug, the mortuary staff will be informed of the potential for exposure to body wastes containing antineoplastic drugs by the attending physician. This will be accomplished by marking the autopsy request form.

(14) Patients' specimens of urine, blood, sputum, etc. obtained within 24 hours of receiving antineoplastic drugs will be labeled (in the remarks sections of the lab slips and on the specimen container) "CHEMOTHERAPY PATIENT" prior to delivery to the laboratory. Employees responsible for the care of the patient and requesting the specimen will ensure that the laboratory request forms are properly labeled.

(15) During chemotherapy and for 24 hours afterward the patient's inpatient chart and room will be labeled "Chemotherapy Patient" in order to notify employees to take appropriate precautions.

Source: Chemotherapy Sourcebook, 1992. McPerry, Editor.

CHAPTER 10

Medical Waste Management: The Problem and Solutions

Michael L. Garvin

INTRODUCTION

During the 1980s the task of waste management was made increasingly more difficult for hospitals. Potentially infectious waste (PIW), generated by every hospital, clinic and doctor's office in the country received intense scrutiny by the media and regulators. The main reason for this attention can be traced to the increase of AIDS cases in the early 1980s and the general lack of understanding of how the HIV virus could be transferred. The public became concerned that the HIV virus could be spread through medical waste placed in landfills. Legislation was passed that banned PIW from the landfill. At the same time, states rewrote definitions thereby classifying a significantly larger percentage of hospital waste as potentially infectious.

Just as the 1980s ended with the passage of the overly comprehensive and unnecessarily costly Medical Waste Tracking Act (MWTA), the 1990s opened with the voice of reason in the form of a congressional mandated report from the Agency for Toxic Substance and Disease Registry (ATSDR). This report[1] concluded that potentially infectious waste presents no more risk to the environment and the public than general household waste. It went on to say that a hospital is better advised to focus its resources on reducing occupational exposures to the waste. The Occupational Safety and Health Administration (OSHA) has picked up that theme in writing a new Bloodborne Pathogens Standard.

Hospitals and health care associations would be wise to seize this opportunity and assess waste management options in the light of scientific data and reason. The objective of this report is to further a rational discussion on the future of potentially infectious waste management.

OVERVIEW OF MEDICAL WASTE

In the course of daily activities, a hospital produces waste. Some of that waste is discharged into the sanitary sewer system and some of it is released in

1-56670-083-3/94/$0.00+$.50

gaseous form through laboratory hood vent ducts, but most of it is "solid waste." Technically, solid waste comprises the largest percentage of hospital-generated waste and includes such waste types as general office trash, food service waste and even the fastest growing waste type, recycled waste. In addition, solid waste includes three types of waste which fall under federal or state regulation: radioactive, chemical and potentially infectious. These three types of waste comprise "regulated medical waste." Simply put, they are medically generated waste which are governed by regulation. In the Medical Waste Tracking Act of 1988, the term "regulated medical waste" was used loosely to apply to those items identified by the federal Environmental Protection Agency (EPA) as being potentially infectious. While those EPA identified items fall under the general term of regulated medical waste, they are by no means the only medical wastes regulated at the present time. Figure 1 shows the seven different medical waste types of combinations which are governed by regulations. The combination waste types will become the hot topic of future regulations.

The focus of this report is potentially infectious waste. A review of the literature also found a number of measurement units for PIW generated. The most useful unit of measurement is "pounds per patient day" as it considers utilization as well as bed capacity. It is difficult to provide exact statistics on how much of this waste is generated. The difficulty arises due to the wide range of definitions hospitals use for potentially infectious waste. The American Hospital Association estimates that an average hospital will produce approximately 20 pounds of solid waste for every patient day. Of that solid waste produced, approximately 10 to 12% is considered potentially infectious. So on the average, a hospital would generate about 2.0 to 2.4 pounds of potentially infectious waste for every patient day if the facility includes only those categories of waste identified by the AHA as potentially infectious. Table 1 provides a comparison of which categories of waste are included in definitions supported by different regulatory agencies. The statistic most often quoted in the literature is that PIW comprises 15% of all hospital waste.

A study[2] conducted by the University of North Carolina of 441 randomly selected hospitals reported the following:

- Hospitals with less than 100 beds generated 1.5 pounds per patient day
- Hospitals with 100–300 beds generated 2.4 pounds per patient day
- Hospitals with 300–500 beds generated 2.7 pounds per patient day
- Hospitals with more than 500 beds generated 2.5 pounds per patient day

The range of this data is probably due to a combination of different definitions and levels of medical care. One recent study of a large university tertiary care facility found that 14.8% of all waste is considered potentially infectious. For every patient day, 3.9 pounds of PIW is generated. This study more accurately assesses the amount of PIW generated at UIHC relative to total waste. Depending on changes in PIW definition, procedures and products purchased, the statistic may continue to be dynamic.

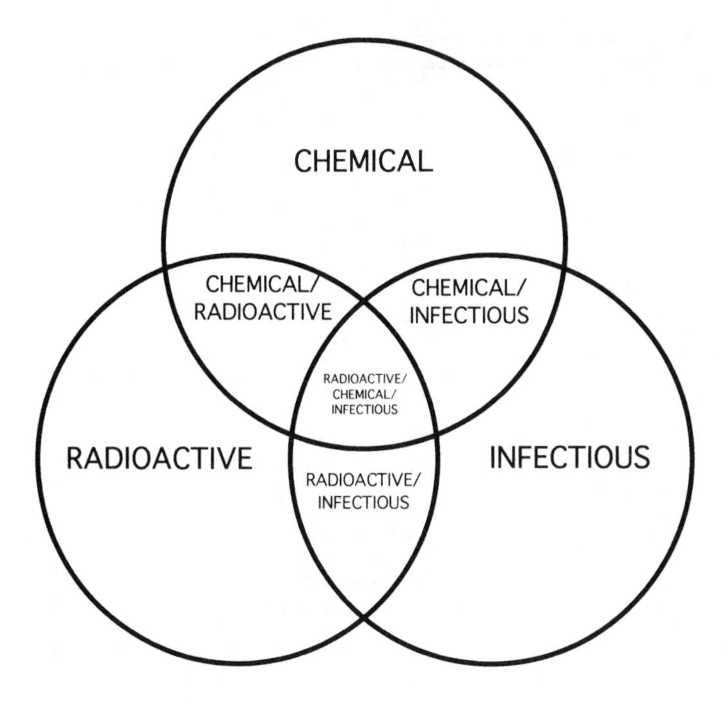

Figure 1. Regulated medical waste.

In a literature search, no data could be found on the impact that clinic activity has on the generation of potentially infectious waste. With clinics playing a larger role in health care delivery, such data would be very helpful. Another issue is the effect that intensity of care has on potentially infectious waste. A tertiary care hospital will probably produce more of this waste than a primary or secondary care facility.

THE LACK OF STANDARD TERMINOLOGY HINDERS THE DISCUSSION OF THE SUBJECT

In reviewing the literature in preparation for this report, one major observation was the lack of standard terminology used when discussing this topic. That portion of regulated medical waste which presents a potential infectious risk is called medical waste, regulated medical waste, biohazard waste, isolation waste, infectious waste and potentially infectious waste. Since it is hospital-generated waste which, if not contained and managed properly, *could* result in the transfer of infection, the phrase "*potentially* infectious waste" is the most accurate term.

"Regulated medical waste" has become the common phrase used by the federal EPA. If the Medical Waste Tracking Act is expanded this year, we may

Table 1. Comparison of categories included in definitions of infectious waste.

TYPES OF WASTE	AHA	CDC	IOWA	EPA	MWTA
1. Microbiological (e.g., stocks and cultures)	Yes	Yes	Yes	Yes	Yes
2. Blood and Blood Products	Yes	Yes	Yes	Yes	Yes
3. Pathological (e.g., tissues)	Yes	Yes	Yes	Yes	Yes
4. Sharps (used)	Yes	Yes	Yes	Yes	Yes
5. (Communicable Disease Isolation (CDC Class 4)	No	Yes	Yes	Yes	Yes
6. Contaminated Animal Carcasses	No	No	Yes	Yes	Yes
7. Other Isolation	No	No	No	Yes	Yes
8. Contaminated Lab Waste	No	No	No	Optional(1)	Yes(2)
9. Surgery and Autopsy Waste	No	No	No	Optional(1)	Yes(2)
10. Dialysis Waste	No	No	No	Optional(1)	Yes(2)
11. Contaminated Equipment	No	No	No	Optional(1)	Yes(2)
12. Sharps (Unused)	No	No	No	Optional(1)	Yes

Not strongly recommended but EPA suggests that these items be considered items saturated with blood as well as intravenous...

all soon be using that term. Chemical waste was called just that until the federal regulation started using the term "hazardous waste."

The most accurate term for this waste is "potentially infectious." Realistically, a great stride forward would be achieved if all parties, including regulators, would agree on either "potentially infectious" or simply "infectious."

A glossary of other terms is included in Appendix 1 to this chapter.

WARNING SYMBOL

For decades the biohazard symbol had served as a strong warning that a serious biological hazard existed with a labeled substance. Extremely concentrated strains of biological agents as well as biological weapons warranted the symbol. During the waste crisis which followed the increase in AIDS cases, the biohazard label was adopted for infectious waste. Now that same symbol is placarded on doors leading to biological weapons as well as those leading to holding areas for potentially infectious waste. There is absolutely no comparison in the degree of risk involved with the two different items. The adoption of this symbol has overstated the risk involved with handling potentially infectious waste while diluting the effectiveness of its original purpose, yet a suitable replacement has not been designed. If a warning symbol is to be used, a new symbol needs to be developed.

DEFINING POTENTIALLY INFECTIOUS WASTE

In the early 1980s the problem was that no one had properly defined potentially infectious waste. In the early 1990s, the main problem is that there are too many definitions. This waste has now been defined by the Centers for Disease Control (CDC), the American Hospital Association (AHA), most state department of natural resources, the EPA and by Congress through the Medical Waste Tracking Act. These definitions can include any or all of the following hospital-generated waste items:

1. Microbiological waste (stocks and cultures)
2. Blood and blood products
3. Pathological waste (tissues, etc.)
4. Sharps (used)
5. Communicable disease isolation waste (exotic diseases known as class 4 by CDC)
6. Other isolation waste
7. Contaminated animal carcasses
8. Contaminated lab waste
9. Surgery and autopsy waste
10. Dialysis waste
11. Contaminated equipment
12. Sharps (unused)

Table 2 shows a comparison of the different potentially infectious waste definitions. The manner in which this waste is defined directly impacts the volume generated by a facility, and volume directly relates to cost. Later this year, Congress will receive information on the cost of the MWTA. Preliminary reports[3] have indicated that the program has increased waste disposal costs dramatically. With a solid majority of states now having potentially infectious waste regulations, the MWTA is seen as a costly and needless regulation which hospitals, facing enough financial pressures, should not be required to bear.

As the case in states such as Iowa, hospital associations and state regulatory agencies have come to agreement on a reasonable definition for this waste. This definition parallels closely the CDC recommendation definition.

Table 2. Waste management summary.

Waste Produced Annually		Amount (lbs.)
Landfill		4,995,360
Incinerator (PIW)		946,782
Recycled Cardboard*		293,350
Recycled White Paper		96,00
Chemical Waste		30,000
Radioactive Waste		4,000
	Total	6,365,492

Total amount of waste generated per patient day: 26.3 lbs
Percentage of PIW to total waste: 14.8%
Pounds of PIW per patient day (based on 1989–1990 data): 3.9 lbs per patient day

* $13.75 per ton for recycled cardboard and avoids a charge of $39 per ton to landfill it.

1. Microbiological waste
2. Blood and blood products
3. Pathological waste
4. Used sharps
5. Class 4 communicable disease isolation waste
6. Contaminated animal carcasses

DISPOSAL TECHNOLOGY

The activities of the last ten years have placed hospital waste disposal programs in a precarious position. As mentioned previously, the concern surrounding the increase in AIDS cases resulted in banning potentially infectious waste from landfills. With most hospitals having easy access to an incinerator, they simply burned the waste. Studies[4] have shown nearly 80% of all hospitals depend on incineration for disposal of at least part of their potentially infectious waste. With the dawn of the 1990s, clean air regulations are being written on both the state and federal levels. Some researchers[5] believe that unless these regulations are reasonable and consider the financial impact on hospitals, on-site incineration will be financially beyond the reach of most facilities. Some rural hospitals may suffer the most. One study[6] indicates that air emission standards can increase incineration purchase and operation cost by a factor of 10.

So where will that leave facilities? The full impact of new air emission regulations will be felt by hospitals in the mid-part of this decade. Their options will include:

1. Developing a hospital consortium-owned regional disposal facility
2. Hospital or associations contracting with a commercial disposal firm

3. Paying for an incinerator upgrade
4. Purchasing an alternative disposal technology

Depending on the severity of the new regulations and the particular characteristics of the hospital's situation, any of these options could be the best option.

Option 1: *Developing a hospital consortium-owned regional disposal facility.* Madison, Wisconsin area hospitals have run a consortium-owned incinerator for five years. At the present time, there are very few of these types of facility in the country. The Volunteer Hospital Association of Iowa is forming a group which will run a regional unit for about 20 hospitals in eastern Iowa.

Option 2: *Hospital or associations contracting with a commercial disposal firm.* At least for the short term, hospitals and hospital associations are choosing this option. When dealing with reputable firms, hospitals receive reasonable rates and favorable contract conditions. Dealing with less than reputable firms or companies having a monopoly on such service can cost hospitals dearly. There is a need for research on the commercial disposal rates in different sections of the country. The research could consider the cost impact on rural versus urban hospitals.

The Iroquois Health Care Consortium based in Albany, New York is supporting the construction of a steam sterilization/grinding facility by Browning-Ferris, Inc. (BFI). The Maryland Hospital Association has encouraged a commercial firm to site a large incinerator in the Baltimore area. The Milwaukee Area Hospital Council is in the process of doing the same in eastern Wisconsin.

A study[7] showed that 35% of Iowa hospitals are choosing (or are forced to choose) to contract waste disposal to a commercial firm for at least part of their potentially infectious waste.

Option 3: *Purchasing an alternative disposal technology.* Hospitals which want to control advantages of an on-site technology but can no longer afford incineration will most likely choose an alternative disposal technology. This has been the case in California where air emission regulation has forced most hospitals away from incineration.

Table 3 shows a comparison of on-site waste disposal technologies which provide an alternative to incineration.

The following is a discussion of alternative disposal technologies.

Steam Sterilization

Except for incineration, steam sterilization is the most common form of infectious waste treatment used by hospitals. This technology requires that the waste

Table 3. Comparison of potentially infectious alternative waste treatment technologies.

Technology	Advantages	Disadvantages	Investment	Operating Cost	Volume Reduction
Steam sterilization and compaction	• Low cost • Small space requirement • Easy to install • Simple to operate • Some volume reduction • Simple to permit	• Not suitable for pathological waste • Requires special containment bags • May create odor problem • Waste appearance is not altered • Weight is not reduced	• 50–500 bed facility • $50,000–$150,000	• 2.5–3.5 cents per pound	• 70%
Grinding and chemical disinfection	• Medium cost • Small space requirement • Easy to install • Simple to operate • Some volume reduction • Alters appearance • Simple to permit	• Not suitable for pathological waste • Weight is not reduced • Liquid discharge may require special permit	• 50–500 bed facility • $50,000–400,000	• 6–10 cents per pound	• 80%

Grinding and microwave treatment	• Low cost • Small space requirement • Easy to install • Simple to operate • Volume reduction • Alters appearance • Simple to permit • Reliable performance	• Not suitable for pathological waste • Weight is not reduced	• 400–1,000 bed facility • $400,000–$500,000	• 7–14 cents per pound	• 80%

be subjected to steam, in some cases under pressure, for a certain amount of time. Liquids produced in the process can be drained and the decontaminated solid waste can be placed directly in a landfill or it can be compacted or shredded prior to final disposal.

Advantages

The major advantages for steam sterilization include the fact that hospitals are familiar with the technology. It has been used to treat infectious waste in hospitals for many years and has well-developed quality control protocols. The technology requires little floor space and is simple to operate. It has one of the lowest "purchase" and "operating" costs of all infectious waste technologies. Steam sterilization produces no air pollution emission which allows for easy siting of a unit, and, when coupled with either compaction or shredding, the technology can reduce waste volume by as much as 80% and render the waste unrecognizable.

Disadvantages

The disadvantages of steam sterilization include the need for special waste packaging. In most cases, the technology requires that the waste be placed in bags which are opened to allow steam penetration. In other cases, bags "melt" when exposed to the heat steam thus requiring secondary containment. Steam sterilization does not reduce the consideration if the local landfill charges by weight. Steam sterilization is also not recommended for wastes such as pathology wastes. Some studies indicate that staff who work around medical waste autoclaves may be exposed to volatile organic compounds in excess of safe levels.

It is suggested that a hospital which chooses this technology clearly explain its plans to the condition of the waste when it reaches the landfill.

Purchase costs for a steam sterilization/compaction unit to service hospitals ranging from

50–500 beds — $50,000–150,000
Operating costs — 2.5–3.5 cents per pound

Chemical Disinfecting/Shredding

The disinfecting/shredding technology grinds the waste in the presence of a hypochloride solution. The hypochloride solution disinfects the waste while the shredding process reduces total volume and renders the waste unrecognizable. The disinfectant solution is then drained from the waste and, in most cases, disposed of to a sanitary sewer. In some cases, pre-treatment may be necessary. The remaining solid waste can be placed in a landfill.

Advantages

The disinfecting/shredding technology can reduce the volume of infectious waste by as much as 80% as well as make the waste unrecognizable. The technology requires little floor space and is simple to operate and produces no toxic air emissions.

Disadvantages

In some cases, wastewater discharged from the chemical disinfecting/shredding unit may have to be permitted by local water treatment authorities. Depending on state emission standards, the cost of a large unit may be as high as a small incinerator. This technology does not reduce the weight of the waste.

Purchase costs for a chemical disinfecting/shredding unit:

Smaller unit — $45,000
Larger unit (1,500 pounds per hour) — $390,000
Operating costs — 6–10 cents per pound

Grinding/Microwaving

The grinding/microwave technology has been imported from Europe. It was developed by a company in Germany, has had over three years of experience in that country and is now in operation in France, Switzerland, Italy and the United States. The technology shreds the waste in a HEPA-filtered chamber and then subjects that waste to microwaves. Steam is ejected into the treatment chamber to aid in disinfection. This process differs from steam sterilization in that the heat is generated from the center of the waste mass.

Advantages

The grinding process reduces the waste volume by as much as 80% and renders the waste unrecognizable. The siting process for a grinding/microwave unit is greatly simplified by the fact that there are virtually no air emissions for the process. The equipment has demonstrated the ability to be very reliable.

Disadvantages

Like the steam sterilization option, this technology will not reduce the weight of the waste and hospitals will have to provide evidence to the landfill that the waste has been disinfected. This technology is not recommended for pathology waste.

Purchase costs for a grinding/microwave unit to service hospitals ranging from:

400—1,000 beds — 375,00–650,000
Operating costs — smaller unit — 9–14 cents per pound
larger unit — 5–7 cents per pound

Incineration: (for comparison with alternative technologies)

Incineration technology uses carefully controlled high temperature combustion to destroy infectious waste. There are a number of designs available within the incineration technology including multiple-chamber, rotary kiln and controlled air systems. Developments in pollution abatement hardware have brought incinerators in compliance with the strictest air emission standards, and, although not as cost effective as ten years ago, waste heat recovery systems can produce energy in the form of steam.

Advantages

Hospitals are comfortable with incinerators. The technology has been the preferred choice for infectious waste treatment in the past. Incineration reduces both volume and weight by approximately 95%. The technology can treat all infectious waste items including pathology wastes.

Disadvantages

Because of air emission concerns, it is becoming increasingly difficult to site an incinerator, especially in urban areas. New sophisticated designs and pollution control equipment required by many states make this technology have the most expensive "purchase" and "operating" costs of all technologies. New legislation on ash management and required operator training is driving operating costs still higher. Public opposition seems to be highest against incineration in comparison to the other treatment alternatives.

Purchase costs of an incinerator which would service hospitals ranging from:

200–1,000 beds — $100,000–3,000,000 (depending on pollution control equipment required)
Operating costs — 12–70 cents per pound

Table 4 shows the recommended disposal or treatment method for different types of potentially infectious waste.

A number of new technologies are currently under development. Gamma Wave Irradiation and Electrothermal Disinfection uses a sealed cobalt source and

Table 4. Recommended methods for disposal of potentially infectious waste.

Type of Waste	Recommended Disposal Technology
1. Microbiological	• Incineration • Disinfection and landfilling (disinfection technology might involve steam, microwave, electrothermal or chemical treatment) Grinding is required by the MWTA and some state regulation
2. Blood and blood products	• Sewer disposal • Disinfection/solidification/landfilling (powder is added to liquid waste—waste is solidified and disinfected)
3. Pathological waste and contaminated animal carcasses	• Incineration
4. Sharps	• Incineration • Disinfection/grinding/landfilling

low frequency radio waves, respectively, to treat PIW. Experiments in burning the waste with lasers and melting it with a welding-type arc also are underway. None of these technologies are available for on-site installation.

Much of this information comes from three technology assessment reports.[8–10]

REGULATIONS THAT AFFECT POTENTIALLY INFECTIOUS WASTE MANAGEMENT

Both federal and state regulators have taken a more aggressive approach to overseeing how hospitals manage their potentially infectious waste. Attachment 7 shows the major regulation generated by the EPA for both air quality and solid waste management. It also details the federal and state OSHA regulation affecting waste management. Even though both the EPA and OSHA have had authority to produce regulations involving potentially infectious waste management for over 20 years, there had not been any reason to develop rules until concern arose over the increase in AIDS cases.

1994 and 1995 will be important years for new regulations. The EPA will develop air quality rules specifically for potentially infectious waste incinerators in late 1992. Congress will decide whether the Medical Waste Tracking Act will be re-authorized and if it should be expanded to the rest of the country in the summer of 1995. The Iowa State Legislature is considering bills which would detail how the waste is managed. The legislature is also asking the Department of Natural Resources to develop air emission standards for incinerators. OSHA and NIOSH are updating regulations which would require that hospitals provide staff with an even higher level of safety concerning bloodborne pathogens.

SUMMARY OF RECOMMENDATIONS

The public concern over the management of potentially infectious waste continues to grow. Unfortunately, a rational discussion is hindered by a lack of consensus on such critical issues as terminology and unit of measurement. The implementation of the following recommendations would assist in addressing these needs:

1. *Standard Terminology* - The phrase "potentially infectious waste" (PIW) most accurately describes the portion of medical facility-generated waste which has the *potential* of conveying disease. Further discussion and regulation should avoid the misnomer "regulated medical waste."

2. *Standardize Unit of Measurement* - The amount of PIW generated is best measured by "pounds per patient day." This unit of measurement incorporates both bed capacity and utilization. Such measurement tools as "pounds per bed" are misleading without a utilization factor.

3. *Standardize Definition* - Because it has a foundation in scientific assessment, the definition recommended by the CDC should be accepted as standard. Such a definition would include:

    a. Microbiological waste
    b. Blood and blood products
    c. Pathological waste
    d. Used sharps
    e. Class 4 communicable disease isolation waste
    f. Contaminated animal carcasses

    Further state and federal regulation should acknowledge this standard and not needlessly add financial burden to health care facilities.

4. *Need for Further Research* - More research needs to be conducted concerning who is most affected by PIW regulations. One study[10] indicates that when regulation increases waste volumes and/or limits disposal technology, smaller rural hospitals are most negatively affected. A survey conducted by the Iowa Hospital Association found that of the 37 hospitals which have contracts with disposal firms, seven state that they had only one firm to provide such a service. More research needs to be done on the impact of including doctors' offices under regulation as well as the impact of health care's shift from in-patient to outpatient services.

Armed with the data such research would provide, hospitals and hospital associations could strengthen their communication with state and federal legislators. Should there not be air emission in standard exemptions for small or rural facilities? Should a one ton-a-day incinerator in rural Iowa be required to have the same pollution equipment as a 200 ton-a-day unit in Los Angeles?

Knowing now that landfilling potentially infectious waste does not pose an environmental risk, would it not be prudent to allow small generators, such as doctors' offices, clinics and small hospitals, to once again landfill that waste as long as occupational risks while transporting are addressed?

5. *Review of Management Practices* - Hospitals, in agreement with OSHA and ATSDR assertions that potentially infectious waste constitutes an occupational hazard, need to fully address that hazard. Hospitals should conduct comprehensive reviews of their waste management programs in light of the new safety standards. Complete assessments of waste separation, packaging, handling and disposal procedures need to be made. Hospital product evaluation and safety committees need to design cost/benefit analyses on new safety-enhancing products such as sheathable syringes, sharps disposal boxes and powders that disinfect and solidify potentially infectious liquid waste. The present cost of treatment for sharps injuries and other PIW exposures needs to be assessed.

6. *Implementation of Waste Reduction Program* - Hospitals can reduce the amount of waste generated. Medical supply manufacturers should be requested to provide products and packaging which can be recycled or which lend themselves to "low polluting" incineration or biodegradation. In addition, hospitals can encourage practices which shift waste from the disposal to the recycle bin. The state of Iowa requires that 50% of all waste generated be recycled trash by the year 2000. This will not happen without waste reduction.

The standardization of potentially infectious waste terminology, units of measurements and definitions can only hasten the resolution of this waste management problem as it assists facilities in evaluating current programs and deciding on cost effective options for the future.

REFERENCES

1. Rettig PC: ATSDR: Medical Waste Poses no Threat to Public. *Health Facilities Management*. June, 1990.
2. Rutala W: Management of Infectious Waste by US Hospitals. *JAMA*. September, 1989.
3. Sedor P: Costs Soar Under EPA's Waste Tracking Program. *Health Facilities Management*. June, 1990.

4. Rutala, *op. cit.*
5. Brodsky R: Strict Incinerator Regulations May Make It Too Expensive For Hospitals To Go It Alone. *Modern Healthcare.* March 18, 1991.
6. Garvin M: Waste Disposal Costs Show Wide Variation from State to State. *Modern Healthcare.* March 18, 1991.
7. Iowa Hospital Association. *Survey on Infectious Waste Management in the State of Iowa.* 1990.
8. Doucet L: *AHA Technical Document Series,* Infectious Waste Treatment and Disposal Alternatives. October, 1989.
9. Cross F: Evaluation of Alternative Techniques for the Treatment of Solid Medical Waste.
10. U.S. Congress Office of Technology Assessment, Medical Waste Treatment Technologies. OTA N3-2045.0. Washington, DC: U.S. Government Printing Office. March, 1990.

APPENDIX 1

Recommendations for Standardized Medical Waste Terminology

Airborne Waste - particles and gases discharged to the air by way of an incinerator stack or a facility exhaust flue.

Biohazard Waste (Biological Waste) - often used synonymously with infectious waste. Traditionally, the term was applied to waste with very high concentrations of infectious agents. Such waste might be found in research or clinical laboratory settings as well as biological defense installations.

Chemical Waste - any waste which includes toxic substances as defined by the Environmental Protection Agency (EPA).

Disinfection - a reduction of populations of disease-producing microorganisms. PIW regulation requires that the waste be disinfected but not sterilized.

General Waste - waste categories such as office and food services trash which present no potential hazard and require no special handling procedures.

Hazardous Waste - term used by the EPA to refer to any waste which may pose a hazard to humans or the environment. The term is used synonymously with "chemical waste."

Infectious Waste - a slightly looser term than "potentially infectious waste" but is used synonymously. The term "infectious" incorrectly implies that the waste has been positively identified as having the capability of transmitting infection.

Isolation Waste - all waste generated by a patient who has been placed on isolation precautions.

Medical Waste - often incorrectly used as a term for "potentially infectious waste," medical waste is simply all waste generated by a medical facility.

Pathological Waste - waste generated by pathological services usually containing tissue, gross specimen and limbs.

Potentially Infectious Waste (PIW) - the most accurate term for waste which has the "potential" for transmitting infectious. The term is used synonymously with "infectious waste."

Radioactive Waste - waste with radioactive properties greater than normal background radiation level.

Regulated Medical Waste - wastes generated by a medical facility which are subject to regulation. The waste includes radioactive, chemical and potentially infectious. The term has been used by the EPA to refer to potentially infectious waste.

Sewage - general term used to refer to all waste discharged into a sanitary sewer.

Solid Waste - defined by state and federal regulators as all wastes generated by a facility which are disposed of by any means except discharge to the air. The term includes wastes placed in landfills and waste incinerated but not waste discharged from an exhaust flue or an incinerator stack.

Sterilization - the complete destruction of microbial life.

CHAPTER 11

Training Issues in Health Care

PART 1

Occupational and Environmental Health Training for Hospital House Staff

Wendy E. Shearn and Kathleen Kahler

Why does the occupational and environmental health training for house staff need to be any different from any other health care workers' training? This is the question we were asked by many occupational and environmental health professionals. Clearly, we must establish a data base of occupational and environmental exposure issues for this group of health care providers. By not providing hospital-specific occupational and environmental health training, a significant risk of financial and criminal penalties is created for medical center administrators and the corporations which own such institutions.

Given traditional extreme hours of work often in excess of 70 hours per week, there are increased risks for health care professionals due to sleep deprivation, including depression, substance abuse, suicide and accidents. They may be exposed to chemicals and physical hazards.

Chemical hazards which place the house staff at risk include antineoplastic drugs, formalin, anesthetic gases, ribavirin, ethylene oxide, nitrous oxide and lasers. Use, possible routes of exposure, health effects, engineering controls and protective equipment need to be addressed in the circumstances in which they are encountered by house officers.

Historically, the house staff has been identified as students, and as such house staff has been underpaid for the number of hours actually worked. This situation has resulted in confusion about the house staff's status as employees. OSHA has assisted medical centers in clarifying the status of house staff by defining physicians as employees. If a physician is employed by a health maintenance organization or a corporation, all OSHA standards apply to them, and compliance with regulations, including training, is required.

1-56670-083-3/94/$0.00+$.50

TRAINING ISSUES

Ideally, occupational and environmental health and safety information should be provided to residents at their orientation. Compliance with OSHA's Bloodborne Pathogen Standard and the Hazard Communication Standard should be stressed. The initial orientation may be the only opportunity to provide this type of information to the residents, because it may be the only time they are together and available.

Some programs rotate the house staff through several departments, while others concentrate on only one. Some residents rotate through different hospitals. This variation in programs makes it difficult to identify who has the responsibility for the house staff's occupational and environmental health and safety training. Interdepartmental and interfacility policies and practices often vary.

Frequently, types of equipment are worn to protect against body fluid exposures. Safer needle devices require special training. Systems for handling biohazardous and hazardous waste should be reviewed by the department or a facility.

House staff should be trained on proper disposal of specimens, needles, bodily fluids and common chemicals such as formalin and alcohol. Violations of environmental and occupational health regulations carry fines, penalties and potential criminal liability.

Clinical decisions and practices of the house staff impact the health and safety of other staff members. For example, the house staff needs to know potential health effects of hospital-administered aerosolized pentamidine on other staff such as nurses and respiratory therapists and what precautions are needed. If they were informed about this, they might order alternate therapy.

Procedures such as chest drainage may create "mixed waste" when body fluids and mercury are combined. The resulting waste requires special decontamination to avoid exposing support staff to bodily fluids and mercury vapors.

Needles left on patient tables, in bed linens, waste baskets, sleep rooms, or floors, or on top of lockers or in pockets of laboratory coats can be fatal since transmission of hepatitis B or AIDS to health care workers by needlesticks has been documented.

NEEDS ASSESSMENT

Facility specific information on common problems should be obtained through a survey of various hospital departments including nursing, pharmacy and environmental services.

Statistics could be reviewed to determine the type of injuries or illnesses incurred by residents in the recent past. Records of health and safety complaints filed by residents should be reviewed.

Based on experience and needs assessment, a hospital house staff orientation program should be developed.

HAZARDS IN THE HEALTH CARE ENVIRONMENT

Cytotoxic drugs are sometimes corrosive and present a reproductive hazard. Because house staff may at times be required to administer them, they need to be educated regarding protective equipment, spill cleanup and management of contaminated materials.

When handling specimens and at autopsies, formalin exposure may occur. Formalin must be disposed of properly according to environmental regulations.

Glutaraldehydes used for sterilization of instruments have been left in open containers in procedure rooms, emergency departments and obstetrics and gynecology for quick turnaround of instruments. Handling of glutaraldehydes in uncovered containers could result in exposure or sensitization.

Nitrous oxide is used as a cryogen in ophthalmology and obstetrics and gynecology. Exposure of the house staff, patients and other staff can occur when the nitrous oxide is vented into the room or into a recirculating ventilation system.

Mercury poses occupational health and environmental hazards. Many facilities still use mercury to weight gastrointestinal tubes. In this circumstance the risk of a mercury spill is significant. In addition, mercury should not be disposed of in the sewer because this method violates environmental regulations. A list of common equipment and locations where mercury is found in a medical center, along with alternative products or work practices is in Appendix 2.

Alcohols and xylenes, which are found in pathology laboratories, should be disposed of in a specific manner and not via the drain.

"Prevention Point," a videotape designed for physicians in training (Altschul Group Corporation, Evanston, IL), has been used to demonstrate safe needle management. The importance of reporting an exposure and receiving appropriate follow-up care is stressed in the video tape.

New needle devices or "safer" needles should be addressed as part of orientation. Many new needleless systems and "safer" devices are being introduced at medical centers. Some require new skills or techniques. House staff require training using these devices.

Ergonomics and physical hazards must be addressed. Back injuries may result or be aggravated by improper height of an examination table. Other hazards include stress and sleep deprivation.

Emergency preparedness needs to be included in orientation. Local hazards such as earthquake or flood should be addressed.

Respirators are now required for house staff working with suspected or diagnosed tuberculosis patients under certain circumstances. Annual training and fit-testing for respirators is a requirement.

House staff should be reminded of their right to refuse to perform hazardous work and to ask for training in a procedure for their own protection as well as for that of the patients.

HOW IS THIS INFORMATION PROVIDED?

Many of the new regulations affecting health care require the physical presence of a trainer to ensure that employee questions are answered. To ensure accuracy and consistency of practices and procedures related to occupational and environmental health and safety, utilizing local experts from the departments within the medical center is important. Orientation for house staff should be presented in accordance with their educational background.

Managers are traditionally used in orientation to discuss procedures for providing or accessing services from their clinical departments. Managers must communicate to those departments that information on health and safety is essential.

Key questions when planning the training should include: 1) topics relevant to the medical center; 2) time required and time available; and 3) who provides the training. See *Essentials of Modern Hospital Safety,* Volume 1, Chapter 20, "Education for Action: An Innovative Approach to Training Hospital Employees" is an excellent resource for developing this type of training.

Monitoring compliance with policies and procedures is the key to ensuring their practice. In addition, mechanisms for addressing noncompliance with occupational and environmental health and safety procedures should also be established.

All training records, including attendance records, and the goals and the objectives of the training should be maintained.

SUMMARY

Occupational and environmental health and safety training should be required for house staff orientation. As new risks arise and present risks are illuminated, additional training will be needed.

We anticipate further development of resources through continued educational research concerning house staff. Education is a critical part of an effective prevention plan.

APPENDIX 1

"Safer" Medical Devices: Preventing Needlesticks and Sharps Exposures Fact Sheet

Increased awareness of occupational exposures to bloodborne pathogens, specifically hepatitis B and HIV, has increased interest in the development of safer medical devices to prevent needlesticks and sharps exposures. Additionally, the Federal Occupational Safety and Health Administration (OSHA) recently introduced the Bloodborne Pathogen Standard. In this standard, OSHA concludes that exposures can be minimized or eliminated by using provisions that include engineering controls (e.g., use of self-sheathing needles), work practices (e.g., universal precautions) and personal protective clothing and equipment.

Selecting new devices which utilize engineering controls is a difficult process for all health care facilities because of the following issues:

1. Design
   a. Many designs are in their initial stage and therefore may not be the *best* design.
   b. Many initial designs were directed at high-use products, not at high-risk products.

2. Protection
   a. Advertised "safety" devices do not always consider exposures to ancillary staff.
   b. New products do not have statistical data to support their use.

3. Availability
   a. Manufacturers have been unable to supply quantities needed for larger facilities.
   b. Products may not be directly exchanged with nonsafety devices.

4. Education
   a. Staff may need new skills to use product correctly.

5. Regulation
   a. CDC and OSHA encourage engineering controls, yet no regulating agency has specifications for new safety designs.

The following safety devices, which utilize engineering controls to prevent needlesticks/sharps injuries, have been selected for the Medical Center with the assistance of the Needlestick Prevention Subcommittee and the Product Selection and Evaluations Subcommittee:

1. Lancets
   a. Ames Glucolet 2™ Automatic Lancing Device
   b. Tenderlett™

2. IV Tubing/Medication Connections
   a. Baxter Protective Needle Lock™
   b. Baxter Needle Less IV Access System - INTERLINK™ (In-service to begin May/June)

3. IV Catheters (Stylets)

   a. Critikon PROTECTIV™ IV

4. Winged Infusion (Butterfly)

   a. Ryan Medical SHAMROCK™ (Currently being piloted)

5. Safety Syringes

   a. Becton Dickenson SAFETY-LOK™ Syringe (Currently being piloted)

6. Vacuum Collection Systems

   a. Ryan Medical SAF-T CLICK® (Currently being reevaluated)

Future evaluations of safety devices which provide engineering controls will be targeted at high-risk procedures for which work practices alone have made it difficult to control exposures (e.g., blood culture collection, blood gas kits, scalpels and suture needles).

Source: Kaiser Permanente Medical Center - San Francisco, Safety Office, May, 1992.

Appendix 2. Mercury in Health Care Settings. Kaiser Permanente. Wendy Shearn, Kathleen Kahler, Jennifer McNary.

Sources of Mercury	Locations Used	Alternative Work Practices	Alternative Products
1. Bougie/Red Maloney esophageal dilator (up to 3 pounds)	Operating Rooms Gastrointestinal Labs Endoscopy Procedures	Inspect tubes for damage or deterioration before use. Remove from service if damage is noted.	Use stainless steel balls instead of mercury to weight. Other suggestions are tungsten or barium.
2. Cantor gastrointestinal tube (6–9 grams of mercury)	Medical Surgical Units	Process for handling mercury and waste disposal should be specific.	No alternative product available.
3. Chest drainage unit (i.e., Pleur-Evac)	CVICU/ICU/Medical Surgical Units	Use water and portable suction units to achieve greater suction.	Many products can achieve adequate suction by utilizing water or by using a vent plug with wall suction.
4. Feeding tubes	Medical Surgical Units	Substitute/replace with tungsten weighted tube.	New products are weighted with tungsten instead of mercury.
5. Thermometers • Room • Older incubator hood • Body 1.5 grams • Laboratory 3–4 grams	 All locations Nursery/ICN All locations Laboratory	Substitute/replace with electronic thermometers.	Electronic thermometers.
6. Sphygmomanometers (200 grams of mercury)	All locations	Substitute/replace with aneroid gauges for mercury sphygmomanometers.	Aneroid gauges.

Appendix 2. Continued.

Sources of Mercury	Locations Used	Alternative Work Practices	Alternative Products
7. Hematology (some analyzers (hydraulic systems used) to draw mercury samples through counting chamber)	Laboratory	Replace with newer model.	Newer models do not use mercury in the hydraulic system.
8. Strip Chart Recorder used in Cardiac Ultra Sound Dry Silver Type (7 milligrams of mercuric bromide)	CV/ICU, ICU	Switch to thermal head recording system.	Thermally sensitive paper.
9. Red mercuric oxide	Laboratory	Substitute	Not required for laboratory use.
10. Mercuric chloride	Pathology (preparing tissue specimens)	Substitute reliable analytical methods.	Zinc can be used, but many pathologists state that results lack clarity.
11. Fluorescent tubes (40 milligrams of mercury)	All locations	Correct handling in the disposal process. No rushing.	No alternative product available at this time. Designers are working to develop tubes with less mercury.
12. Barometers	Laboratories	Substitute electronic barometers for mercury containing devices.	Electronic barometer.
13. Alkaline batteries and mercury batteries	All locations	Consideration should be given to disposal process. Recycle products.	Low-mercury batteries.

All products containing mercury need to be disposed of as either hazardous or extremely hazardous wastes. New products contaminated with biohazardous material need to be decontaminated before disposing of material as hazardous waste.

APPENDIX 3

Tuberculosis (TB), Exposures, and Health Care Workers (HCWs)

TB is on the rise in the United States today. This fact sheet will address the concerns of HCWs with an explanation of what TB is and how it is spread. It will also discuss the detection and prevention of TB.

HOW IS TB SPREAD?

TB is a disease caused by bacteria and is spread by coughing. You do *not* "catch" TB by sharing a glass or by touching something handled by a person with active TB. (Active TB generally means TB in the lungs.) Coughing people with active TB expel bacteria into the air in very small numbers. When a patient with active TB coughs directly in the face of a HCW, that worker can become infected if a closed area is shared for a prolonged period of time.

WHAT KINDS OF PRECAUTIONS ARE TAKEN AT KAISER MEDICAL CENTER TO PREVENT THE SPREAD OF TB?

The Kaiser Foundation Hospital uses 100% fresh air in its ventilation system. This means that the air from a patient's room or any room is vented to the outside. Air is not reused or recycled in the hospital. Therefore, HCWs not working in an enclosed area with a patient with TB are at minimal risk of contracting TB.

Respiratory precautions are another method of protecting HCWs from exposure to TB. A sign stating "STOP—See nurse before entering" may be posted outside the door of a patient with TB. Anyone entering the room of a patient with TB should wear appropriate personal respiratory protection. In actuality, the *patient* should wear a mask, although compliance with this seems difficult.

Respiratory precautions are recommended even though the air in the hospital rooms "gets changed" by the local exhaust ventilation 2–3 times an hour. Here's why: it still takes from 2 to 3-1/2 hours to remove 99.9% of the air, including the airborne bacteria.

WHAT IS A TB EXPOSURE AND HOW CAN IT HAPPEN?

Usually a TB exposure occurs when someone has prolonged contact with a patient who has TB but who has not yet been diagnosed with it. The lack of a diagnosis is usually not due to an oversight or error: TB can be hard to diagnose. The physicians have to make an evaluation of all the illnesses (including TB) that might cause the same symptoms that TB causes before they can make the diagnosis. Although a chest X-ray can be helpful, the preliminary diagnosis of TB is made from a positive sputum smear. Depending on the patient's condition, it may be difficult to get a good sputum smear. All of these factors may delay the correct diagnosis of TB.

HOW WILL I KNOW IF I HAVE BEEN EXPOSED?

You and your co-workers will be notified of the TB exposure by the normal means of communication in your department (e.g., posted memo, communication log, staff meeting, etc.)

WHAT SHOULD I DO IF I HAVE BEEN EXPOSED?

The Infection Control Committee recommends that routine annual TB skin testing (or semi-annual testing for certain designated areas) is sufficient for most exposures. Exposed employees requesting testing/evaluation before their routine testing date should contact Employee Health. Employee Health recommends that employees who had significant prolonged contact and/or who are immunocompromised (e.g., HIV+, taking anticancer drugs or steroids) should contact their health care practitioner or the Employee Health Department to discuss their exposure. Depending on the circumstances of your exposure and your state of health, you may not need follow up or you may need to get a TB skin test 12 weeks after exposure to see if your test becomes positive.

WHAT WILL HAPPEN IF I DEVELOP A POSITIVE TB SKIN TEST?

A positive skin test probably means that you, like 10 million other people living in the U.S., are infected with TB, but are not contagious or sick with it. To see if you have active TB you will need to get a chest X-ray. If your skin test was positive but your chest X-ray does *not* show TB in your lungs and if you are not coughing, having drenching night sweats, losing weight or spitting up blood, then you have dormant or latent, not active, TB. You cannot infect anyone else.

You may need to get medication for the latent infection to reduce your chances of developing active TB in the future. You will need to discuss the results of your test with your health care practitioner or with the health care practitioner in Employee Health to decide whether or not you will need medication.

PREVENTION

To prevent or minimize exposures, the Infection Control Committee recommends that you:

1. Teach all patients to cover their mouths when coughing
2. Follow universal precautions
3. Observe postings on patient doors

Prepared by: G. Denton, NP, Employee Health, K. Kahler, MPH, Safety Office, and W. Shearn, MD, Chief of Occupational Medicine and Employee Health; Kaiser Permanente Medical Center: San Francisco, Employee Health Center.

APPENDIX 4

Scabies Fact Sheet

WHAT IS SCABIES?

Scabies is an infectious disease caused by a human parasitic mite, *Sarcoptes scabies.*

WHAT ARE THE SYMPTOMS OF SCABIES?

Symptoms of *classic* or *typical* scabies include:

Itching: more intense at night
Skin lesions: particularly around finger webs, sides of toes and fingers, wrists, elbows, knees, ankles, abdomen and thighs.

In people with depressed immune systems, scabies may not cause itching or lesions. Rather, the skin may appear scaled and crusted, most often on the hands, feet, elbows and knees. These symptoms are characteristic of *crusted* or *Norwegian scabies.*

WHAT IS THE INCUBATION PERIOD?

In a previously unexposed individual, 2 to 6 weeks may elapse between exposure and onset of itching. In people who have had scabies in the past and are sensitized to the mite, re-exposure may produce itching as soon as 48 hours after exposure.

HOW IS SCABIES TRANSMITTED?

Scabies is transmitted by skin-to-skin contact with "infested persons" or, less commonly, by contact with their freshly infested clothing or other personal objects.

WHAT CAUSES THE ITCHING AND THE BUMPS?

The mite and its secretions, feces and eggs (all foreign protein material) stimulate the body's immune responses, causing the symptoms of scabies.

HOW IS SCABIES DIAGNOSED?

The diagnosis of scabies is made by history taking and physical examination. Sometimes a microscopic examination of a skin scraping can demonstrate the presence of the mite, its eggs or its fecal pellets.

HOW IS SCABIES TREATED?

There are various creams or lotions that may be prescribed for scabies: Permethrin (Elimite), Lindane (Kwell) or Crotamiton 10% (Eurax). In general, all medication should be applied after a brief shower or bath. The skin should be dry and the medication applied to the body from the neck down with special attention to the areas around the finger webs, wrists, elbows, axillae, breasts, buttocks, and genitalia. It should be left on for 12 hours. If the medication is removed—such as by hand washing—it should be reapplied. After 12 hours, a cleansing bath or shower should be taken.

Sheets, pajamas, towels and any clothes worn during the previous week should be washed on the hot cycle and dried in a clothes dryer. Nonwashable items should be sealed in plastic bags and not opened for 2 weeks. It is *not* necessary to wash your carpets, furniture and all the clothes in the closet. For asymptomatic individuals, a single treatment is recommended. For those who are symptomatic, treatment should be repeated in 1 week.

CAN I STILL WORK?

For asymptomatic personnel, no work restrictions are necessary. If an exposed employee is symptomatic, he or she may return to work after one treatment with medication has been completed. Family members or close contacts generally do not need to be treated unless the employee is symptomatic was exposed to a patient with Norwegian or crusted scabies.

PREVENTION

Hand washing before and after patient contact may prevent infestation. The mite roams freely on the skin for up to 4 hours before burrowing, so washing with soap and water can prevent it from making your body its home. When caring for patients with Norwegian scabies who have not yet been treated with Elimite, keep the ends of your gown sleeves tucked in the ends of your gloves. Complaints of itching and/or even a minimal rash should be reported to your supervisor so that a physician can examine the patient to make the diagnosis.

Source: Kaiser Permanente Medical Center - San Francisco.

APPENDIX 5

Hepatitis and Health Care Workers

INTRODUCTION

Hepatitis is an inflammation of the liver that can be caused by a number of agents, including viruses, toxic chemicals (e.g., alcohol), certain drugs and parasites. This fact sheet, which was written for health care workers, (HCWs), describes hepatitis B and hepatitis C, the most likely forms of *viral* hepatitis to which HCWs could be exposed. This fact sheet briefly addresses hepatitis A as well.

TRANSMISSION OF HEPATITIS A

Viral hepatitis A is spread primarily through the "fecal-oral" route, which means, for example, by eating contaminated food, particularly shellfish. Being an HCW does *not* make a person at increased risk of contracting this illness.

TRANSMISSION OF HEPATITIS B

Viral hepatitis B is not spread like hepatitis A (by food or feces), nor is it transmitted through everyday contact, as with a cold. It is transmitted through contact with blood and other body fluids (saliva, semen, vaginal fluids, etc.). Hepatitis B is most commonly spread through contaminated needles, sexual contact and perinatal transmission (from mother to infant around the time of delivery). HCWs are potentially exposed when they work around blood and body fluids.

Symptoms

A person who contracts hepatitis B may not have any symptoms or he/she may have any or all of the following: loss of appetite, loss of taste for cigarettes, fatigue, headache, stiff or aching joints, low grade fever, nausea, vomiting, abdominal pain, jaundice (yellowing of the skin and eyes), dark urine and light colored stools. Similarly, a person with hepatitis A may or may not have the same symptoms. Blood tests can determine if someone currently has or has had either hepatitis A or B.

Incubation and Contagious Periods

The hepatitis B incubation period (the time from exposure to the onset of symptoms) varies from 6 weeks to 6 months. A person infected with hepatitis B is contagious as long as the virus remains in the blood: 1 to 6 months for most people but indefinitely for some.

Complications of Hepatitis

For hepatitis B, 90% of the patients recover spontaneously and develop lifelong immunity to the virus. However, 5 to 10% continue to have symptoms for more than 6 months and may develop some degree of chronic hepatitis, which can cause cirrhosis or liver cancer. Mothers can pass the virus on to their babies at delivery.

WORK RELATED EXPOSURES TO HEPATITIS B

Every year, 200 to 300 HCWs die from hepatitis B or related illnesses. HCWs can be infected by hepatitis B if they are stuck by a needle or another sharp instrument that is contaminated with infected blood. Workers can also be infected through a splash of blood or other infectious body fluids to the eyes, nose, mouth or open skin (cuts, sores, rashes, etc.). Bites that penetrate the skin can also transmit hepatitis B. Statistically, about 6 to 30% of HCWs who are exposed to hepatitis B infected blood or body fluids will contract the disease. After an exposure, HCWs may need gamma globulin or a special hyperimmune globulin.

Prevention

The most important ways to protect yourself from hepatitis B are the following:

1. Get the hepatitis B vaccine. *Take all 3 doses.*
2. Always follow universal precautions. Treat all blood and body fluids as potentially infectious and protect yourself.
3. Wash your hands before and after physical contact with any body fluids. Even if you were wearing gloves, you should wash you hands after removing the gloves.
4. Wear gloves and other protective equipment (goggles, masks, face shields) when you expect to come in contact with blood and other body fluids.
5. Dispose of needles and other sharps safely in the sharps container. *Never* bend, break or cut needles before disposing of them. If you need to recap them, use the one-handed scoop technique.

TRANSMISSION OF HEPATITIS C

Hepatitis C is thought to be transmitted in a manner similar to that for hepatitis B: by contact with blood and body fluids. Intravenous drug use and sexual contact with someone who is hepatitis C positive may place a person at risk of contracting hepatitis C. At the present time there are conflicting reports about perinatal transmission of hepatitis C.

Transfusions also place the people receiving them at risk of contracting the hepatitis C virus. Transfusion blood is tested for both hepatitis B and C, but hepatitis B is much easier to detect and the test for it is currently more accurate than the test for hepatitis C.

Symptoms and Incubation

The symptoms of hepatitis C are similar to those of hepatitis A and B. Hepatitis C's incubation period of approximately 60 days is longer than hepatitis A's but shorter than hepatitis B's.

Chronic Hepatitis C

Like hepatitis B patients, many hepatitis C patients develop chronic infection; in fact, half the people infected with hepatitis C develop chronic hepatitis.

WORK RELATED EXPOSURES TO HEPATITIS C

HCWs may become infected if they are stuck by a needle or another sharp which is contaminated with infected blood. Splashes of body fluids to the eyes, nose or mouth or open cuts or rashes seem less likely to transmit hepatitis C than hepatitis B. About 3% or 1 out of every 27 HCWs exposed to hepatitis C positive blood will contract the virus.

Prevention

Although no vaccine is currently available for hepatitis C, gamma globulin may be given to HCWs after blood or body fluid exposure to protect them from contracting hepatitis C. In addition, the same prevention methods listed for hepatitis B should be followed by HCWs to protect themselves from work related exposures to hepatitis C.

If you think you have been exposed to any kind of hepatitis, you should notify your supervisor and the Employee Health Center at x2636. If you have any questions regarding hepatitis, please call the Employee Health Center.

PART 2

Nursing Student Health and Safety Training in the Hospital Workplace: Risks and Responsibilities

Jolie Pearl and Marian McDonald

INTRODUCTION

How well prepared are nursing students to protect themselves form occupational hazards when they enter the hospital workplace? How prepared are hospitals to provide a safe environment for nursing students? What are the potential consequences of not providing a safe workplace for nursing students—for students and institutions alike?

This chapter discusses hazards encountered by nurses in hospital settings and the special risks and problems faced by nursing students during clinical practicums, as well as issues faced by hospitals that have nursing students on site. The concerns raised are also relevant for new nursing graduates, many of whom may not be prepared to avoid hazardous exposures in the hospital environment. Current nursing school approaches and curriculum content regarding occupational hazards faced by nursing students are described and deficiencies addressed. This section then offers an overview of what an approach to educating nursing students about workplace hazards should include.

Hospital-based nurses continue to be at high risk for exposure to occupational hazards and workplace injuries and illnesses. Nurses-in-training (e.g., nursing students) are at *especially* high risk for exposure to workplace hazards on the job, because of the following factors:

- Widespread lack of appropriate education about avoiding exposure to workplace hazards
- Inexperience
- Their role as patient providers
- Lack of or limited protection under existing regulations because of student status

Nursing students are not the only ones at risk, however. Hospitals are vulnerable as well, both because of their legal responsibility to protect patients and the potential legal ramifications of the provision of a clinical environment for student practicums with access to patients and exposure to hazards. Hospitals' vulnerability to litigation is exacerbated by the fact that most nursing students receive little or no formal health and safety education or training, and that which they do receive is rarely monitored or evaluated by the hospitals in which students do their practicums.

1-56670-083-3/94/$0.00+$.50

Hospital departments of nursing, as well as other departments with responsibility for staff education and/or compliance with health and safety training, have a critical role and responsibility in the prevention of occupational illnesses and injuries in hospitals and in the training, education and protection of nursing staff and students. Hospital nursing departments are also responsible for coordination and oversight of on-site nursing student clinical practicums. It is to these hospital departments, as well as educators interested in the health and safety training of nursing students and future nursing staff, that this chapter is addressed.

HOSPITAL WORKPLACE HAZARDS FACED BY NURSING STUDENTS

Nurses are exposed to a wide variety of occupational hazards in the hospital work environment and suffer high rates of injuries and illnesses on the job. Nurses encounter virtually every type of hazard present in the hospital environment. These include:

- Infectious hazards, such as tuberculosis, hepatitis B, HIV, and CMV
- Chemical hazards, such as ethylene oxide, formaldehyde, and anesthetic gases
- Pharmaceutical hazards, such as antineoplastic agents, ribavirin, and pentamidine
- Physical hazards, such as lifting, radiation, and assault
- Psychological hazards, such as stress and shiftwork
- Reproductive hazards, such as ethylene oxide, infectious agents, and ribavirin

There is no doubt that nursing is a hazardous profession. Unfortunately, the health care industry has often dragged its feet in responding to health threats to its employees and been slow to act upon its responsibility to provide a safe and healthy work environment.

A case in point is occupational exposure to hepatitis B. Hepatitis B has been a serious and well-documented occupational health hazard to health care workers for many years.[1] As many as 200 health care workers die each year from occupational exposure to hepatitis B, with many others suffering other serious health effects.[2] Health care workers with both frequent and infrequent blood contact were known to be at risk for acquiring hepatitis B infection.

Despite the availability and proven effectiveness of a vaccine against the virus, many health care institutions did not offer it to staff at risk, or they required staff to cover the cost of the vaccine themselves. It was not until the promulgation of the OSHA Bloodborne Pathogens Standard (BBPS) in 1991 that provision of the vaccine at no cost to health care employees was required by law.[3]

For nursing students, workplace hazards include those faced by fully trained and practicing nurses. The likelihood of workplace exposure may be even greater among nursing students, however, due to lack of training and experience, and the medical and financial implications more dire.

Little data are available on injuries and illnesses sustained by nursing students during hospital practicums. In addition, few studies of occupational injuries and illnesses among nursing or other health care staff include nursing students in their study cohorts. Consequently, the true magnitude of occupational injuries and illnesses among nursing students remains unknown.

One study of blood and body fluid exposures among health care workers in a large teaching hospital did include nursing students in its cohort. The study found that these students sustained the third highest overall incidence rate of needlestick and blood and body fluid exposures among all staff categories in the institution. Nurses in the "General" and "Respiratory/Rehabilitation" areas sustained the first and second highest incidence rates, respectively. Nurses and nursing students sustained 78.8% of all reported exposures in this study.[4]

Nursing students are at high risk for acquiring hepatitis B. However, hepatitis B vaccination rates among nursing students are low. The percentage of nursing schools requiring students to be vaccinated was as low as 4.2% in 1988,[5] despite Centers for Disease Control and Prevention recommendations that students in the health care professions be vaccinated during their training period, prior to workplace contact with blood.[6]

The impact of the Bloodborne Pathogens Standard on hepatitis B vaccination rates among students in schools of nursing is not known. Under the standard, hospitals are not required to provide the vaccine to nursing students.

SPECIAL RISKS AND PROBLEMS FACING NURSING STUDENTS IN HOSPITAL SETTINGS

Nursing students today face a number of challenges. Nursing students perform many of the same job functions and provide almost as many patient care activities as do nurse employees. They face the same workplace hazards as staff nurses. Yet nursing students may be at even greater risk than staff nurses, for the following reasons:

- As students, they do not benefit from required health and safety training that staff nurses are entitled to receive
- They lack the protection of regulatory bodies such as the Occupational Safety and Health Administration (OSHA) and, in many states, Workers' Compensation
- Occupational surveillance and screening data on nursing students are not being gathered on a systematic basis.

Historical Background

The special risks and problems facing nursing students during clinical practicums are not new. Historically, nursing students were seen as low cost or free labor. Nursing students were expected to function fully as staff, facing widespread and dramatic rates of occupational illness and injury.

For example, in Minnesota in the 1920s, very high rates of tuberculosis were detected among nursing and medical students. They were particularly high among those who worked in tuberculosis sanitoria, where rates of active disease among nursing students ranged from 5 to 19%, and TB infection rates reached 100% in some instances.[7]

In that period, nursing students were required to train in tuberculosis wards. It was not until the institution of the "contagious disease technique" in 1933 that a "serious effort was made to set up a barrier between patients and students for the protection of the latter against tubercle bacilli" in the Minnesota University Hospitals.[8] Advocates of this approach also understood the importance of education: "It has been shown that no matter how cognizant the student nurse is of the presence of tuberculosis among her (sic) patients she cannot protect herself against tubercle bacilli unless she is provided with the necessary equipment, encouraged to use it, and taught every necessary step in carrying out contagious disease technic."[9]

Exemption from Regulation

The OSH Act (1970) covers only those people who are employees of a specified employer.[10] Nursing students are not employed and therefore are not covered. Because of this, nursing students are not covered under OSHA. This means nursing students do not benefit from a number of the OSH Act's provisions.

Under OSHA's Hazard Communication Standard (1983) and the Bloodborne Pathogens Standard (1991), hospitals are required to provide training to employees on the various occupational hazards they will encounter in the workplace. Nursing students are likely to be excluded from these trainings.

When nursing students do receive education about hazards in the hospital work setting, generally it is at the discretion of the nursing school and individual nursing instructors. Hospitals that provide student practicums on site may also request or require student education on specific workplace hazards prior to coming to their clinical practicums. More routinely, the focal point of hospital concern is on prevention of disease transmission from nursing students to patients, with documentation of freedom from childhood diseases and active tuberculosis often required of students prior to patient contact.

Since OSHA standards do not cover unpaid students, the likelihood of nursing students being assured of routine and on-going in-service training on health and safety is not high. Without adequate training, these students are at risk of inappropriate use of equipment, which could lead to injury to themselves and/or others. Devices which might seem routine to staff in a hospital can be potentially dangerous to the untrained.

A recent episode in a Bay Area hospital provides an example. A nursing student who had not been properly trained in the correct disposal of contaminated needles sustained a needlestick injury (and exposure to HIV-infected blood) while disposing of a needle in a "sharps" container.[11] While it could be argued that the cause of the event was the design of the container, the student was at a disadvantage. She had not received training on the recommended usage of the device, nor had she practiced needle disposal techniques with a less hazardous substance.

Traditionally, hospitals and nursing schools have provided a certain amount of site-specific training to nursing students on workplace practices for infection control and, more recently, on practices designed to prevent infection transmission to employees, in the form of universal precautions. This site-specific training is often provided by the nurse preceptor in conjunction with a nursing instructor from the nursing school, who is on site with the students. Often during clinical practicums, a nursing student will be assigned to work with a staff nurse from the institution, who will be responsible for demonstrating to the student techniques and practices of patient care. The quantity and quality of those demonstrations will be reflective of the abilities and knowledge of the individual nurse. The likelihood is not high that a staff nurse will be fully versed in the theoretical or conceptual underpinnings of all potential health hazards and the correct ways to work safely with known hazards in the work environment; and such an expectation places an undue burden on the individual nurse. Nurses are not routinely trained as teachers in this area and are often overburdened with patient care duties as a result of understaffing.

Nursing students are exempted from OSHA protections such as provision of the hepatitis B vaccine, which hospitals must now provide at no cost to employees at risk. As mentioned above, only 4.2% of nursing schools required hepatitis B vaccination of their students in 1988. Of these 26 schools, 77% required students to cover the cost of the vaccine themselves.[12] The high cost of the vaccine may be prohibitive to some students.

Hospitals do not routinely include student nurses in on-going staff screening programs. Hospitals may require that student nurses be free from active TB and have various vaccinations. Students are expected to get these services from their own private practitioners or from a student health service. If a student sustains an injury during a clinical practicum, she/he generally must seek consultation and treatment outside the facility. The injury may or may not be reported to the health care facility; and since follow-up treatment for the injury or illness is generally not done on site, it is unlikely student illnesses and injuries will be included in statistical data gathering on occupational injures or illnesses for any given worksite.

Exemption from Workers' Compensation

The definition of who is covered by workers' compensation varies from state to state. In California, for example, nursing students are considered to be volunteers under state labor law, and they are not covered by the disability provisions of workers' compensation laws. If a student nurse were to sustain and appropriately document an injury or illness during the clinical practicum s/he could, in theory, be eligible for only medical coverage under Workers' Compensation.[13]

Nursing students may, on the other hand, exercise their tort law rights to sue a hospital which has been negligent and therefore contributed to a nursing student's illness or injury, as will be discussed below.

Exclusion from Surveillance

It is worth noting that interns and residents (physicians-in-training) are much more likely to be included in hospital surveillance cohorts than are nursing students. The more frequent inclusion of physicians-in-training may be due to the extended period of time which they spend in clinical settings, or to the fact that during training, they are paid and therefore afforded employee status. Inequities in status are certainly a factor as well.[14] In the hospital hierarchy, physicians are afforded greater status than nurses; nursing students are on yet a lower rung of the prestige ladder. The marginalization of nursing students is perpetuated by their exclusion from these study cohorts.

For more insidious problems such as exposure to TB, the exclusion of nursing students from routine hospital staff screening ensures that new conversions among students are never linked to a specific worksite or patient exposure. With certain types of exposures, such as HIV, immediate evaluation and prophylactic treatment are warranted. Yet because the BBPS does not apply to students, they are not guaranteed either appropriate follow-up evaluation or prophylaxis. Additionally, without the immediate establishment of a baseline negative HIV test, students might be ineligible for medical coverage under workers' compensation and in danger of losing evidence for future legal recourse if they were to seroconvert.

Schools of nursing or student health centers may compile data on injuries and illnesses among nursing students, but these data are not easily accessible, nor are aggregate data available for the full range of workplace hazards encountered by student nurses.

These problems are compounded by lack of communication and clear protocols between health and safety programs and on-site student nursing programs. For example, recently a nursing student at a San Francisco Bay Area hospital did an extensive (three hour) intake interview with a newly admitted patient who had an undiagnosed cough. The patient was later placed on respiratory precautions to rule out active tuberculosis. The nursing student was not informed, as staff caregivers were, of potential exposure to TB, but discovered the information inadvertently.[15] This situation illustrates two critical points. First, the student probably had not been adequately educated about potential symptoms of active tuberculosis, nor about the protocol for initiating respiratory isolation for the patient. Secondly, she was not notified of her potential exposure, and so was not alerted to the potential need for appropriate follow-up evaluation.

SPECIAL ISSUES AND PROBLEMS FACING HOSPITALS

Hospitals that allow nursing students on-site for clinical practicums need to be concerned about providing appropriate health and safety education of these students for several reasons:

- Possible legal liability if students have occupational exposures on site
- Hospital concern about compliance with Joint Commission on the Accreditation of Healthcare Organizations (JCAHO)
- The legal obligation to provide a safe work environment
- The accurate collection of screening and surveillance data on occupational illnesses and injuries on site

Hospital Liability for Occupational Injuries and Illnesses Sustained by Nursing Students

In the fall of 1992, a California nursing student sustained a needlestick injury while disposing of a needle contaminated with HIV-infected blood. As a result of that incident, the student has brought a civil lawsuit against the hospital where the event occurred, as well as against the nursing school, the nursing instructor and the manufacturer of the needle disposal box.[16] Although there has been no determination made in this case as of this writing, the arguments against the hospital are compelling and raise the question of whether other hospitals are vulnerable to similar lawsuits if they do not change their practice regarding health and safety training and protection for nursing students.

The arguments being used against the hospital address three important issues. The first is the hospital's liability, because it provided the clinical environment in which the student was functioning and in which the injury occurred. The second is the hospital's failure to properly ascertain the competency of the nursing instructor who had oversight of the nursing student at the institution. The third issue concerns the lack of guidelines from the hospital (or from the nursing school) to address how students are to be protected from exposure to communicable diseases while they are learning new skills, such as the proper handling and disposal of needles and other hazardous equipment.[17]

The first two arguments are straightforward and are similar to the liability a hospital has for the safety and appropriate care of its patients. For example, if a patient was harmed by a nursing student or instructor, the hospital would be held liable for the harm done to the patient (despite the fact that neither student nor instructor are employees of the institution), because the institution allowed them access to a patient who was under its care.

The third argument, regarding policies about student access to patients with communicable diseases, raises critical issues. If guidelines were developed which allow student choice regarding working with patients with communicable diseases, implicit in such a policy would be the identification of patients who have specific diseases—namely AIDS or infection with HIV. Such a policy could easily entail the revelation of confidential information about patients, and could be a breach of state law, ethical guidelines (including providers' duty to treat), existing hospital policy and patient trust. Because of the "window period" in which the AIDS virus may be present without antibodies being detected, there is no way to ensure that any given patient is *not* infected with HIV and therefore free of the infection at the time a student may work with that patient. In short, such a policy would be unethical, if not illegal, and essentially unworkable.

This case demonstrates a range of complex issues regarding hospital liability which can arise when a nursing student is injured during a practicum. Failure to address these issues can leave a hospital open to lawsuits.

Compliance with JCAHO

Despite student nurses' exclusion from worker health and safety regulations, guidelines exist which may apply to students, for which hospitals could be responsible. The guidelines for JCAHO apply to all hospitals accredited by that organization and may eventually be modified to address students specifically.[18]

Under the 1994 JCAHO guidelines on "Orientation, Training, and Education of Staff," the term "individual" (*not* "employee") is used to describe those for whom the institution is responsible for providing appropriate "orientation."[19] "Orientation" may include discussion of specific safety hazards and related policies applicable to the individual's assigned duties. The hospital's responsibility for provision of this orientation includes individuals from off-site agencies who provide nursing care to patients. "Whether done by the hospital or the off-site agency, the hospital is responsible for assuring that each individual from an off-site agency has completed an adequate and timely orientation to the hospital."[20] It is possible that "off-site agency" could be interpreted to include nursing schools.

Providing a Safe Work Environment

A hospital could be in violation of OSHA's requirements for providing a safe and healthy workplace, if an illness or injury sustained by an employee could be shown to have resulted from the actions of a student nurse. For example, a student nurse who had not been properly trained in the safe handling of contaminated needles could endanger hospital staff.

Accurate Surveillance

Finally, hospital-based occupational exposure screening and surveillance data will be incomplete and misleading if they do not include nursing and other students who have patient contact and are at risk for exposure to communicable diseases. It behooves hospitals to develop systems for accurately gathering exposure data on students. Such data would be extremely valuable to students and institutions alike.

WHAT NURSING STUDENTS ARE CURRENTLY TAUGHT ABOUT HOSPITAL WORKPLACE HAZARDS

Material for this section was developed through interviews with faculty and students in San Francisco Bay Area and East Coast nursing schools.

With the advent of HIV/AIDS, a new awareness has developed about occupational hazards found in the health care setting. Nursing practice and education have begun to change to reflect an increased understanding of hospitals as potentially hazardous environments, and new standards of practice have been developed which are designed to better protect nurses on the job.

Despite important advances in teaching about bloodborne hazards, the education nursing students receive is still inadequate and does not address the full range of hazards they will face during their clinical practicums. The current overall approach to this aspect of nursing education does not ensure that students will know how to adequately protect themselves in the clinical environment. No specific standards for health and safety curriculum in nursing schools exist; the content, methodology and evaluation of the health and safety education nursing students receive varies considerably. Criteria are lacking to evaluate existing curricula on the attitudes, knowledge and behavior of nursing students regarding occupational hazards.

With the growth in the theory and practice of universal precautions, nursing students have benefited from a greater emphasis in their education on avoiding needlestick injuries and exposure to patients' bodily fluids. Universal precautions have become a primary focus of health and safety education in nursing schools. However, education about other workplace hazards and other health and safety concerns are generally not taught as a separate or specific content area.

When nursing students are taught about the hazards of an infectious agent, it is in the context of the education they receive about a particular disease, and the related prevention of nosocomial infections, rather than as part of a health and safety curriculum. For example, when students are taught about the pathophysiology of tuberculosis, they may also receive information about the importance of preventing its spread to other patients and themselves. Relatively few links may be made between the study of the disease and the appropriate protective measures students should take. To compound this problem, many nursing schools have students begin clinical rotations during their first semester, before having studied disease pathophysiology and associated protective measures.

Because of current general concern about HIV/AIDS and tuberculosis, students benefit from relatively greater attention to the prevention of these diseases. Additionally, prevention of back injuries is usually covered in nursing school curricula. Education is minimal about most other occupational hazards encountered in the hospital. Protection from exposure to chemical and reproductive hazards may be absent altogether.

KEY COMPONENTS OF AN EFFECTIVE APPROACH TO EDUCATING NURSING STUDENTS ABOUT HOSPITAL WORKPLACE HAZARDS

Responsibilities regarding nursing student health and safety education and training are different for nursing schools and hospitals. The general parameters of what each institution should address are outlined below.

Responsibilities of Nursing Schools

The focus of nursing school education about hazards should be the general and applied principles of occupational health and safety in the health care workplace. It should delineate the following:

- The full range of hazards, including where and under what circumstances and work procedures they will be encountering
- Health risks of exposures
- Protective measures, including the roles and limitations of engineering and administrative controls and personal protective equipment
- Medical and reporting post-exposure protocols
- Legal rights and protections, as well as resources, including clarification of what protections are *not* extended to nursing students

Such training can be based upon health and safety curricula and materials developed by nursing professionals, occupational health professionals and unions. A number of such articles, training materials and handbooks are available.[21–26]

Responsibilities of Hospitals

Hospital health and safety education and training of nursing students should address the specifics of the workplace and should be as comprehensive as that provided to all hospital staff nurses. It should accomplish the following:

- Address the institution's policies and procedures for infection control and prevention of infectious disease transmission to staff and patients.
- Provide specific instruction on both existing and new procedures or devices utilized or practiced by nursing staff in all patient care areas where nursing students rotate.
- Review the protective measures in use and/or recommended or required by the institution.

- Delineate post-hazard exposure protocols of the institution, including screening and reporting requirements.
- Include all workplace hazards nurses may encounter, and review as well fire and electrical safety, and personal and hospital security.
- Evaluate the effectiveness of the education and training done on site, as well as determine the baseline health and safety knowledge students have received from their school.

Hospitals and nursing schools will need to work together to ensure that nursing students are prepared to enter the hospital workplace with sufficient knowledge and skill to protect themselves and others.

CONCLUSION

Nursing students are tomorrow's healers. Yet current health and safety education and training practices of both nursing schools and hospitals leave nursing students in the dark. As a result, nursing students are uninformed and unprotected during hospital practicums.

Because of these practices, nursing students are being forced to make unacceptable sacrifices which can lead to problems for them, their nursing schools and for hospitals. Nursing students deprived of health and safety training and education are especially vulnerable to occupational illnesses and injuries. Their patients and co-workers may be disadvantaged as well. Hospitals offering practicums may face liability issues.

These problems can be addressed if the key institutions involved—nursing schools and hospitals—take heed and act upon their distinct and interconnected responsibilities for providing nursing students a safe learning environment. This can be done by adding workplace health and safety to the nursing school curriculum, by including nursing students in hospital health and safety training, screening, and surveillance and by strengthening inter-institution communication and evaluation.

Such policy and training changes would benefit all parties involved. These changes would serve to strengthen the health, quality and commitment of future generations of nurses while enhancing patient care and the overall safety of the hospital environment.

REFERENCES

1. Nelson Kenrad E: Prevention of Hepatitis in Health Care Workers. In Emmett Edward A (Ed.). Health Problems of Health Care Workers. *Occupational Medicine: State of the Art Reviews.* Volume 2, Number 3, July-September, 1987.
2. Centers for Disease Control: Guidelines for Prevention of Transmission of Human Immunodeficiency Virus and Hepatitis B Virus to Health-Care and Public-Safety Workers. *MMWR* 38(S-6):5, 1989.
3. OSHA, Final Rule, Occupational Exposure to Bloodborne Pathogens. *Federal Register,* 64145–64182, December 6, 1991.
4. Yassi Annalee, McGill Myrna: Determinants of Blood and Body Fluid Exposure in a Large Teaching Hospital: Hazards of the Intermittent Intravenous Procedure. *American Journal of Infection Control,* 19(3):129–130, June, 1991.
5. Goetz Angela, Yu Victor: Hepatitis B and Hepatitis B Vaccine Requirements in Schools of Nursing in the United States: A National Survey. *American Journal of Infection Control,* 18(4):243, August 1990.

6. Centers for Disease Control: Protection Against Viral Hepatitis: Recommendations of the Immunization Practices Advisory Committee (ACIP). *MMWR*, 39(RR-2):14, 1990.
7. Myers Arthur J: *Invited and Conquered - Historical Sketch of Tuberculosis in Minnesota.* Minnesota Public Health Association, St. Paul, Minnesota. pp 602–603, 1949.
8. *Ibid*, p 604.
9. *Ibid*, pp 612–613.
10. 29 USCS § 652, n 9.
11. Nursing Student Sues USF Over Tainted Needle. *San Francisco Chronicle*, September 25, 1993.
12. *Ibid*, Goetz, p 243.
13. Berg Gary: The Noetics Group, San Francisco, CA. Personal communication. December, 1993.
14. Butter Irene, et al.: *Sex and Status: Hierarchies in the Health Workforce.* Washington, DC: American Public Health Association, 1985.
15. Burgel Barbara: Associate Clinical Professor, School of Nursing, University of California, San Francisco. Personal communication. November, 1993.
16. *San Francisco Chronicle, Ibid.*
17. Silver Melvyn: Attorney for University of San Francisco nursing student. Personal communication. November, 1993.
18. Dubor Gayle: Nursing instructor, Samuel Merritt College School of Nursing. Oakland, CA. Personal Communication. November, 1993.
19. Joint Commission on the Accreditation of Healthcare Organizations. *1994 Accreditation Manual for Hospitals - Volume II Scoring Guidelines. Section 4: Orientation, Training and Education of Staff.* p 2. J.C.A.H.O.: Oakbrook Terrace, Illinois, 1993.
20. *Ibid*, p 2.
21. California Nurses Association. *Health and Safety Issues for Nurses.* San Francisco, CNA, 1986.
22. California Nurses Association. *If it Happens to You. What You Need to Know About Occupational Exposure to HIV, HBV, and Other Blood-borne Pathogens.* San Francisco, CNA, 1990.
23. Service Employees International Union (SEIU). *Health and Safety Manual.* Washington, DC. SEIU, 1987.
24. Labor Occupational Health Program (LOHP). *Hospital Hazards: A Union Guide to Inspecting the Workplace.* Berkeley, CA. Labor Occupational Health Program, 1991.
25. Weinger Merri, Wallerstein Nina: Education for Action: An Innovative Approach to Training Hospital Employees. In Charney W, Shirmer J (Eds.), *Essentials of Modern Hospital Safety, Volume 1.* Lewis Publishers: Chelsea, MI, 1990.
26. Coleman Linda, Dickinson Cindy: The Risks of Healing: The Hazards of the Nursing Profession. In Chavkin Wendy (Ed.), *Double Exposure Women's Health Hazards on the Job and at Home.* New York: Monthly Review Press, 1984.

The authors would like to extend heartfelt thanks to the many people throughout the country who granted time and helped think through the different aspects of this chapter. Without their expertise and interest this chapter would not have been possible.

CHAPTER 12

Case Studies of Health Care Workers in the Compensation System

Lenora Colbert

My education about workers' compensation began in 1986 when I started working for 1199, National Health and Human Service Employees' Union. As Occupational Safety and Health Director, I suddenly became aware of issues I had previously overlooked as a worker. I could now relate the dangerous incidents I had personally witnessed to issues of occupational health and safety.

I have worked in the health care field for 26 years in several different capacities including a nurses aide, central supply packaging and sterilizing equipment worker, file clerk, messenger and driver, and also a radiology technologist. I have experienced many work-related injuries that went unreported.

My real education about workers' compensation began when I came on staff at 1199. I could connect the incidents I had personally experienced to these issues. At 1199, the need for strong advocacy in the prevention of injuries, illness and diseases attributed to the work place are understood.

This essay examines the experience of health care workers with the state-based workers' compensation systems by focusing on certain salient and fairly typical problems encountered by workers sustaining workplace injuries and, especially, work-related illnesses. The scenarios described are based on case studies selected from the files of the New York workers' compensation law firm of Pasternack, Popish and Reiff, which for more than two decades has represented the majority of workers' compensation claimants from the 1199 National Health and Human Service Employees' Union. Other scenarios are case histories from the 1199 National Benefit Fund Disability Department.

While these case histories are from New York State, I want to provide a generic treatment of how the state systems too often fail to adequately and promptly compensate health care workers for job-related ailments, particularly in relation to occupational illness rather than workplace accidents.

BACKGROUND

The workers' compensation system is the country's earliest forms of social insurance, introduced in most states over three quarters of a century ago. Since no federal minimum standards exist, the laws, their administration and the ability of workers to receive adequate compensation vary from state to state. Generally,

1-56670-083-3/94/$0.00+$.50

state systems impose "no fault" liability on employers for job-related disabilities. Workers relinquish their common-law rights to legal action. In return, employers promise workers prompt and adequate compensation, complete medical care and rehabilitation services for work-related injuries and illness. In most states workers are compensated at a rate pegged to two thirds of their gross average weekly wages up to a maximum generally at or near the state average weekly wage (SAWW) for the period in which they are temporarily totally disabled and some proportionate amount thereof if they are classified as "partially disabled."

Medical evaluations of the degree of workers' disabilities are the basis for both the duration and amount of wage replacement benefits. These are sometimes referred to as the "indemnity" portion of the benefits and provide entitlement to medical and rehab care at the employer's expense. For claims of occupational disease, medical evidence of the work-relatedness of illness is the basis for determining whether or not the illness will be treated as compensable under workers' compensation.

The Health Care Industry and Workers' Compensation: Cost-Containment Through Preventing Injuries and Illness vs. Cost-Containment by Preventing Compensation of Claims

Since employers' liability for the cost of both medical care and wage replacement benefits are at stake, as we shall see below, employers or their insurers relentlessly pursue cost-containment policies which make medical expenses and wage replacement benefits the subject of litigation. They challenge (controvert) the work-relatedness of occupational disease claims or, in respect to both accident and illness claims, seek to cut off or reduce wage replacement benefits as soon as possible, while workers who may be too disabled to return to work must struggle in the system to maintain wage replacement weekly payments at a level which can sustain themselves and their families.

Similarly, medical costs are supposed to be paid in full under workers' compensation systems, but as medical costs have increased in recent years they too are the focus of employers' efforts to cut or shift such costs. Workers in need of diagnostic tests, surgical procedures or palliative treatment are often forced to litigate to obtain needed medical care or face having such costs borne by the union's health benefit plan with workers picking up the costs of co-payments and deductibles. This economic conflict is manifested on the medico/legal terrain in the form of disputes about the work-relatedness of occupational illness claims, medical treatment and the degree and duration of workers' disabilities.

While health care workers, particularly those employed at hospitals with employee health services and emergency rooms close at hand, have less difficulty accessing medical care than many other workers, it is by no means assured that the work-relatedness of their ailments will be recognized, diagnosed and recorded so that their entitlements under workers' compensation systems will be established without hardship or impediments. Doctors employed by health care institutions are often the first to see health care workers presenting symptoms of illness, either in employee health service offices or emergency rooms. Like their counterparts in other practice settings they usually lack training in occupational health, fail to take thorough occupational histories and are primarily interested in diagnosis and treatment rather than in determining whether the work or work environment has been the cause of, or a significant contributory factor in the origins of the illness or injury. Such shortcomings play a role not only in the failure to recognize occupational diseases attributable to health care workers' chemical exposures and infectious diseases but also in failing to connect certain chronic diseases of the musculoskeletal system to previous incidents of workplace acute trauma or cumulative trauma due to repetitive tasks involving heavy lifting, grasping and the like.

In the case of claims for occupational disease, employers—and health care institutions are not exceptions—have controverted a high percentage of claims on the issue of compensability, frequently by exploiting medical and scientific uncertainty surrounding the etiology of occupational disease. While the claim is contested and awaiting adjudication or settlement, sick and disabled health care workers do not receive workers' compensation wage benefits and might not receive medical benefits since health insurers usually do not cover work-related illness.

These problems are compounded by the marked trend in recent years toward the corporatization of health care. The "new" medical-industrial complex is characterized by the advent of profit-making health care institutions and the impact they have had on so-called non-profit health care providers which emulate their profit maximizing orientation and methods. Hospital administrators are not interested in establishing systems or structures to ensure that occupational injuries and illness are properly identified and recorded since costs associated with such injuries and illnesses will have to be borne by the institutions. Indeed, the lack of such systems will facilitate the shifting of such costs onto union health insurance funds, state disability systems partly funded by workers, social security disability systems partly funded by workers, social security disability and public assistance systems funded by taxpayers or onto the injured workers and their families. And the ability of the health care institutions to externalize such costs will tend to undermine efforts toward prevention since less incentive will exist to reduce such costs and health and safety measures cost money, at least in the short term. So health care institutions can be seen as reflecting contradictory missions. Within the institutions are professional care givers who are dedicated to treating their own health care workers when they are injured and ill, as well as administrators who are dedicated to minimizing costs, including labor costs like those associated with compensation and treatment of the institutions' employees.

The Politics of Workers' Compensation

It should be noted, moreover, that the state-based nature of workers' compensation has worked to the disadvantage of workers and to the advantage of employers. Unlike the Social Security system which covers everyone in the nation and makes its entitlement structure virtually immune from the political depredations of those politicians who might pursue cost-cutting in the name of deficit reduction, when it comes to the state-based workers' compensation systems, business forces often threaten to move out of state, pointing to cheaper workers' compensation costs in other states or regions to induce nervous state legislators to enact changes in the law which have the effect of eroding workers' entitlements. Thus, workers' political power to defend their interests is fragmented state by state rather than concentrated as it is in a uniform federal system. Accordingly, most of the problems outlined below are likely to get worse rather than better, just as declining real wages for most segments of workers in the face of rising worker productivity may be attributable, in part, to declines in unionization, historically the means by which workers exercised some power in defense of their interests and secured a rising portion, during the post-war expansion, of their contribution to production.

The Health Care Work Environment Is Among the Most Hazardous

According to the Industry Risk Index (IRI) which ranks industries by degree of hazard, hospitals and health care facilities (SIC codes 8062 and 8099) are among the fifty most hazardous.[2] Health care workers are exposed to a wide array of illness-inducing hazardous substances. Autopsy workers are exposed to formaldehyde, x-ray technicians to radiation hazards, respiratory therapists to pentamidine. Instruments are sterilized with glutaraldehyde, contaminated needles may transmit bloodborne infectious diseases like hepatitis B or HIV, housekeepers may be exposed to biohazardous medical waste. Approximately one fourth of all workers' compensation indemnity expenditures in eight states were for back injuries. Nurses and patient aides appear to be at high risk for work-related back injury. Multi-state analyses of back injury risk among worker groups revealed that nursing aides and garbage collectors ranked first or second in each state. Nurses aides, porters and laundry workers suffer low back injuries and chronic diseases of the musculoskeletal system after long term employment involving heavy lifting, bending and transporting tasks, while clerical workers at ergonomically unsound computer keyboard work stations are turning up with repetitive stress injuries like carpal tunnel syndrome, tendinitis, de Quervains' syndrome, tenosynovitis and other upper extremity disorders.

Workers' Compensation and the Health Care Worker with an Occupational Disease

While health care workers may contract a variety of occupationally induced diseases, there is a significant gap between the incidence of such illness and the number of claims filed and compensated. The recognition that the system, historically designed around workplace accidents and "no fault" principles, has broken down in relation to occupational diseases is widely shared by diverse observers who have examined it.[3–4]

No discussion of health care workers' experience in the workers' compensation system would be complete without mention of the fact that many workers who contract occupational diseases do not file claims for compensation and therefore never enter the system. A variety of factors are responsible:

1. Relatively little is known about the potential health effects of most synthetic chemicals. No information is available regarding the toxicity of many chemicals in the health care setting.
2. Doctors are not trained to consider work as a cause of disease. Surveys show that adequate work histories are reflected on less than ten percent of hospital charges.[5] Accordingly, many illnesses of occupational origin are mistakenly diagnosed and attributed to their causes, such as smoking, aging or lifestyle factors.
3. Doctors receive little or no training in occupational medicine in medical school—an average of only four hours in four years of medical school.[6]
4. Health care workers are typically exposed to more than one hazardous substance and are often not informed that they have been exposed to hazardous chemicals or substances. Symptoms may become manifest only after many years from the onset of exposure.

Controversions and Delay

Those health care workers who know or suspect that their illness is work-related are confronted with a high incidence of controversions and delay. (Interim Report) "Individual employer liability appears to provide a strong incentive for employers (or their agents) to adopt a defensive litigation strategy which results in extensive litigation within a no-fault system."[7] (ibid.)

Workers' Compensation for Occupationally Contracted Tuberculosis

Health care workers who contract tuberculosis through workplace exposure are often the victims of dual system failures—a failure of occupational health policy to prevent disease transmission—often because of inadequate infection control procedures and lack of proper ventilation in treatment facilities, and a failure of the compensation system to identify and compensate claims for work-related tuberculosis.

The failure of the workers' compensation system should be understood not only in terms of the legal barriers which may in the end defeat a worker's claim for compensation, but also as a failure of the system to educate workers and require health care institutions responsible for diagnosing and treating those who contract the disease to take occupational histories and determine whether the disease was "to a reasonable degree of medical certainty" the outcome of a workplace exposure.

The Case of Sharon B.

Sharon B. was, at the time of her illness in 1992, a 42-year-old lab technician who assisted in the performance of autopsies, including cutting open cadavers, removing, weighing and bagging body organs where the cause of death was determined to have been tuberculosis. She carried out such work without any protection other than a dust mask. After becoming ill and missing work intermittently for some time, she was laid off for excessive absences. About six months after she was discharged from work she was admitted to a hospital with symptoms diagnosed in June, 1992 as multi-drug-resistant tuberculosis.

She was treated in hospital for three months before being discharged and underwent an extended period of convalescence. At no time during her stay in the hospital was any attention given to whether she might have contracted the disease through her occupation. Doctors are focused principally on the diagnosis and treatment of disease not on which insurers have liability for the payment of the bills. All bills were presented to Medicaid. In hospital, understandably, the main objective was to restore the patient to health.

But from the perspective of the worker who will not necessarily be able to return to work upon discharge from the hospital, and who will not necessarily have been paid while sick and unable to work, the issue of income maintenance is critical. Sharon B. was reduced to being a recipient of public assistance.

When her plight came to the attention of the union that had formerly represented her during her employment, she was assisted in obtaining legal representation and filed a claim for compensation in January, 1993. Her case was controverted by her former employer. The doctor who diagnosed and treated her in the hospital provided a report which indicated the diagnosis but expressed no opinion regarding whether her disease was likely caused by her work and work environment.

When the doctor's opinion regarding the work-relatedness of Sharon B's illness was solicited by her attorneys, the physician communicated to the firm his view that she could not have contracted tuberculosis by assisting in autopsies in as much as tuberculosis is an airborne infectious disease which could not be transmitted even by infected cadavers since they "don't breathe."

The doctor was clearly ignorant of the literature regarding nosocomial transmission of tuberculosis during procedures such as autopsy.[7–9] (Lundgren, 1987, Tuberculosis infection transmitted at autopsy. Tubercle, 1987; 68:147–150; Kantor, 1988. Nosocomial transmission of tuberculosis from unsuspect disease. American Journal of Medicine, 1988; 84:833; See also Federal Register, Tuesday, October 12, 1993, p. 52812: "Nosocomial transmission of TB has been associated with close contact with infectious patients of health care workers, and during procedures such as ... autopsy").

The case of Sharon B. will have to be litigated at a hearing in which an expert physician trained in occupational medicine will be required to overcome the uninformed opinion of the doctor who diagnosed and treated Sharon B. It is nearly one year since she filed her claim. It is still pending. Her illness has left her deaf in one ear and with permanent nerve damage. In the past six months all 21 of her co-workers have been tested and ten have tested positive for the first time.

The Case of Ethel T.

Ethel T. is a hospital coding clerk who has performed data entry work for over twenty years. Over the course of that period of time she has worked on several computer keyboards and work stations. Recently, she was diagnosed with bilateral carpal tunnel syndrome after seeing a hospital-affiliated doctor to whom she was referred by the hospital's employee health service to which she had reported with complaints of persistent, severe pain in her hands, wrists and arms.

As in the case of Sharon B., the doctor who diagnosed Ethel failed to render an opinion regarding whether or not the disease was the result of Ethel's long years of keyboard work. Under the New York workers' compensation law, Ethel's work need not be the sole cause or even the main cause of the disease. The law requires that her work be a significant contributing factor in the cause, or aggravation of her condition. Nonetheless, her claim was controverted. This means that she is not eligible for wage replacement benefits and has been refused authorization for a surgical procedure known as a carpal ligament release to relieve the pain caused by compression of the median nerve in each wrist by inflamed tendons. Her claim may take one or two years to resolve before the Workers' Compensation Board.

If doctors wish to treat such patients conservatively in the hope that a period of several weeks off the keyboard together with physical therapy might reverse the progress of the disease and prevent the need for surgery and permanent impairment, the employer's controversion of the claim usually makes such a course impossible. First, the worker usually cannot stop working without a prospect of wage replacement benefits. Second, physical therapists will not treat the worker because they are not assured of payment by the hospital or its carrier or agent. Even if they might be paid when the claim is ultimately established, they are usually not willing to wait a year or 18 months until the claim is established. It is ironic that an employee of a health care institution is having compensation and medical treatment blocked by the hospital.

Toward a Reform Agenda Linking Workers' Compensation to Prevention of Workplace Injuries and Illness

Barth and Hunt recommended the dissemination of information on workplace hazards to both workers and employers as a means to contribute to the prevention of occupational disease. In New York, the state legislature, responding to an initiative of the New York Committee for Occupational Safety and Health (NYCOSH), established a $5 million grant program, funded by an assessment on workers' compensation premiums, to train workers concerning occupational hazards, right to know and workers' compensation in 1985. NYCOSH led a coalition effort in which labor union, COSH groups and occupational health professionals strongly lobbied for the program.

New York's Training and Education Grant Program Funded by Workers' Compensation Premiums

Now three years old, this program elicited worker-training proposals totaling more than $11 million in its first year, attesting to both the need for such training and the strong interest on the part of unions, COSH groups, employers and health professionals in the academic sphere.

Training courses and curricula, films, and other resource materials specifically targeted to worker populations at risk of occupational disease have been produced and disseminated. Such programs constitute an approach to the prevention of occupational disease that follows up on the right-to-know movement of the early 1980s.

A mere legal "right to know" can hardly make a substantial contribution to preventing workplace exposures and illnesses, without the training that educates workers to the hazards they face and the means to protect themselves. This program is likely to have an important educational impact on employers, particularly smaller employers, who as Barth points out, "[i]n some instances... know, little more than their employees about the hazards to which they are exposed."

Occupational Health Diagnostic Clinical Centers

Under New York law, the workers' compensation system does not pay the medical costs of occupational disease screenings unless such screening results in a positive diagnosis of an occupational illness. Even if an individual or group of workers has clearly been exposed to dangerous levels of toxic substances in the workplace and a screening is medically indicated, no compensation for the costs of such evaluations will be made.

The statewide Occupational Health Diagnostic Clinical Center network is a promising efforts toward the prevention of occupational illness, and it's reasonable that the law should be amended to require that the workers' compensation system contribute to the cost of such efforts through payment for occupational disease screenings for workplace exposures associated with disease.

An ancillary and closely related element in New York's approach to preventing occupational illness will be the development of a statewide data collection system incorporating workers' compensation claims information to identify hazardous occupational exposures and diseases and thereby target enforcement and preventive efforts toward hazardous industries with worker populations at high risk.

As the Interim Report put it, "individual employer liability appears to provide a strong incentive for employers (or their agents), to adopt a defensive litigation strategy which results in extensive litigation within a no-fault system." That report made four important recommendations addressed to the state workers' compensation system's treatment of occupational disease claims (pp. 99–100).

1. Establish legal presumptions to reduce the difficulty of proving the cause of occupational disease.
2. Establish an employer-and/or producer-financed trust fund to pay benefits.
3. Eliminate artificial barriers to occupational disease claims in the law.
4. Establish a neutral administrative body to administer the compensation of occupational disease claims.

To these I would add the following provision: insurer-mandated payment of medical expenses for claims. To achieve a more equitable and expeditious payment of medical expenses incurred by individuals filing for workers' compensation benefits, a system should be developed in which health insurers are required initially to pay the cost of medical care for individuals seeking care for a suspected work-related illness. If an individual's claim is sustained by the workers' compensation system, the health care insurer would be reimbursed at the time of settlement or award. If no award is made and the worker is not financially responsible, the insurer could be reimbursed through a special fund. This system would remedy the present situation in which often neither medical insurance nor workers' compensation coverage is available to pay the cost of health care for individuals. As a result of this lack of coverage, individuals often defer a much-needed medical evaluation, and secondary prevention interventions are often foregone.

1199 National Benefit Fund

The 1199 National Benefit Fund for Hospital and Health Care employees grew out of a small medical plan whose coverage was limited to basic surgery and hospitalization. This plan which initially covered pharmacists was expanded to include hospital workers in 1962. This plan, moreover, which began in 1945, today covers over 78,000 members employed in health care and human services.

The fund provides disability benefits for its members. Included in the Disability Benefits Plan is an Intervention Program to assist members during the initial stages of their disability and to help them successfully return to work at the earliest possible date. This is best accomplished through early intervention.

Early intervention in its broadest sense may be viewed as that initial contact with the patient and medical provider when the first disability claim is submitted. This intervention is essential in assisting the patient to return to employment at the earliest and most feasible date possible. The benefits of early intervention are implicit in the proven belief that disabled workers overwhelmingly want to return to the workplace forthwith for both financial security and for their own sense of esteem and well being.

An important component of the Intervention Program is identification of Workers' Compensation claims to ensure that the costs associated with such claims are not shifted into our fund.

The following are two case histories which can be used as examples of what members sometimes experience when engaged in the process of obtaining Workers' Compensation benefits. Although the data that was collected were from actual claims that were handled in the Disability Benefits Department, the names have been changed to maintain confidentiality. In both instances, the role that the Fund played was as a result of our Intervention Program which is part of the initial processing, upon receipt of a disability claim or notice of a potential disability claim. The only exception would be a claim of very short duration, particularly where the member has already returned to work at the point the claim is being processed.

Case Study #1

A. Jones, a 55-year-old respiratory therapist, submitted a claim to the fund on January 23, 1992. A. Jones is a male hospital worker employed at a hospital in New York City that treats AIDS patients. The initial diagnosis on the claim form was pulmonary tuberculosis. The member had last worked on October 16, 1991 and shortly thereafter filed a claim for Workers' Compensation with the employer's plan. When no benefits were forthcoming after approximately two months the member filed for disability benefits. Upon review of the claim form at the fund office it was noted that the member, the doctor and the employer failed to indicate that the illness may have been job related. It appears the member did not acknowledge the possibility of Workers' Compensation due to his experience with the compensation carrier up to that point. The employer failed to indicate the possibility of a job-related illness for obvious reasons. The doctor failed to make a connection since there may have been a relationship between the doctor and the hospital.

As part of our intervention program the claims examiner having connected the member's illness with the type of job performed contacted the member to verify the type of employment and to raise the question of Workers' Compensation. As a result, the claims examiner was able to convince the member that to pursue the Workers' Compensation claim was in his best interest. The member filed a claim with the carrier and the board. After several months a hearing was scheduled.

In the meantime, the Benefit Fund paid 26 weeks in disability benefits and filed an appropriate notice with the carrier and the board. The member received $5,590.00 in disability benefits from the fund. However, it took more than six months before the member's case was finally adjudicated by the board. The Workers' Compensation Board rendered a decision that the member's illness was job related and made an award. The fund was subsequently reimbursed by the carrier and the member received Workers' Compensation benefits beyond the 26-week period which the fund had paid.

Cast Study #2

B. Roberts, a 36-year-old female nursing assistant employed at a hospital in Brooklyn, New York was diagnosed with carpal tunnel syndrome with chief complaints of pain in the right wrist radiating to the elbow and numbness in the right hand involving all fingers. The member filed a claim for disability benefits on May 26, 1992. An initial payment was made by the fund on June 10, 1992 at which time an intake was completed by the claims examiner which includes pursuing the issue of Workers' Compensation.

It was determined, with the assistance of a Workers' Compensation attorney that this member was a good candidate for the Mount Sinai Occupational Health Clinic to determine if the member's condition was job related. The member was very reluctant at first due to her perception of how she would be treated by the Workers' Compensation carrier and the possibility of being left without funds for a substantial period of time. After being assured that the fund would continue the payment of disability benefits while her case was being adjudicated by the board, the member agreed to be examined by the Mount Sinai Clinic. Accordingly, a claim was filed with the Workers' Compensation board on or about September 10, 1992. It was subsequently determined that the member's condition was job related and evidence of this was submitted to the board and the fund was reimbursed for 26 weeks of disability benefits paid in the amount of $5,590.00. The member subsequently returned to work in November, 1992. The period of disability paid by the Fund was 4/22/92 to 10/20/92. It should be noted that the hearing took place almost one year after the member filed a claim for Workers' Compensation benefits.

Had the member not returned to work after receiving the maximum of 26 weeks in disability payments, the claims examiner would have referred the member to the fund's Members' Assistance Department. The Members' Assistant Department is staffed with certified social workers who among other duties, provide counseling and assistance to members who have received maximum benefits from the fund.

Integrating the Workers' Compensation Medical Component in National Health Care System Reforms

The following comments are regarding Title X-Coordination of the Medical Portion of Workers' Compensation and Automobile Insurance, from the Health Security Act, 1993.

The most significant areas of concern we have about the legislation in its current form are as follows:

An individual with a work-related occupational disease or injury should be guaranteed the right to choose the provider for both the diagnosis and treatment of his/her occupational condition.

Since both the diagnosis and the disability determination are used to decide on the benefit level for an affected worker, it is not appropriate for workers to be evaluated by physicians hired either by an employer or by any other health care provider who may not be able to provide an independent assessment of the diagnosis, work-relatedness and extent of disability.

The proposed legislative language speaks only in terms of the treatment of work-related conditions, failing to recognize that the diagnosis and determination of work-relatedness frequently require evaluation by specialized occupational medicine physicians.

Case members should also be responsible for coordination of return to work with work place modification and exposure cessation, as this is the mainstay of treatment of occupational diseases. Finally, the case manager should coordinate treatment via referrals to appropriate providers. However, it is not appropriate that this be done in consultation with the workers' compensation carrier; rather, it should be done in consultation with the occupational medicine specialist responsible for the affected workers care.

Experience rating of employers' workers' compensation experience needs to be added to provide incentives toward the reduction of these preventable illnesses and injuries. Therefore, we urge that incorporation of wording recognizing the importance of experience rating in assessing workers' compensation premiums and the mechanisms by which this can be linked to disease prevention.

REFERENCES

1. Salmon, J. Warren: The Corporate Transformation of Health Care: Issues and Directions, Baywood Publishing Co., 1990.
2. Pederson D, Young R, Sundin D: *A Model for the Identification of High Risk Occupational Groups Using RTECS and NIOSH Data.* NIOSH Technical Report 83-117. Washington, DC; US Government Printing Office, 1983.
3. An Interim Report to Congress on Occupational Diseases, US Department of Labor, 1980, pp. 54–78; Barth P, Hunt A. *Workers' Compensation and Work-Related Illnesses and Diseases:* MIT Press, Cambridge, MA, p. 163–178, 1980.
4. APHA Policy Statement No. 8329 (PP): Compensation for and Prevention of Occupational Disease. APHA Policy Statements 1948-present, cumulative. Washington, DC. APHA, reprinted in *American Journal of Public Health,* 74(3):292.
5. Rosenstock L: Occupational Medicine: Too Long Neglected. *Annals of Internal Medicine* 95:774–776, 1981.
6. Levy BS: The Teaching of Occupational Health in US Medical Schools: Five-year Follow-up of an Initial Survey. *American Journal of Public Health* 75:79–80, 1985.
7. Interim Report, Op. Cit. pp 98–99.
8. Lundgren: Tuberculosis infection transmitted at autopsy. *Tubercle* 68:147–150, 1987.
9. Kantor: Nosocomial transmission of tuberculosis from unsuspect disease. *American Journal of Medicine,* 84:833, 1988.

10. Federal Register, Tuesday October 12, 1993, p 52812. "Nosocomial transmission of TB has been associated with close contact with infectious patients or health care workers, and during procedures such as...autopsy."
11. Harber P, Billet E, Gutowski M, Soottook, Law M, Roman A: Occupational low back pain in hospital nurses. *J. Occup. Med.* 27:518–524, 1985.
12. Cust G, Pearson JDG, Mair A: The prevalence of low back pain in nurses. *Nurs Rev* 19:169–179, 1972.
13. Spratley, Samuel, Assistant Director, 1199 National Benefit Fund, 1199 National Health and Human Service Employees Union, 310 West 43rd Street, New York, New York 10036.
14. Tuminaro, Dominick, Pasternack, Popish & Reiff, P.C., 111 Livingston Street, 22nd Floor, Brooklyn New York 11201.

APPENDIX 1

Sample Protocol: Hospital Safety Office Program

The following is a sample protocol for hospital safety office activities developed by the Employee Health Service, Safety Officer and Health and Safety Committee of a medium-sized acute care hospital. Periodic activities are scheduled after initial walk-through and industrial hygiene inspections have determined the extent of hazards in various units within the hospital. Therefore this protocol reflects the distribution of hazards in one hospital; it should not be used in other hospitals without completing the same initial investigation to identify problem areas and special needs. Infection control activities are coordinated with the infectious disease committee, and these are not included in the protocol below. If a department other than the safety office will perform a monitoring activity this is indicated by a (c) for the industrial hygiene contractor or an (s) for the security department.

Section/Activities	Schedule
Environmental Health	
Anesthetic gases - Surgery (c)	Every 3 months
Formaldehyde - Dialysis (c) - Pathology (c)	Every 3 months
Ethylene oxide - Central supply (c)	Every 3 months
Solvents/reagents - Clinical lab (c) - Eye pathology (c)	Every 3 months
Mercury - Cardia cath lab	Every 3 months
Carbon monoxide - Garage - Engineering	Every 3 months
Noise - Engineering - Food services - As requested	Every 3 months
Fume hoods - Clinical lab - Eye pathology	Every 3 months
Microwave ovens - Each location	Every 3 months
Ambient radiation - As requested	Every 3 months
Chemical disposal - Each location	Every 3 months

1-56670-083-3/94/$0.00+$.50

Section/Activities	Schedule
Safety	
New employee orientation	Every 2 weeks
Nursing orientation	Every 2 weeks
Area inspections	Every month
Departmental consultations - As requested	
Departmental safety training - As requested	
Policy review and update	Every year
Accident investigations - As requested	
Statistical monitoring report	Every month
Safety bulletin board poster change	Every 2 months
Transport safety training	Every 6 months
Personal protective equipment inventory	Every 3 months
Fire	
Fire drills	Every month
Sprinkler valve inspection (s)	Every week
Fire extinguisher - Check (s) - Maintenance (c)	Every month Every 6 months
Building inspection	Every year
Fire equipment/systems check	Every 2 months
Meetings	
Safety Committee	Every month
Risk Management Committee	Every month
Infection Control Committee	Every month
Radiation Safety Committee	Every 3 months
Occupational Health Department (=Employee Health Service)	Every month
Education and Intervention Subcommittee (=Worker Education and Program Participation)	Every 3 months
Facilities Maintenance Department (with Safety Office representatives)	Every week
Administration and Supervisors	Every 2 weeks
Other	
Disaster drill	Every 6 months
Review disaster plan	Every year

APPENDIX 2

The Top Offenses

The top 11 most frequently cited OSHA standards for hospitals are listed below. Do any of these problems exist at your facility?

1. 29 CFR 1910.151 (c). Eyewashes or showers for emergencies not provided where employees could be exposed to hazardous materials.

2. 29 CFR 1910.132 (a). Personal protective equipment was not provided whenever it was necessary because of processes or environment.

3. 29 CFR 1910.133 (a)(1). Protection for eyes and face was not provided where there was a reasonable chance of injury.

4. 29 CFR 1910.22 (a). Passageways, storage rooms and service rooms were not maintained in a clean, orderly and sanitary condition.

5. 29 CFR 1910.22 (a)(2). Floors were not kept clean and dry. Where wet processes were used, drainage and mats were not provided.

6. 29 CFR 1910.1200 (e)(1). A written hazard-communication program has not been developed, *implemented and maintained.*

7. 29 CFR 1910.1200 (f)(5)(i). Containers of hazardous chemicals were not labeled, tagged or marked with the names of the chemical(s).

8. 29 CFR 1910.1200 (f)(5)(ii). Containers of hazardous chemicals were not adequately labeled with the hazard posed by the chemical.

9. 29 CFR 1910.1200 (g)(1). A Material Safety Data Sheet was not available for each hazardous chemical use.

10. 29 CFR 1910.1200 (g)(8). Copies of required MSDSs were not maintained in the *workplace,* were not readable or were not accessible during each work shift.

11. 29 CFR 1910.1200 (h). Employees were not provided with information and training concerning hazardous chemicals in their work area at the time of their initial assignment and whenever a new hazard was introduced into their work area.

INDEX